D1518513

MALAYAN FOREST PRIMATES

Ten Years' Study in Tropical Rain Forest

Siamang

Agile gibbon

Dusky leaf monkey

Long-tailed macaque

PRIMATES OF PENINSULAR MALAYSIA

MALAYAN FOREST PRIMATES

Ten Years' Study in Tropical Rain Forest

Edited by

David J. Chivers

University of Cambridge
Cambridge, England

PLENUM PRESS • NEW YORK AND LONDON

Library of Congress Cataloging in Publication Data

Main entry under title:

Malayan forest primates.

Includes bibliographical references and indexes.
1. Primates–Malaysia–Krau Game Reserve, Pahang. 2. Primates–Behavior. 3. Mammals–Behavior. 4. Mammals–Malaysia–Krau Game Reserve, Pahang. 5. Krau Game Reserve, Pahang. I. Chivers, David John.
QL737.P9M275 599.8'09595 80-25181
ISBN 0-306-40626-8

A Division of Plenum Publishing Corporation
227 West 17th Street, New York, N.Y. 10011

Printed in the United States of America

To the memory of

Murgatroyd

Paragon of monogamous fidelity and fatherhood

FOREWORD

The primates that provide the central theme of these studies by David Chivers and his colleagues are the dominant large herbivores of the tropical evergreen rain forest. To this extent, they are the ecological counterparts of the great herds of ungulates inhabiting the savannahs of tropical Africa (and the monsoonal plains of Asia in their pristine state). Both groups comprise the chief primary consumers of living vegetable tissue in their respective environments. Members of each show appropriate anatomical adaptations for such a diet. As efficient exploiters of a dispersed but generally abundant food source, each group collectively forms the main vertebrate component of animal biomass in the environment.

Yet, despite superficial convergence, there are important differences in the biology and behaviour of members of these two groups of herbivores. Of greatest practical moment to the enquiring biologist are the ready visibility of most plains-dwelling ungulates, the ease with which the researcher can travel over (or above) their habitat by motor transport (or light aircraft) and the facility for near approach without causing disturbance that a closed vehicle has proved to offer. Given the additional attractions of wide, open views and stupendous scenery, generally invigorating climate and easy life-style, it is perhaps not surprising that in past decades much research effort has focussed on the larger herbivorous mammals of the tropical savannahs.

Contrast the problems of fieldwork in the immensely tall evergreen rain forests of the humid tropics. The primates range freely in three dimensions through a vegetation which obtrudes and obstructs the line of sight at every point and limits an unrestricted view to a few yards. The canopy foliage is liable to conceal even a large animal from any but partial and intermittent observation. To stay with his study subjects, earth-bound man must follow on foot amid the mud and thorns and leeches. To fix an eye on his quarry he must either contort his neck in a fashion that strains every cervical ligament, or adopt more aberrant postures that vary from the uncomfortable to the absurd. So far, no primatologist's equivalent to the portable bird-hide has been developed, and no subterfuge has succeeded in concealing the observer more than momentarily from the astute vision of the creatures he would investigate. It is true that there are some hours guaranteed in every 24 when the forest assumes an inspiring beauty, but there is no seasonal remission from the drenching rain. Good shelter is essential for efficient functioning; tropical forest camp-building is an arduous, time-consuming skill, acquired only with experience. Books, papers, notes, films and tapes must always be protected from the pervading wet. Is it surprising that for so long no student came forward to study these animals in the unrestrained state in their natural environment?

It has been my pleasure - initially when employed in the agreeable ambience of the University of Malaya and dwelling with all essential comforts amid the Ulu Gombak Forest Reserve, and subsequently at greater remoteness in rural Suffolk - to become acquainted in turn with many of the scientists whose work is displayed in these pages. In the progression of primatologists in Peninsular Malaysia, I acclaim as trail-blazer Dr John Ellefson. After trials in other areas, he devoted his main effort to the study of a small, isolated population of gibbons on a forest-clad promontory on the East coast. He showed that human determination can ultimately overcome the wayward shyness of wild gibbons, so that they accept the presence of the attentive watcher. Through his responses, too, I took note of the physical and psychological strains imposed by the difficult working conditions and long hours of solitary life involved in research of this nature. John wisely summoned Judy, who then also found time to study the facial expressions of Long-tailed Macaques in Singapore Botanic Garden (a locality from which, I understand, civic authority has since extirpated the species).

The Ellefsons were succeeded chronologically by the Chivers, who arrived already married and with experience of the rain forest environment in another hemisphere. It is to David Chivers' determination to conduct an exhaustive preliminary survey of the entire Peninsula that we owe the selection of the Kuala Lompat site in the Kerau Game Reserve, location of so many of the related studies that have followed. It is worth remembering that the regular grid of traces (*rentis*) that assisted David in his initial assessment of the plot had been cut by a survey party investigating the potential of the area for rice-growing. Kuala Lompat has since assumed an importance that must be measured on an international scale. Yet the Malaysian authorities are under constant pressure to expand agricultural resources or otherwise develop natural areas, and the site cannot be regarded as immune from such a threat.

The results of the research described in these pages can be

evaluated from many different aspects. For example, the understanding of natural behaviour of non-human primates can be used comparatively in the interpretation of human behaviour, both normal and pathological. Such comparisons have been made by (among many others) Suzanne Ripley who also, in the course of a brief visit, studied Malaysian primates (e.g. S. Ripley, 1980. Infanticide in langurs and man: adaptive advantage or social pathology? In M.N. Cohen, R.S. Malpass and H.G. Klein, eds., *Biosocial mechanisms of population regulation.* Yale U.P.).

For primatologists, the ethology of these species of cercopithecine and colobine monkeys and the three gibbons has inherent significance in context in the wider scholastic field. Few studies have detailed so meticulously the time and activity budgets of any higher primate in its natural environment. New insight has been gained into the structure of the social units of these species. The relative roles of male and female in the monogamous gibbons have been elucidated to a great degree. In so doing, some conventional beliefs have been up-turned. For instance, in siamang as among the gibbons, the 'great call' is uttered by the female and it is the male (not the female) who then utters 'a series of high-pitched staccato cries that sound rather like applause' (G. Durrell, 1966, *Two in the bush*, p. 223. Collins).

The family life of the hylobatids proves to be stark but not unappealing. Most human societies would regard as meritorious the persistent monogamy, the coherence of the family during the long immaturity of the offspring and the assiduity with which each parent plays its role in support of the dependant infant. The value of the Kuala Lompat study area in academic terms increases exponentially as time passes, allowing repeated glimpses of the evolving family group as successive members are born, mature, depart or die.

Among the most remarkable of discoveries, in fact, has been the extraordinary stability of the social organisation, in siamang in particular, as far as can be judged from the small sample. It is somewhat astonishing that one family group should have remained, recognisably continuous despite changes in composition, static in one area for almost a full decade. In relation to conservation requirements, this feature of siamang territoriality is a factor of obvious major significance. It must be viewed in the context of the paramount need of the group to know in the greatest possible detail all the available resources of the vicinity. This necessity stems from the fact, perhaps not previously demonstrated with such force, that the tropical rain forest is not an environment of high productivity, and that only a comparatively small proportion of the total plant biomass offers exploitable resources at any one instant. The vital requirement of the gibbon family group is to obtain (and, for efficiency, to retain) a detailed familiarity with the plants of a sufficiently extensive area of forest, and subsequently (and continuously) to defend this against other exploiters with similar needs.

Evaluation of the conservation potential of a given trace of forest, ideally, would consist of a similar survey in equivalent detail. With the information provided by David Chivers and his fellow authors, such a task begins to be possible. They are to be congratulated most heartily on the determination and devotion with which all have applied themselves to the elucidation of these details. Some of the information published here has appeared previously. It is nonetheless extremely valuable to have this review of the programme of research brought together between one pair of covers. The resulting work is erudite, comprehensive, impressive in its coverage, and provides an exemplary model for future projects of this nature. It stands as a monument to the dedication and industry of those who participated.

It also stands as an example of international cooperation. Although no Malaysian appears among the authors, the project has only been possible because of the support of Malaysians, both personal and official. Malaysians have contributed to the work and Malaysians will now benefit from the publication of the combined results in a single volume. I welcome the book as a landmark in primatology and as a key source for conservation management.

Great Glemham CRANBROOK

August 1980

CONTENTS

LIST OF PLATES

LIST OF CONTRIBUTORS

Dr Pelham Aldrich-Blake
BBC Natural History Unit,
Broadcasting House,
Whiteladies Road,
Bristol BS8 2LR, U.K.

Dr David J. Chivers,
Sub-dept. of Veterinary Anatomy,
University of Cambridge,
Tennis Court Road,
Cambridge CB2 1QS, U.K.

Dr Sheila Curtin,
942 Shevlin Drive,
El Cerrito,
California 94530, U.S.A.

Dr John G. Fleagle,
Dept. of Anatomical Sciences,
Health Sciences Center,
State University of New York,
Stony Brook,
New York 11794, U.S.A.

Dr Paul Gittins,
Dept. of Applied Biology,
University of Wales Institute of Science and Technology,
King Edward VII Avenue,
Cardiff CF1 3NU, U.K.

Dr John and Dr Kathy MacKinnon,
F.A.O., d/a P.P.A.,
Jalan H. Ir. Juanda No. 9,
Bogor, Indonesia
<u>and</u> 86 Aldreth Road,
Haddenham, Cambridge, U.K.

Dr John B. Payne,
Forest Department, Wildlife Branch,
P.O. Box 311,
Sandakan, Sabah, East Malaysia
<u>and</u> 14 Boston Road,
Haywards Heath, Sussex, U.K.

Dr Jeremy J. Raemaekers,
Department of Biology,
Mahidol University,
Rama 6 Road,
Bangkok 4, Thailand

ACKNOWLEDGEMENTS

It is impossible to do justice here to all those who have helped each of us, in Malaysia or our home country, during the last 10 years. I hope here to acknowledge our main debts, knowing that each of us has been able to go into more detail in our separate publications. Each of us has come to love Malaysia, through the warm hospitality extended to us and the profound experience of working in the tropical rain forest.

The Federal and State Departments of Wildlife and National Parks (formerly Game Departments) have been the constant contact throughout our studies in Peninsular Malaysia. Our first siamang study, including a country-wide survey, was developed with the assistance of the Chief Game Warden, Mr Bernard Thong, subsequently a good ally in the State of Perak. For the last nine years the Director-General has been Encik Mohamed Khan bin Momin Khan, and we are indebted to him for his continued friendship, advice and assistance. The studies at Kuala Lompat have depended on the approval and help of successive Game Wardens in the State of Pahang: Encik Abdul Jalil bin Ahmad (whose vision in the '60s led to the development of the Ranger Posts around the Krau Game Reserve, with special treatment for Kuala Lompat), Mr Cecil Rajah, the late Encik Musa bin Long, and Encik Hassan bin Hussein. We have valued also the friendship and help of the Federal Research Officer, Mr Louis Ratnam.

This series of studies started at Universiti Malaya, under the guidance of Lord Medway (now the Earl of Cranbrook), with help given to us by David Labang and Yong Gong Chong (Dennis). Sheila Curtin, Pelham Aldrich-Blake and Jeremy Raemaekers have also been based there, and we are all very appreciative of the enthusiastic interest shown in our studies by Professor José Furtado, Dr Enke Soepadmo, Dr Yong Hoi Sen and Dr David Wells. On my second visit, in 1972, I was sponsored by the Institute for Medical Research in Kuala Lumpur, as were the MacKinnons and John Fleagle subsequently, and we thank the then Director, Dr Bhagwan Singh, Dr Roy Sirimane, and the Head of the Medical Ecology Division, Dr Lim Boo Liat, for their help and advice. (Most of us have spent many happy hours with

Lim on convivial nocturnal expeditions around Kuala Lumpur!) Paul Gittins was sponsored by Universiti Sains Malaysia in Penang, and he is indebted to the Dean of Biological Sciences, Professor C.P. Ramachandran, and Dr Joe Charles, as well as to the Forest Departments in Perak and Kedah and the Department of Irrigation and Drains in Kedah for his accommodation in Ulu Mudah. In 1974 and 1976 I was sponsored by the Wildlife Department as was John Payne, and in 1977 by Universiti Pertanian Malaysia (along with Universiti Kebangsaan Malaysia, the base for our current studies), and I thank them for their enthusiastic interest and help, especially the Dean of Veterinary Medicine and Animal Science, Professor Omar bin Abdul Rahman, the Head of Animal Science, Dr Syed Jalaludin, and Dr M.K. Vidyadaran (Menon).

We have received considerable assistance with plant identification and discussions of the vegetational effects on animals from the Forest Research Institute at Kepong, initially from Dr T.C. Whitmore, and subsequently from Mr K.M. Kochummen, Dr Francis Ng, Y.C. Chan and H.S. Loh; and from Dr Soepadmo, Professor E.J.H. Corner and Jack Putz. Various Government Departments have been very generous in supplying essential information, such as weather and vegetation records and maps, and the Ministry of National Unity has processed all our research proposals and sanctioned our studies. John Fleagle thanks the management of Lima Blas Estate, especially Mr M.R. Chandran, for assistance in his work there. World Wildlife Fund (Malaysia) have shown continued interest in our studies, and we thank Mr Ken Scriven for his friendship and help.

We have all worked at Kuala Lompat, and during this time we have received invaluable field assistance from (in chronological order): Sarah Chivers, Ali bin Draman, Shaharil bin Abdul Jalil, Kalang anak Tot, Greysolynne Fox, Barbara Sleeper, Caroline Harcourt, Patricia Crofton (now Raemaekers), Dr Chris and Margaret Spencer, Eamonn Barrett, Glyn Davies and Surender Sharma. Without casting any aspersions on the rest, we owe a special debt to Kalang, who, over the years, has generously shared with us his encyclopaedic knowledge of the plants and animals in the Reserve and a great deal of his time; his energy, skill and enthusiasm in the forest have been crucial to the success of these studies.

I have been funded by a Malaysian Commonwealth Scholarship, a Science Research Council (SRC) Overseas Studentship, a Goldsmiths' Company Travelling Studentship, grants from the Boise Fund (Oxford), the Emslie Horniman Fund of the Royal Anthropological Institute (London), the New York Zoological Society and the Merchant Taylors' Company (London), and by the University of Cambridge Department of Anatomy and Travelling Expenses Fund. Since 1972 vehicles have been made available by Royal Society Government Grants-in-Aid.

Pelham Aldrich-Blake received grants from the John Spedan

Lewis Trust, the Boise Fund and the Explorers Club (New York), and an SRC Post-Doctoral Fellowship. Sheila Curtin received a National Institutes of Health (U.S.A.) training grant (GM-1224) and a grant from the G.W. Hooper Foundation for Medical Research (San Francisco). John Fleagle received grants from the National Science Foundation (U.S.A.) (GB-42033, BNS-7683114, BNS-7724921) and the Hooton Fund (Harvard University). John and Kathy MacKinnon received grants from the James Poulton and Boise Funds (Oxford).

Jeremy Raemaekers was funded by a Medical Research Council Studentship, with grants from Peterhouse, Cambridge and the H.E. Durham and Worts Funds (Cambridge). Paul Gittins received an SRC Overseas Studentship, and grants from the New York Zoological Society, Bartle Frere Fund (Cambridge) and St John's College, Cambridge. John Payne received an SRC Studentship and grants from the Boise Fund (Oxford), the Smuts Memorial and Bartle Frere Funds (Cambridge) and from Sidney Sussex College, Cambridge.

In 1972 the Fauna Preservation Society (London) assisted with the development of the Kuala Lompat Post as a research and conservation area, and in 1977 the B.B.C. Natural History contributed to the Wildlife Department's costs in building a tree ladder overlooking two fig trees at Kuala Lompat, and the World Wildlife Fund (Project 1530) financed a survey of fauna and flora in the Krau Game Reserve for the formulation of management plans; surveys conducted by myself and colleagues and by Drs Ng and Soepadmo respectively.

The writing of this book was helped immeasurably by a Working Party Grant from the Cambridge Philosophical Society, which aided the inter-change of manuscripts between authors and allowed five of us to gather for five days in August 1979 to discuss and revise the contributions. While wishing to preserve the individual style and, in some cases, different approach of each author, chapters were edited and recast to establish some degree of conformity to help the reader. We are very grateful to those who read through the whole volume and supplied many constructive comments: Dr Alison Jolly, Lord Cranbrook, Jules Caldecott, Elizabeth Bennett and Yarrow Robertson - and to those who read certain chapters: Kathy Wolf (chapter 4), Mah Yoon Loong (chapter 5), Drs Peter Rodman and Jack Stern (chapter 7) and Ron Tilson (chapter 8). Dr David Wells and Mah Yoon Loong contributed important comparative data to Appendix I.

The incredibly complex task of typing this manuscript for printing was executed superbly by Pat Reay (who had already typed doctoral dissertations for three of the contributors). Figures were drawn by Jeremy Raemaekers (chapters 2 and 3), John MacKinnon (some of chapter 6), Lazlo Meszoly (chapter 7), John Payne (chapter 9 and one each in chapters 1 and 2) and myself (the rest). Photographs for the plates, and the larger figures, were prepared for publication by the Visual Aids Unit, Department of Anatomy, and we thank Tim

Crane, Roger Liles and Andrew Tiley (and formerly Chris Burton and Don Manning) for their efficiency, tolerance and good humour. We thank Priscilla Barrett for her initiative and skill in blending the key features of the book into the cover design.

We are very grateful to Professor Richard Harrison, Head of the Department of Anatomy, for the support he has given us in so many ways during this decade of study. In addition to those mentioned previously, Mike and Pat Goh, Al and Liz Laursen in Malaysia, Dr Ivan Polunin and his wife Siew Yin in Singapore, and Professor Robert Hinde and Dr Michael Simpson in Cambridge, have been important friends and mentors to many of us.

To finish on a personal note, I am very grateful to Professors John Corner and John Napier, and Drs David Pilbeam and Alison Jolly, for the encouragement they gave me at the start of my studies, at a time when field work was not very respectable among academics. My wife Sarah and I remember with affection those with whom we shared the research laboratory at Universiti Malaya - Kiew, Tho, Pala and Ming Pow. We are indebted to the Earl and Countess of Cranbrook for the hospitality and friendship they extended to us during our initial studies, and I owe much to Gathorne for the precise guidance he has given then and since; it is most appropriate that he should introduce this volume, since, as a leading force in Malesian zoology and ecology, he has been a source of inspiration to us all. I am very grateful for the cooperation of the authors, throughout our association, which has culminated in this volume relatively painlessly; it has been good fun and very stimulating. I can never repay Sarah for the crucial and unselfish contribution she has made to so many aspects of my professional life; I am sure that there would be no material for this book without her, and I know how appreciative the other contributors are of the help she has given them.

Great Eversden D.J.C.

August 1980

INTRODUCTION

David J. Chivers

Sub-Department of Veterinary Anatomy
University of Cambridge

PRIMATE SOCIO-ECOLOGY

Background

Interest in Man's closest mammalian relatives - the primates - has escalated dramatically during the last twenty years. The layman has become as enthusiastic as the scientist, whether it be to marvel at the complexity and cleverness of their behaviour or to protest at abuses such as habitat destruction and trading evils. It is perhaps understandable from an anthropocentric viewpoint that no other group of animals has attracted such intense multi-disciplinary interest that numerous national and international societies have been established to promote their protection and study. This is an unique situation, in the academic world at least, where teaching and research are otherwise organised strictly according to scientific discipline.

During the nineteenth century the main interest was in their natural history and anatomy. It was not until the 1930s that intensive studies of primate behaviour began. The first phase was dominated by psychologists investigating mental processes, and thus they concentrated on man's closest relatives, the apes. Claims emanating from these times, that the sexual bond was the basic cohesive element of primate groups, whose structure was determined by social dominance, have had a stifling effect on studies of primate behaviour until surprisingly recently. C.R. Carpenter pioneered the field study of primates, observing howling monkeys and spider monkeys in Central America, and gibbons in Thailand; his observation and perception led to an uncanny understanding of primate society, and he often predicted accurately (as revealed by subsequent studies) the occurrence of behaviours that he did not himself observe. He was inspired, as were Nissen and Bingham in their searches for African apes, by Robert Yerkes, whose own important contribution was in the laboratory study of primates.

Further progress in defining the frameworks of primate society was delayed by the Second World War. It was not until the late 1950s that studies were resumed, mainly in Japan and by Americans in Africa. In the West at least, the way was led by anthropologists searching for models of human evolution. Much information was amassed on the ecology and behaviour of wild primates (DeVore, 1965; Altmann, 1967; Jay, 1968), with the surprising discovery that behaviour patterns were not species specific, that considerable variation in behaviour occurred within species.

Since behaviour could be determined only partly by genes, the next major step was to establish what role the environment played in shaping the observed behaviours, adaptive or non-adaptive. This brings us to the present phase of quantitative description and analysis, to which we hope to contribute. Under the influence of zoologists the exciting questions about the origin of man and his behaviour have had to be deferred until a more secure primate base is constructed. In recent years several stimulating reviews (Kummer, 1971, Jolly, 1972; Rowell, 1972; Michael and Crook, 1973; Hinde, 1974) have produced some explanations and many hypotheses, which are now being tested by our greatly refined techniques. These syntheses have led to increased efforts on ecological, sociological and physiological aspects of behaviour (e.g. Clutton-Brock, 1977a; Chivers and Herbert, 1978). The critical states which many primate populations in the wild are now reaching have been highlighted by Rainier and Bourne (1977).

Approaches

Among those with ecological interests, a main theme has been efforts to relate differences in social organisation to differences in the environment. Initially this involved social comparisons with gross habitat differences, such as forest and savannah, or gross dietary differences, such as frugivore or folivore (Crook and Gartlan, 1966; Eisenberg et al., 1972; Jolly, 1972), but success was limited and correlations were weak (Crook, 1970). There are two main explanations for these poor correlations (see Gittins, 1979). Firstly, species living together in the same habitat have to exploit different features of that habitat to coexist; there is no reason to suppose, therefore, that social organisation should respond in the same way to these different selection pressures. Secondly, social organisation is the outcome of a complex set of inter-related factors, and changes in one behaviour are likely to cause changes in others (Crook, 1970; Hinde, 1974); the interaction of these variables may differ between phylogenetic groups, and different combinations of traits may evolve in response to the same selection pressure.

Such problems led to a greater emphasis in identifying more precisely those environmental features affecting social structure (e.g. Clutton-Brock and Harvey, 1977). The most critical factor is now thought to be the abundance of food and its precise pattern of distribution at times of scarcity, and recent field work has concentrated on defining the feeding niche of different primate species, and their strategies of ranging, in relation to social organisation (e.g. Clutton-Brock, 1977a). This has been our approach over the last 10 years in the rich environment of the Malayan tropical rain forest (Plate I).

The simpler environment of the African savannah and woodland, the variety of primates and the better visibility for observation therein, and the closer proximity to the West resulted in the main research effort being in this continent, certainly during the 1960s. It was the very complexity of the Asian forests, and the mystery surrounding them, contrasting with an apparent simplicity of social organisation among the primates residing therein, which drew some of us to the East. Since most primates are arboreal, field workers in Africa also paid more attention to rain-forest species, once the obsession for models of human evolution declined (Chalmers, 1968; Aldrich-Blake, 1970; Gautier-Hion, 1978; Struhsaker, 1978; Gartlan et al., 1978).

Although more remote and complex, with difficult terrain, the forests of South-east Asia seemed to offer exciting opportunities to unravel some of the problems of primate socio-ecology, mainly because of the rich primate fauna - lorises, tarsiers, cercopithecine and colobine monkeys and lesser (and great) apes - occurring in the contrasting habitats of evergreen and deciduous tropical forests. Coordinated long-term studies seemed to offer the best prospects for unravelling the details of ecology and social behaviour of each species, and their interrelationships, and the aim of this volume is to collate the results of numerous field studies in West Malaysia over the last 10 years.

Our belief has been that the most productive results are those where individuals with different interests and training combine to produce long-term community studies. It has become clear that, as the sociologist is dependent on the ecologist for a full interpretation of his results, so is the ecologist dependent on the nutritionist and plant biochemist. Such findings, and their ramifications, are the basis of our second decade of study.

The most urgent pressure for understanding the workings of the tropical rain forests comes from the rapid destruction of such resources, as increasing proportions of the world's human populations damage the various ecosystems with which they used to live in harmony as they come to operate on a global, rather than ecosystem, scale (Dasmann, 1975). If species of mammal are to avoid extinction, and to survive in any numbers, the slowly emerging system of National Parks needs to be augmented by positive action to safeguard wildlife in areas of light disturbance by man. Both features of conservation - preservation and management - require a sound and comprehensive data-base on which to make decisions. It is our hope to rise above our personal academic interests, to contribute to this desperate task, for if we fail to do so we are unlikely to have subjects for future study, or possibly we shall not even survive to study! At last, the 'biosphere people' (Dasmann, 1975) are realising the horrific consequences of increased flooding, soil erosion and imbalance of atmospheric oxygen and carbon dioxide, following further clearance of the 'lungs', 'sponge' and 'filter' of this planet - the tropical rain-forests. We must make whatever contribution we can to resolving such problems, in the hope that it is not too late.

MALAYA

The Malay Peninsula is situated in the heart of Sundaland - the continental shelf of South-east Asia extending towards Australasia from the Asian mainland (fig. 1.1). Today it is known as Peninsular or West Malaysia, to distinguish it from the East Malaysian states of Sabah and Sarawak on the island of Borneo. Since it is the major part of a zoogeographical entity extending north to the Isthmus of Kra, we shall continue to refer to it as Malaya or the Malay Peninsula for convenience.

The Oriental region includes the mainland and islands of tropical Asia; its three main sub-divisions are India, China and Malesia (the continental shelf of South-east Asia, also known as Sundaland).

Sundaland

The present flora and fauna of continental Asia were well established by the Pliocene. It was only at the end of the Miocene, 15-10 million years ago, that the Sunda Shelf was formed, by an uplifting of land in the South China Sea to form islands, which subsequently became joined to the Asian mainland by the end of the Pliocene (van Heekeren, 1972). By this time all the main existing plant genera had spread into the region, dispersing rapidly through the main hill systems as semi-evergreen dipterocarp forest (Ashton, 1972). The mammals followed during the Pleistocene, first from the Indian sub-continent, and then from South China (Medway, 1972).

The Early Pleistocene was moister, with corresponding

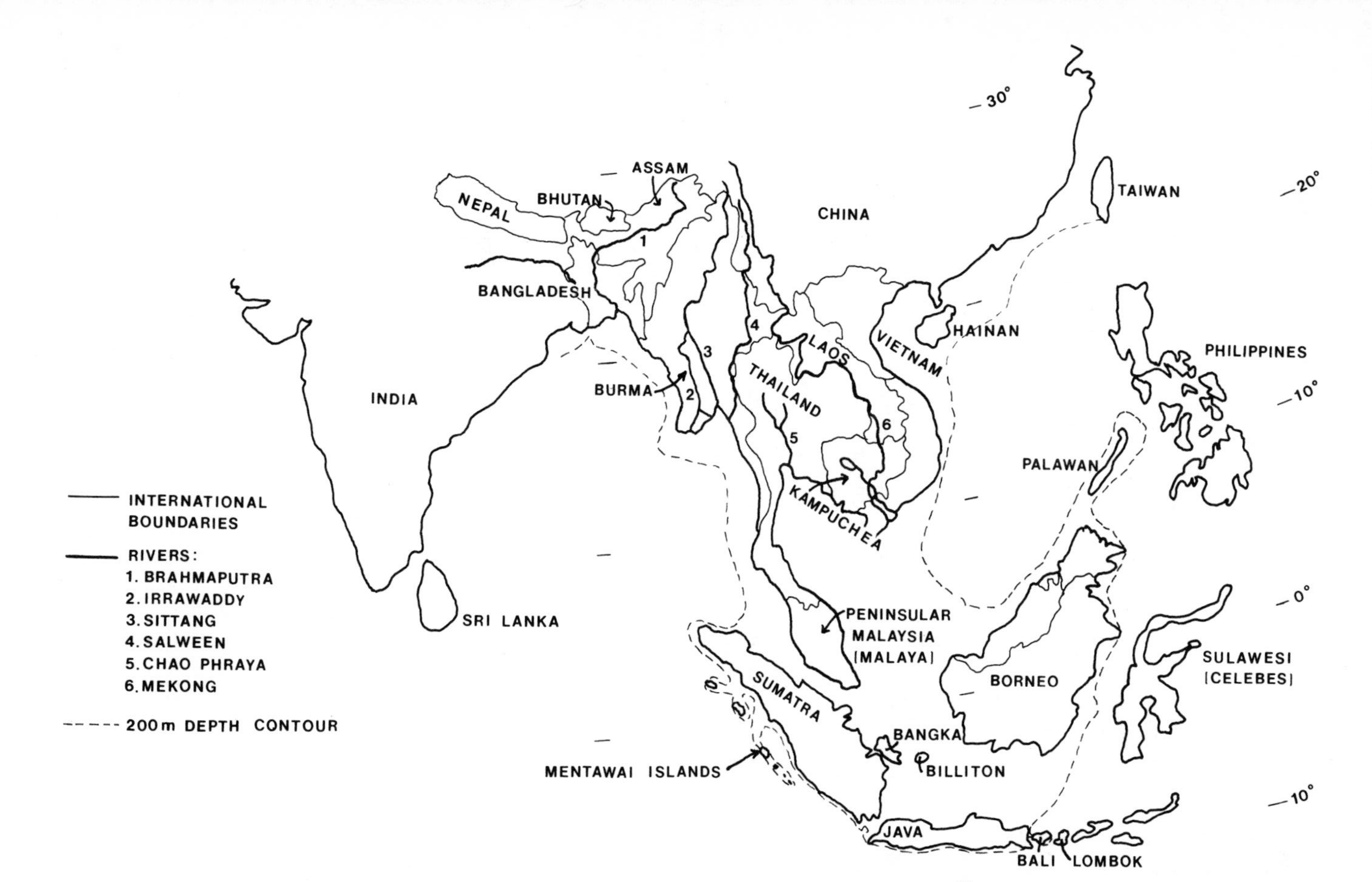

Fig. 1.1. Map of South and South-east Asia, indicating political boundaries, major rivers and the Sunda Shelf (from Payne, 1979b)

evolution of mixed dipterocarp forests. It was also a time of considerable volcanic activity, which produced further changes in land surfaces; these eruptions have continued to the present day. Large mammals, with affinities to the Villafranchian fauna of India and Burma, spread into Sundaland during the Early Pleistocene; subsequent migration of this Siva-Malayan fauna seems to have been blocked by the growth of tropical forest, and later arrivals are Sino-Malayan. The Middle Pleistocene was the driest period, with major glaciations in temperate regions and the largest depression of sea level. It was presumably the time when savannah flora and fauna, including man, spread over large areas of the exposed Sunda Shelf, reducing and isolating tracts of tropical forest. A major pluvial followed, resulting in an extensive spread of tropical forests and the advent of true forest animals into Sundaland, including all the primate genera.

Marked climatic fluctuations continued into the Upper Pleistocene, with alternate depressions and elevations of the sea by 200 feet or more, causing continual changes in land area, with floral zones shifting several hundred feet up and down mountains. The periods of isolation to which most areas were subjected produced new habitats and vegetation, and corresponding adaptations of the fauna. Thus it was this instability of the region which produced such an exciting diversification of plants and animals, so that nowhere else on this planet can we find better opportunities to study evolutionary processes in action, especially among the primates. For example, gibbon populations appear to have expanded when the opportunities arose and then diversified when isolated (fig. 1.2) to produce extant populations of varying status (see below; fig. 1.6).

Forests

The tropical rain-forests of the Far East, extending through the Malay Archipelago (Malesia) from Sumatra in the west to New Guinea in the east, are broken into western and eastern blocks by monsoon forests in the seasonally dry central part - west Phillipines, Sulawesi (Celebes), the Moluccas and Lesser Sunda Islands. These two blocks of evergreen rain-forest correspond to the Sunda and Sahul shelves (of South-east Asia and Australia respectively). The eastern block extends into north Australia and out into the Pacific with increasing floristic poverty, and the western block persists in regions with ever-wet climates; in addition to Malaya, Sumatra, Borneo and west Java, these include parts of the Isthmus of Kra, S.E. Thailand, Cambodia, S.W. Sri Lanka (Ceylon) and the Western Ghats of south India (Whitmore, 1975).

Five climatic climax forest formations have been recognised in the Malay Peninsula (Wyatt-Smith, 1953). The forest canopy up to about 1220 metres (4,000 feet) above sea level is dominated by trees of the Dipterocarpaceae; within this altitudinal range, there is

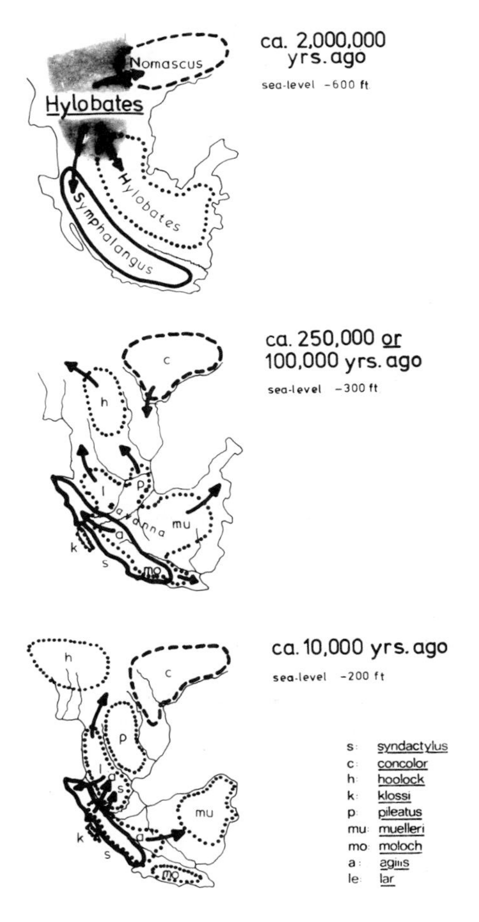

Fig. 1.2. Changes in the Sunda Shelf during the last two million years, indicating the possible spread of gibbon populations at times of low sea level (condensed from Chivers, 1977).

zonation into lowland, hill and upper dipterocarp forests, with the main transition zones at 305 m (1,000 ft) and 760 m (2,500 ft) respectively (fig. 1.3). The first two belong to the tropical lowland evergreen rain-forest formation (Whitmore, 1975), and the last, along with the oak-laurel forest up to 1524 m (5,000 ft), belongs to the world-wide formation of lower montane forest. Above 1524 m the upper montane forests are characterised as montane ericaceous, because of the predominance of species of Ericaceae, Coniferae and Myrtaceae. Such altitudinal zonation is only approximate and it varies from place to place, being more compressed on isolated massifs. Across the Straits of Melaka in Sumatra, the transition from lower to upper montane forest, for example, occurs around

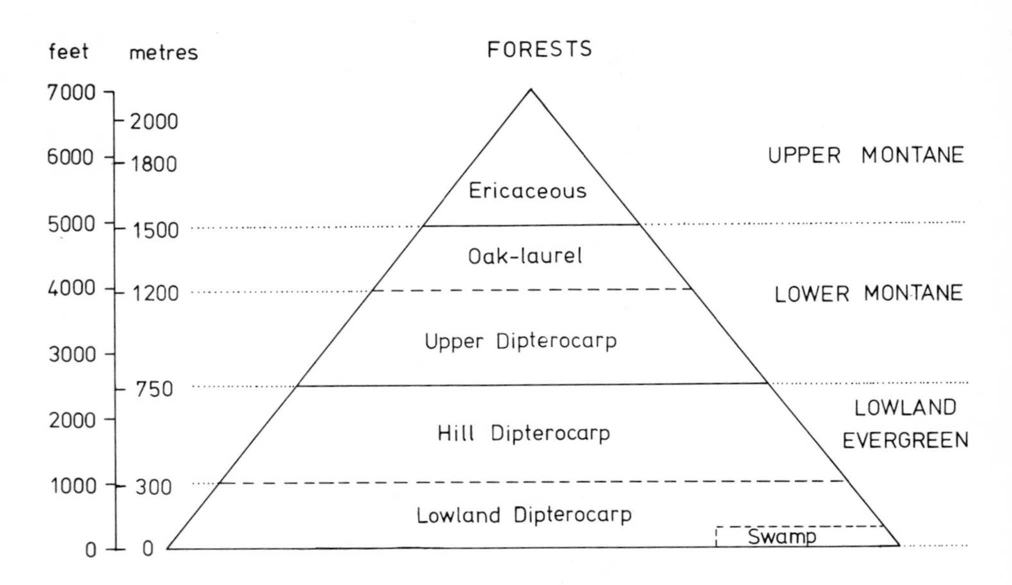

Fig. 1.3. A diagrammatic representation of altitudinal zonation in the rain-forests of the Malay Peninsula.

1830 m (6,000 ft) above sea level. The diversity of tree species and the mass of vegetation decrease markedly with altitude, and it is the lowland forests which provide the richest habitats for wildlife.

It is these dryland forest formations with which we are mainly concerned, but mention should be made of the wetland forests, particularly the peat swamp and mangrove forests of the west coast of Malaya, in which primates also occur.

Climate

The Malay Peninsula extends for 1300 km from the Isthmus of Kra in Thailand, with a maximum breadth of 500 km (fig. 1.4). Intermittent volcanic activity has thrown up several arcs of mountain ranges running much of the length of the peninsula. Conspicuous among these are the Main Range on the west and the Trengganu Highlands on the east, linked partially along the Pahang-Kelantan border by Gunung Tahan (2187 m, 7,174 ft, a.s.l.) in the Taman Negara (National Park) and its foothills. Situated as it is in the rain shadow of these ranges, the Krau Game Reserve, the main site of our studies, escapes the full force of the rain-laden north-east monsoon that is experienced first on the west coast and then on the east at the turn of the year, and the milder south-west monsoon that arrives on the west coast in April or May. Thus the annual rainfall may be less than 1800 mm (70 in) annually in this central region, compared

Fig. 1.4. Map of Peninsular Malaysia, showing states, major towns, roads and railways (Chivers, 1974).

with 5100 m (200 in) in Kedah and Perak on the west, and 3800 m (150 in) in the wetter parts of Kelantan and Trengganu (Dale, 1959).

The equable equatorial climate is modified, therefore, by maritime and monsoonal factors. Such seasonality as occurs is most marked on the east coast, but the prediction of wet and dry seasons is difficult, because of variations from year to year in their timing and intensity; in some years they may be impossible to identify. The driest and sunniest months are usually at the start and in the middle of the calendar year. Since rainfall may be more continuous but lighter in the wet season, compared with the torrential downpours that may occur at other times, this season may be identified more easily by persistent cloud cover and lower temperatures than by the amount of rain falling. The number of rain days each year ranges between 120 and 230 according to location and altitude (Dale, 1960). There is a daily temperature range of about $9\frac{1}{2}$ °C (15 °F), with a mean annual temperature of $29\frac{1}{2}$ °C (85 °F) in lowland Malaya, whereas above 1220 m a.s.l. it may be as low as $18\frac{1}{2}$ °C (65 °F). At Temerloh, the weather station of long standing nearest to the Krau Reserve, there was 5.8 hours of sunshine each day on average, with monthly means varying between 4 and 7 h/day according to the time of year (Dale, 1964).

One corollary of the small seasonal changes, compared with those of cooler climates, is that several years may pass without a critical period of shortage of resources. It is at such a time that competition between coexisting animal species becomes apparent and of consequence to their survival. This has special significance for studies of ecological segregation, which usually last for only two years at the most.

MALAYAN PRIMATES

Among the great variety of birds and mammals inhabiting the forests of Sundaland are four kinds of primate:- (1) nocturnal prosimians (strepsirhine and haplorhine); (2) the opportunistic cercopithecine macaques; (3) the leaf-eating colobine monkeys; and (4) the acrobatic lesser apes or gibbons, with the sluggish, "solitary" great ape, the orang-utan, in Borneo and Sumatra.

Today Peninsular Malaysia is inhabited by ten primate species (table 1.1). The nocturnal slow loris, *Nycticebus coucang*, has so far eluded systematic study. The large troops of long-tailed macaques, *Macaca fascicularis*, occur mainly in riverine forest and disturbed habitats, with those of pig-tailed macaques, *M. nemestrina*, in higher, more remote habitats. Of the three species of langur or leaf monkey, the silver leaf monkey, *Presbytis cristata*, is confined in Malaya to the mangrove forests along part of the west coast, and so is outside the scope of this volume.

Table 1.1. The primates of the Malay Peninsula.

Scientific name	Common name	Local names
Hylobates syndactylus	siamang	siamang; ungka
Hylobates lar	lar gibbon	wa-wa; wak-wak; mawa
Hylobates agilis	agile gibbon	ungka; wa-wa
Presbytis obscura	dusky leaf monkey	cengkong (chengkong); lotong berchelak
Presbytis melalophos	banded leaf monkey	ceneka (cheneka); kekah
Presbytis cristata	silvered leaf monkey	lutong (lotong); lotong kelabu
Macaca fascicularis (irus)	long-tailed macaque	kera
Macaca nemestrina	pig-tailed macaque	beruk (berok)
Macaca arctoides (speciosa)	stump-tailed macaque	berok kentoi

The dusky or spectacled leaf monkey, *P. obscura*, and the banded leaf monkey, *P. melalophos*, occur throughout the inland forest. There are three species of lesser ape: the largest, the siamang, *Hylobates syndactylus*, is based on the Main Range and confined by the Perak river in the north, the Pahang river in the east and the Muar river and Tasek Bera in the south; the lar gibbon, *H. lar*, occurs throughout the forests south of the Perak and Kelantan rivers (with the Thai sub-species to the north of the Mudah river on the west); between the Mudah and Perak rivers is found the agile gibbon, *H. agilis*.

The same genera are represented throughout Sundaland, and at lower densities, in the deciduous monsoon forests of Assam, Bangladesh, Burma, north Thailand, Laos, Kampuchea (Cambodia), Vietnam (Indo-China) and southern China. These forests are composed of two main formations, according to the extent and intensity of the dry season - tropical semi-evergreen and moist deciduous forests (Whitmore, 1975).

The two macaque species are distributed widely throughout South-east Asia (fig. 1.5); on the Asian mainland they are sympatric with other species, of which the stump-tailed macaque, *M. arctoides*, just reaches into the extreme north of Peninsular Malaysia (Medway, 1970). The silver leaf monkey occurs widely through the inland forests of Borneo and Sumatra, and the dusky leaf monkey extends north into Thailand; otherwise there are numerous "island"

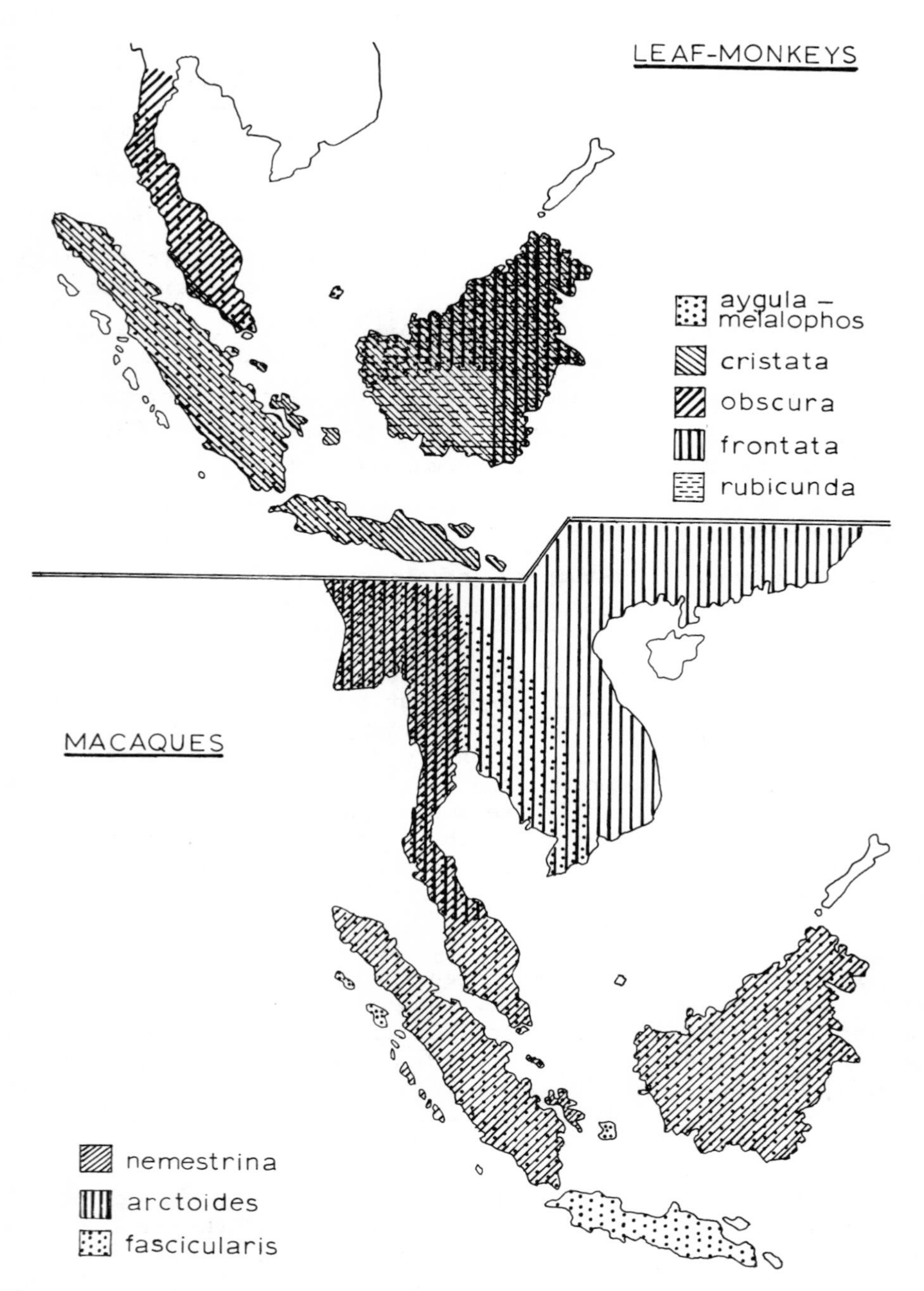

Fig. 1.5. Distribution of leaf monkeys and macaques in South-east Asia (from Medway, 1970).

Plate I. (a) tropical rainforest in Peninsular Malaysia (DJC)
(b) hill forest in Ulu Gombak in the Main Range (DJC)

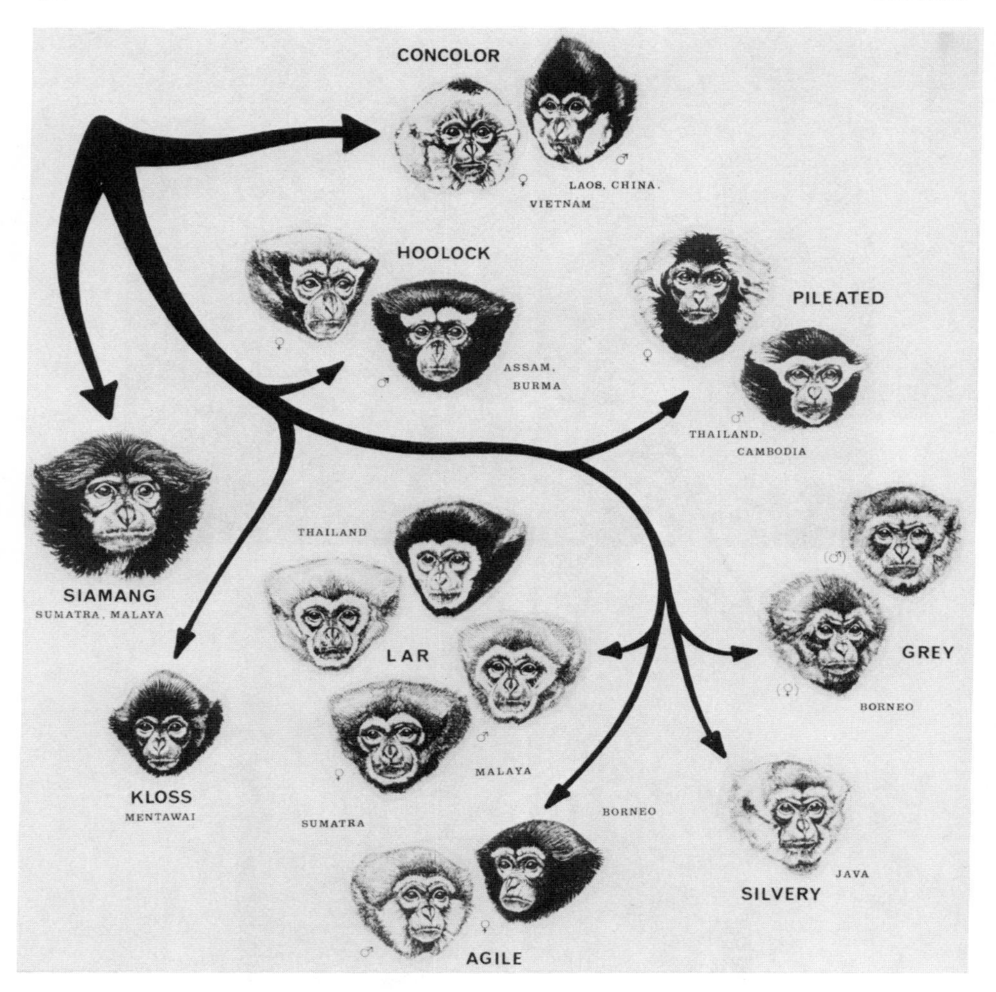

Fig. 1.6 (a)

Fig. 1.6. Distribution of gibbons in South-east Asia, illustrating affinities (drawings by Priscilla Barrett).

congeners, representing the great diversification undergone by the genus *Presbytis* (fig. 1.5; Medway, 1970). There are moves to return to the generic distinction advocated by Pocock (1934) between *Trachypithecus* (including *P. obscura* and *P. cristata*) and *Presbytis* (including *P. melalophos* and the many other *'aygula'* species)

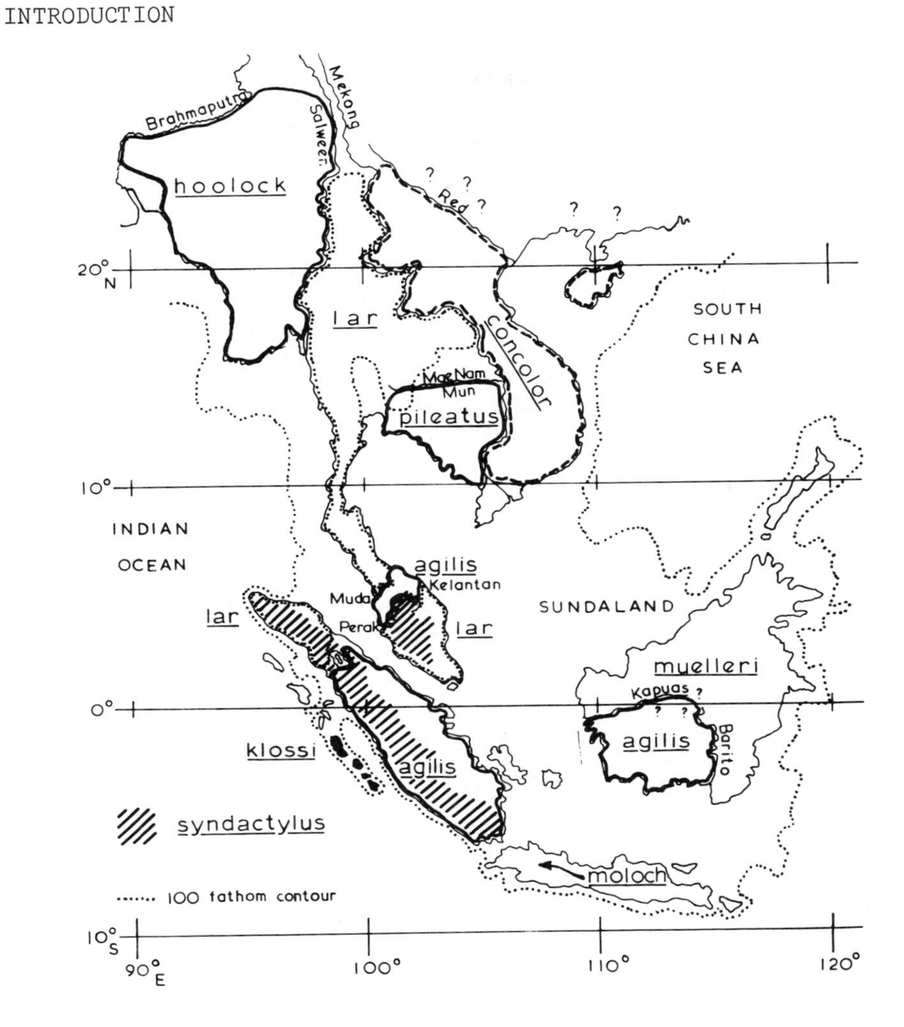

Fig. 1.6 (b)

(Brandon-Jones, in prep.). There are also the curious colobines in the north and east of the region, species of *Rhinopithecus*, *Pygathrix* and *Nasalis* (and *Simias* in the south-west), which show terrestrial adaptations and primitive colobine features. The siamang occurs also in the mountains of Sumatra, from where it seems to have originated (Hooijer, 1960; Frisch, 1971), and the lar gibbon occurs in the north of Sumatra, as far south as Lake Toba. Elsewhere in Sumatra, and in south-west Borneo (Marshall and Marshall,

1976), the agile gibbon is found. There are six other allopatric species of gibbons (fig. 1.6), of varying degrees of divergence between the siamang and lar/agile extremes, from Assam in the north-west, to Java and Borneo in the east and southern China in the north (Chivers, 1977).

Previous studies

Following Carpenter's pioneering study of the lar gibbon in Thailand (Carpenter, 1940), no further interest in South-east Asian primates was shown until the 1960s, when Ellefson (1968, 1974) studied the lar gibbon in Johore, Peninsular Malaysia. Other studies in the peninsula (table 1.2) included those on macaques in Selangor and lower Perak (Bernstein, 1967a, 1967b), and those on silver leaf monkeys at Kuala Selangor (Bernstein, 1968a; Furuya, 1961). Judy Ellefson (1967) and Chiang (1968) studied the long-tailed macaques in the Singapore Botanic Gardens, and Spencer (1975) observed them in the Waterfall Gardens, Penang, while Kawabe (1970) and Koyama (1971) studied siamang at Fraser's Hill in the Main Range, and Southwick and Cadigan (1972) surveyed the abundance of primates in a variety of habitats. Some of the most pertinent observations of forest primates were made by McClure (1964), during his study of bird migration across a pass in the Main Range, from a tree platform in Ulu Gombak, 22 miles from Kuala Lumpur. Kathy Wolf (in prep.) has recently completed a detailed behavioural study of the silver leaf monkeys at Kuala Selangor, including the incidence and consequences of infanticide following male take-overs of groups (Sugiyama, 1965; Hrdy, 1977).

Present studies

The series of studies that constitute this volume started twelve years ago in 1968 (Table 1.2). Following a country-wide survey of gibbon distribution and numbers, David and Sarah Chivers selected three areas for a detailed study of the ecology and behaviour of the siamang (Chivers, 1971, 1972, 1974; Chivers and Chivers, 1975) - the Kuala Lompat Post of the Krau Game Reserve in central Pahang, Ulu Sempam on the eastern slopes of the Main Range near Raub, West Pahang, and Ulu Gombak in Selangor (fig. 1.7); at least 10 days each month were spent in the first two areas, with just a few days of roadside observation in the rugged terrain of Ulu Gombak (Plates I and II). For a variety of reasons, both practical and academic, it was Kuala Lompat which became the key site in our studies. It is a relatively undisturbed site, which abuts a large tract of forest rising up to the summit of Gunung Benom.

The 17-month siamang study ended in mid-1970, and later in the year Sheila Curtin (1976a, b) started her 12-month study of the two leaf monkey species there. The following summers saw follow-up

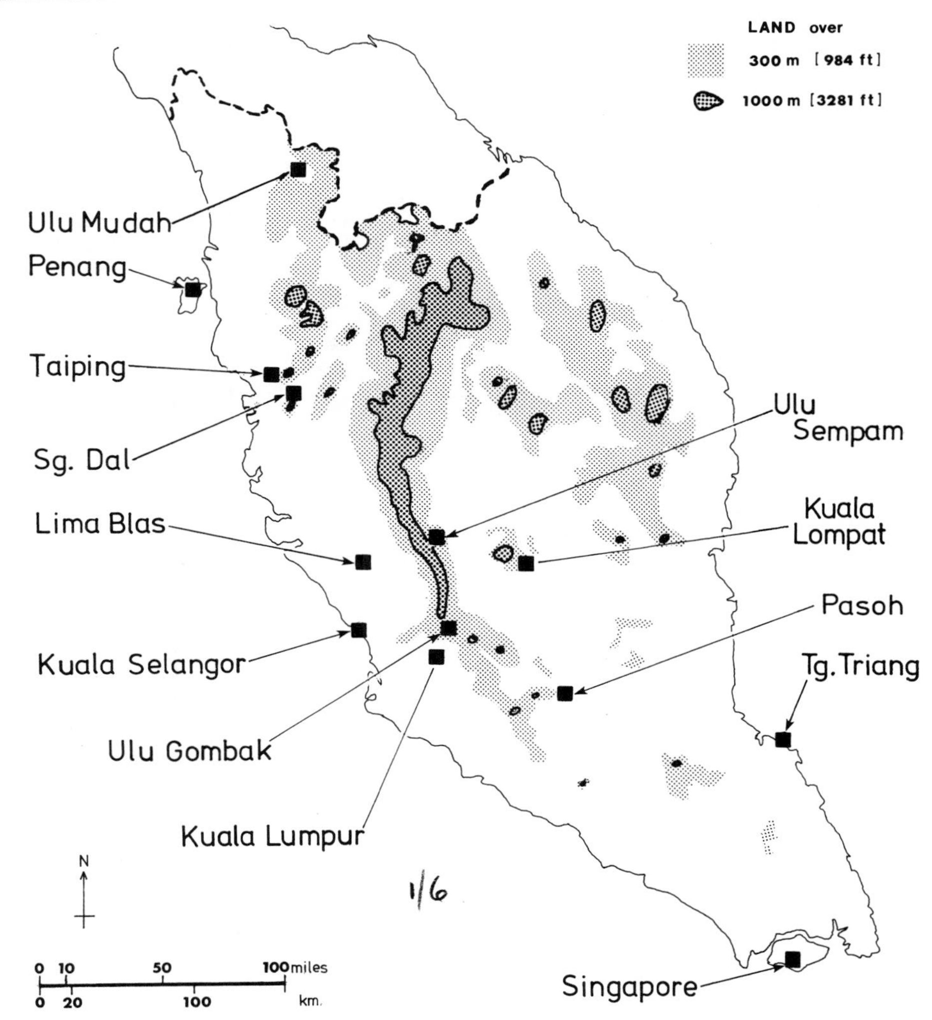

Fig. 1.7. Map of Peninsular Malaysia showing high land and the study sites referred to in this volume.

studies on the siamang, and efforts to habituate the lar gibbons, in 1971 by Pelham Aldrich-Blake (Aldrich-Blake and Chivers, 1973) and in 1972 by Chivers and others (Chivers, Raemaekers and Aldrich-Blake, 1975). John and Kathy MacKinnon spent six months in 1973 assessing comparative habitat use, in terms of feeding and ranging, among the gibbons, leaf monkeys and macaques, with special interest in some aspects of gibbon behaviour (MacKinnon, 1977; MacKinnon and MacKinnon, 1977; Chivers and MacKinnon, 1977). John Fleagle (1974, 1976a, b and c, 1977, 1978) also studied comparative habitat use, in 1972, 1973 and 1974-75, but with special reference to the

Table 1.2. Studies of Malayan forest primates.

Hylobates syndactylus	*H. lar*	*H. agilis*	*Presbytis melalophos*	*P. obscura*	*P. cristata*	*Macaca fascicularis*	*M. nemestrina*
McClure (1964)			Curtin (1976a,1976b)		Furuya (1961)	Ellefson (1967)	Caldecott (in prog.)
Kawabe (1970)	Ellefson (1968,1974)		Fleagle (1976b,1977,1978)		Bernstein (1968a)	Bernstein (1967a,1967b,1968b)	
Koyama (1971)					Wolf (in prep.)	Chiang (1968)	
Chivers (1974)						Spencer (1975)	
Southwick and Cadigan (1972)							
Aldrich-Blake and Chivers (1973)	Vellayan (in prog.)					Aldrich-Blake (this vol.)	
Fleagle (1974, 1976c)							
MacKinnon and MacKinnon (1978)							
Raemaekers (1977)		Gittins (1979)					
Chivers and Davies (1979)							
Bennett et al. (in progress)			Bennett (in prog.)			Mah (in prep.)	
Wilson and Marsh (in progress); Johns (in progress)							

size and type of substrate in relation to postural and locomotor behaviour and anatomy.

In 1974 Pelham Aldrich-Blake started the study of long-tailed macaques, which was continued from 1976 by Mah Yoon Loong (in prep.); in the same year Raemaekers (1977, 1978a, 1978b, 1979) started a two-year comparison of the details of feeding and ranging in the siamang and lar gibbon. In 1977 interest spread to animals that appeared to compete with primates for food, especially fruit, with John Payne's (1979b) two-year study of tree-squirrels, extending considerably the preliminary observations made in 1973 by Kathy MacKinnon. He continued the phenological records of trees, started by Aldrich-Blake and Raemaekers. It still remains to observe more thoroughly birds, such as hornbills and pigeons, although some reference will be made to them herein. David Chivers continued observations of siamang (and lar gibbons) from 1977 to 1979, mostly through the services of the local assistant Kalang s/o Tot, but in 1977 he organised a survey of primates and other wildlife throughout the Krau Reserve (Chivers and Davies, 1979; Payne, 1979a; Caldecott, 1980).

Elsewhere in the Peninsula, studies have been limited to those by Paul Gittins (1978, 1979) of the agile gibbon in Sungai Dal, Gunung Bubu (Perak), and at its border with the lar gibbon in Ulu Mudah (Kedah), where some hybridisation was observed to occur. John Fleagle spent much of his time collecting comparative data from the Lima Blas Estate on the Selangor-Perak border.

SOCIO-ECOLOGY OF MALAYAN FOREST PRIMATES

Approach

The contrasts in (a) social structure, from monogamous pairs to multi-male groups, (b) in ranging, from highly territorial to highly tolerant of intrusion by neighbours, and (c) in feeding, from mostly frugivorous to mostly folivorous, among Malayan primates provides exciting opportunities, that had hitherto been neglected, to study these different adaptations to a similar environment, and to gain an appreciation of inter-relationships within this primate community.

To this end this volume is divided into four sections. This first introductory one is concluded with a description of the Malayan rain forest and its phenology, with special reference to the Krau Game Reserve, by Raemaekers, Aldrich-Blake and Payne (Chapter 2). The second section contains qualitative and quantitative descriptions of the ecology and behaviour of gibbons, by Gittins and Raemaekers (Chapter 3), dusky and banded leaf monkeys, by Curtin (Chapter 4), and long-tailed macaques by Aldrich-Blake

(Chapter 5), based wholly or partly on observations at Kuala Lompat, in an effort to portray the natural history of each of those six species.

We then move on to a variety of comparisons. The MacKinnons combine spot samples and prolonged (dawn-to-dusk) observations of all higher primate species at Kuala Lompat during a short intensive study focused mainly on ranging and feeding behaviour (Chapter 6). Fleagle assesses positional behaviour - locomotion and posture - in relation to habitat use in the various species (Chapter 7). Chivers and Raemaekers present an analysis of changes in behaviour over time, based on about 500 days of siamang observations, spread through 70 months over 10 years, with briefer comment on the other species (Chapter 8). Payne discusses relationships between primates and other animals inhabiting the forest canopy (Chapter 9).

These varying approaches allow a discussion (Chapter 10) of spot samples and prolonged observation, of single-species studies and inter-specific comparisons, of the contribution of functional anatomy to understanding ecology and behaviour, of feeding and ranging behaviour in relation to forest productivity, and of climatic and other long-term effects on behaviour. This collation of results aims to clarify methods and findings, and set these studies in perspective with regard to South-east Asia and to current ideas in behavioural ecology. Finally, we assess the survival needs of each species and suggest directions for further research, with conservation of the tropical rain-forest as the primary concern, and the advancement of primate biology and mammalian socio-ecology (and behavioural biology) as secondary concerns, although such a distinction should not mislead the reader as to the close relationship between these needs.

Methods

Given the financial and logistical problems of field study, and the multitude of variables encountered when observing Nature's experiments, in contrast to the controls that can be achieved in the laboratory, field methods require a broad approach, with a wide range of parameters being sampled over a long time. Difficulties in locating animals and the variability of the environment make short-term sampling (only a few hours each day at different times) impractical, and introduce problems of bias that are usually insuperable within the two-year periods over which individual studies are usually conducted. We have sought to avoid, or modify, techniques devised for tackling specific problems in the laboratory; the field approach has usually to be more general. Implicit in our approach has been the belief that no amount of slaving over a hot computer can remedy deficiencies in the data resulting from inadequate time and effort in the forest.

Plate II. (a) the Krau Game Reserve from a hill just south of the Kuala Lompat Ranger Post looking north-west over lowland forest towards the massif of Gunung Benom in cloud (JBP)

(b) the hill forest of Ulu Sempam in the Main Range (DJC)

The very existence of this book reflects our realisation that the most productive studies to resolve the critical needs of man and primate require the collaborative efforts of several people in the same place over a long time. Furthermore, we recognise the need for balance between studies of social and maintenance behaviours, the environment and digestive, reproductive and other aspects of the individual's biology. Deficiencies in this respect within the present volume are being remedied in the current programme of study (table 1.2), as these seem to present the only hope for a full understanding of the behaviours observed.

Details of methods are given in each chapter, where they differ from our efforts at standardisation to allow comparisons between observers and interpretation over time. Although specific problems require different methods, the initial studies of each species required a quantification of individual behaviour and the environment so as to provide answers to a wide range of behavioural and ecological questions. Thus each observer sought to record the behaviour of his subjects from dawn to dusk on at least five consecutive days each month (Curtin for 2-3 days) for at least one, preferably two, years, combining time-interval sampling with continuous notes as follows:

(1) Daily activity - this is defined as the active or alert period between leaving one night tree and entering the next. Samples of one or more individuals are taken every 5 or 10 min in each hour to deduce patterns of the major activities of rest, feed, travel and call. The frequency and duration of sub-categories, such as play during rest or travel, grooming during rest, movement or immobility during travel or feed, and postures or other details are also recorded. Such scores can be totalled to produce activity budgets for each age and sex class of animal; samples at ½- or 1-min are necessary to quantify the details of rest, feed and travel bouts (and even more frequent for travel).

(2) Feeding - daily feeding time can be divided into bouts of feeding, and within that, the numbers of food sources visited, and the kind of food consumed. In terms of availability and digestibility (structure and composition) these foods are most usefully divided into the vegetative (leaves, stems, shoots, bark, etc.) and reproductive (fruits - seeds and pulp - and flowers) parts of plants, and animal matter (mostly termites, spiders and caterpillars). Time spent feeding on each of these categories (and subcategories) is expressed as number of visits and time each day and as a percentage of the active period, with analyses in terms of the dietary contributions of at least the more important tree species, and of hourly, daily and monthly patterns.

(3) Ranging - measured by (a) the distance actually travelled by one or more individuals or the group, and (b) the direct distance

between successive sleeping positions. These can be analysed in terms of the number of metres covered or the total number, and the number of different, hectare quadrats entered each day, month and so on. Such data permit examination for the stability of limits of ranging and of differential range use, in relation to concepts of home range, core area and territory. Samples of height above the ground at regular intervals allow consideration of preferences for particular levels of the canopy.

(4) Social organisation - the size and composition of the social group, with details of maturation and the identification of different age classes; records of the nature, sequence and quantity of interactions between members of the group, with data on spacing, to allow an assessment of the degree of group cohesion and the coordination of activities (the main interactions involving grooming, play, sex and aggression, as well as overt vocal and visual signals); the details of the spacing of, and interactions between, groups (particularly the timing and frequency of loud calls).

(5) Environment - daily measures of rainfall and temperature, and samples taken every 10 or 15 min of sunshine, cloud cover and rainfall; descriptions of the structure and plant species composition of the habitat, with quantification of the availability of potential leaf, fruit and flower foods on a monthly basis; and details of interaction with other animals in the habitat.

While this has been a convenient sequence in which to describe the main parameters recorded, the reverse order seems more appropriate for presenting the main results of our studies (Chapters 2-5), both to give the reader a feel for each primate species, and to lead more logically into the discussion. The validity of 5-day periods for assessing behaviour over one month (should it be necessary to extrapolate beyond the observation period) was justified by Chivers, Raemaekers and Aldrich-Blake (1975).

Habituation

Information could not be collected systematically on the daily round of a primate social group, until it was fully accustomed to the continuous presence of an observer. Not only does habituation depend on the animals realising that there is nothing to fear from the observer, but the observer has to become thoroughly familiar with the tract of forest in which the study group lives and with some of its habits. It is the ability of the observer to place himself in advance of the travelling group, particularly if it can be close to an important food source, which does most to accelerate the process of habituation.

The time taken to achieve continuous observation from dawn to dusk depends on the species, the habitat and the observer. The observer has to be highly motivated and to show diligent (almost fanatical) and skilful application. The habitat should ideally be of flat terrain with primary forest, in which visibility is better, and a good trail system; dry weather is much better, but rain at night will keep the forest floor damp and allow the observer to move swiftly and quietly. Some features of primate behaviour are helpful, others are unhelpful; it is the net effect which produces differences between species in the ease of habituation. The dramatic vocal and locomotor behaviour of gibbons, and their regular habits about a small territory, are partly offset by their small group size and speed or stealth of movement. The large groups of leaf monkeys and macaques are perhaps more difficult to habituate, because of their less predictable habits, their greater stealth and their tendency to disperse, especially in the presence of an observer; while it may be easier to locate individuals, even in secondary growth, individual recognition is very difficult.

Gibbons seem to be less fearful than monkeys, and most of the behavioural descriptions presented herein are of habituated animals, but ease of habituation is very variable, mainly for environmental reasons. For example, Chivers (1974) habituated a group of siamang at Kuala Lompat after 34 days, 294 hours, in the forest, whereas in the hilly terrain of Ulu Sempam, and with the advent of the rains at a crucial stage, it took 139 days, 1245 hours. The lar gibbons at Kuala Lompat were habituated by Raemaekers (1977) almost as easily as the siamang there had been, but the agile gibbons at Sungai Dal posed Gittins (1979) the same problems as the siamang at Ulu Sempam. The value of the considerable amounts of data that can be collected during habituation are greatly enhanced by comparison with those collected subsequently.

The leaf monkeys were much more difficult. Whereas sustained observations of gibbons resulted from open "pursuit", Curtin (1976a), having tried all approaches, achieved best results over 12 months by stealth - by stalking and observation from hiding. While dusky leaf monkeys became very tolerant of her presence, they were not really habituated; banded leaf monkeys were even more difficult. Aldrich-Blake (pers. comm.) habituated a group of long-tailed macaques over 3 months, by following quietly and openly at a distance that did not alarm them, until he was tolerated at distances of about 15 metres (50 feet).

Study areas

The search for study areas in 1968 (Chivers, 1974) yielded several possibilities according to the desired criteria of accessibility, accommodation, ease of movement and large primate populations. The flat terrain of the lowland forest near the Kuala Lompat Ranger Post of the Krau Game Reserve in central Pahang appeared best in all respects (Plates III and IV), and the choice

Plate III. (a) aerial view of the Kuala Lompat Ranger Post in the Krau Game Reserve in 1969 (DJC)

(b) the rest house at the Kuala Lompat Ranger Post in 1970 (DJC)

(c) the old reservoir-keeper's hut 800 feet above sea-level at Sungai Dal, restored by Paul Gittins during his study of agile gibbons (DJC)

Plate IV. (a) banded leaf monkeys settling for the night in the petai tree by the Kuala Lompat Ranger Post (JBP)

(b) the forest edge, showing the variety of plant forms, just south of the Kuala Lompat Ranger Post (JBP) to make this the main site of all primate studies has been amply justified, except in the cases of the pig-tailed macaque and slow loris. Since siamang are most abundant in the Main Range, the Gombak valley and Fraser's Hill offered as good prospects as one could hope to find in mountainous terrain (fig. 1.7). The terrain in Ulu Gombak was just too rugged (Plate I), however, as suggested by Ellefson (1974), and, although excellent discontinuous observations of all species have been made from the roadside, the choice of Ulu Sempam, on the northern slopes of Fraser's Hill, as the hill forest study area proved to be a good one. Its value was enhanced by the opportunity to study primates in forest that was carefully logged selectively, but increased disturbance and the local security situation have prevented further work there.

The area around Maxwell's Hill in Perak seemed to offer the best prospects for a study of agile gibbons, and Gittins (1979) selected Sungai Dal in the Gunung Bubu Forest Reserve to the south for his main study (Plate III). This 400-ha block of lowland and hill dipterocarp forest, extending from 150 to 875 m (492-2871 ft) above sea level, is being left as a standard with which to compare the surrounding forest that is regenerating after logging. It contains an abundance of primates in its rugged terrain and merits more continuous attention. The other exciting area is Ulu Mudah, in Kedah, but close to the border with Thailand, where lar and agile gibbons meet across the Mudah river. It was possible to work on the shores of the lake created by the construction of a dam for irrigation purposes, and Gittins (1979) did so for as long as the security situation permitted. He worked on a flat, swampy peninsula on the south side of the lake at 105 m (344 ft) above sea level, which had been extensively logged so that the original structure of the forest canopy had been destroyed.

While these areas selected for studies of gibbons have proved, to varying extent, suitable for studies of other primates, success is being achieved with leaf monkeys and macaques in the Lima Blas Estate on the Perak-Selangor border, where adaptations to forest patches among plantations can be documented, and in the Pasoh Forest Reserve in Negri Sembilan, lowland forest at the southern end of the Main Range (fig. 1.7). Sites such as Lima Blas Estate (Fleagle, this volume; Bernstein, 1967a, 1967b), as well as parks in Singapore (Ellefson, 1967; Chiang, 1968), Kuala Lumpur (Bernstein, 1968b), Penang (Spencer, 1975), Taiping and Kuala Selangor (Bernstein, 1968a; Wolf, in prep.) permit closer observation of the details of social and positional behaviour. Nevertheless, this volume is concerned mainly with the adaptation of primates to the forest proper, hence the emphasis on Kuala Lompat (described in Chapter 2), supplemented by work in Ulu Sempam, Ulu Gombak and Sungai Dal. Such studies will hopefully provide the basis for explaining the adaptations of monkeys to disturbed habitats - from selectively logged forest through farmland to urban areas - behaviour that is now being fully documented.

J.J. RAEMAEKERS

THE FOREST

J.J. Raemaekers, F.P.G. Aldrich-Blake and J.B. Payne

Department of Physical Anthropology, University of Cambridge; Department of Psychology, University of Bristol; Sub-Department of Veterinary Anatomy, University of Cambridge

INTRODUCTION

The visitor to the tropical rain-forest is impressed first by its many dimensions: there seems to be so much more going on above ground than there is in a temperate forest. Other features soon impress one - the majestic scale (especially in lowland forest), the tangle of climbing plants festooning the trees, the density and variety of animal life, the variety of sweet-rotting smells, and, above all, the paradox between the baffling diversity of plants and the sense of monotony which the prevailing green imposes on the whole, relieved only here and there by a flush of pink new leaves or pale flowers. Set against this backdrop is an incessant noise of birds and other animals and the discordant buzz of insects.

The forests of the Malay peninsula contain some 2,500 species of trees, 30% of an estimated 7,900 seed plant species present (Whitmore, 1975). In contrast, the British Isles harbour only 35 native tree species, 2½% of 1,430 seed plant species, in over twice the area (Mitchell, 1974). The lowland forests of Malesia are richer in species than any other forests in the world, including the tropical forests of Africa and America (Whitmore, 1975: 6). Stands of single species are rare, excepting for a few species in seasonal swamps, on ridges, along river banks or in disturbed areas. It is not at all obvious why tropical rain-forests should be more diverse than forests elsewhere, nor why the Malesian tropical rain-forests are more diverse than the American or African ones. One theory explaining diversity in tropical rain-forest holds that it is promoted by the selective pressure exerted by host-specific insect pests, with trees speciating to escape pests and pests in turn speciating to catch up with the trees (Janzen, 1970). It has also been suggested that genetic drift might augment this process of speciation, with individuals of already rare species being so isolated from each other that they seldom interbreed, so that drift then creates differences which actually prevent interbreeding (Federov, 1966). It might even be that there is evolutionary advantage to be gained from rarity (Flenley, 1979: 128); for example, to escape insect pests by lowering the chances of being located. Any viable theory, however, faces the problem that the processes it invokes must apply more forcefully in tropical than in non-tropical environments.

The floristic diversity of the lowland tropical rain-forest is matched by structural grandeur and complexity. Emergent trees can reach 80 m (260 ft), exceeded only by some Australian eucalypts and Californian redwoods. A vertical section through the forest canopy shows tree trunks to be very slender in proportion to their heights, by comparison with temperate broad-leaved trees. The taller specimens often have boles clear to 25 m (80 ft) before throwing out spreading crowns; shorter trees, in the shade of these tall ones, grow narrower and more cylindrical crowns (fig. 2.1). The trees are often draped with woody climbers, soft-stemmed creepers, and stranglers which begin life as epiphytes and end up encircling and often killing their hosts. Numerous other sorts of epiphytes, such as orchids, occupy wide forks and branches. Large climbers grow thick stems, and may be as old as the trees which support them. The undergrowth consists mostly of seedlings, saplings, young climbers, palms (often spiny) and gingers. It is generally easy enough to walk through, except where plants have found full sunlight in a gap recently created by a tree fall, and scramble to overtop one another.

These gaps provide the only chance most trees and climbers ever get to grow beyond the seedling stage, for less than one percent of the sunlight striking the top of the forest canopy penetrates to the forest floor (Yoda, 1974). Fall gaps are the forest's means of regenerating itself; fresh ones are colonised by light-loving plants specialised for rapid growth and reproduction and the wide dispersal of their seeds. Among these plants there grow up others, which can tolerate shade in the early stages and which eventually replace the colonisers.

Climate within the forest varies greatly with position (Richards, 1952). Near the floor it is constantly warm and humid. Higher up, exposed canopy is drier and hotter during the day, because it receives direct sunlight, but cooler at night. Rain falls mostly in the evenings in brief, often torrential storms, heralded by a passing rush of wind. Even at sea level, the rain cools the loitering human to shivering point. In the hills, cold, damp mists persist until late into the mornings, shrouding the trees from view.

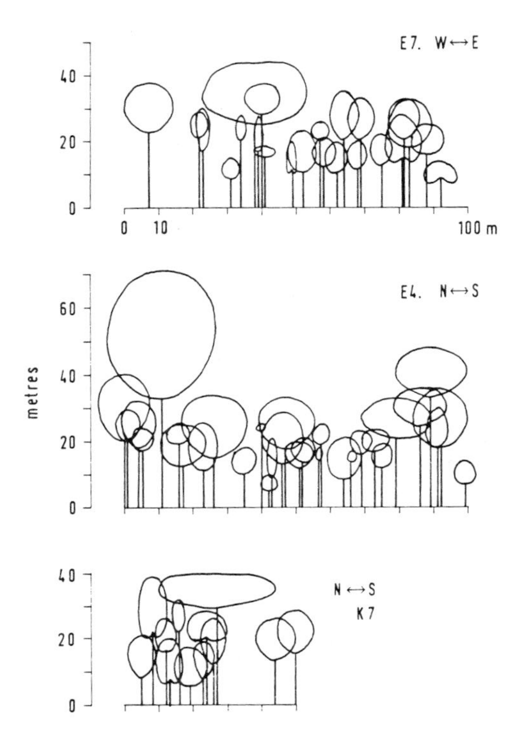

Fig. 2.1. Canopy profiles of strips 10 m deep from 3 plots in virgin forest. Drawn from location of trunk base, height to bottom of tree crown, height to top of same and maximum diameter in the plane above. The crown is assumed to lie directly above the base.

The great stature of the forest, the variety of its structural environments, the great diversity of plant species and the warm, rather stable climate are features which profoundly affect the fauna. Diversity of structure and plant species diversity, separately and in combination, provide a large number of potential niches into which animals have radiated, themselves creating new ones for other animals as they do so. The sheer volume of canopy also makes for high primary (plant) production and hence high secondary (animal) production. Annual dry leaf litter production varies within the narrow limits of 6 to 9 tonnes/ha in all lowland tropical rain-forests investigated in Asia, Africa and America (7.5 tonnes/ha at Pasoh, Malaya). In contrast, temperate forests produce only 1.6 - 3.4 tonnes/ha (Leigh and Smythe, 1978, table 2; A. Hladik, 1978, table 2).

The animal biomasses of tropical rain-forests are correspondingly high. Primates, the commonest large mammals in most such forests, reach live-weight biomasses of 5 - 30 kg/ha (Hladik and Chivers, 1978). Total animal biomass figures do not exist.

Eisenberg and Thorington (1973) calculate 44 kg/ha live weight for all non-volant (i.e. not flying or gliding) mammals on Barro Colorado Island, Panama. Insects are the main contributors to animal biomass and can reach prodigious abundance. Abe (1975) gives a figure of 109 kg/ha live weight for termites alone in Pasoh Forest, Malaya, representing 3,225 termites/m^2. A. Hladik (1978) records a dry-weight production of macroscopic insects in the leaf litter of Ipassa Forest, Gabon, of 23 kg/ha/yr. Total live-weight animal biomasses in tropical rain-forests should therefore be in the order of 150 - 200 kg/ha, as high as some of the exceptionally productive East African savannahs with their huge ungulate populations. These ungulates alone may reach 120 kg/ha without overgrazing (Bourlière and Hadley, 1970, table 2).

The other factor contributing to high animal biomass is that, due to high insolation and humidity, plant growth can continue all year round, with rapid recycling of nutrients ensured by rapid decomposition. Temperature varies little, and the diurnal range exceeds the annual range of monthly mean maxima and minima. Rainfall generally exceeds 1500 mm (59 inches) annually and may exceed 7500 mm (295 inches), and dry periods seldom last longer than three months. Nevertheless, while the climate over tropical rain-forest is more constant than elsewhere, it does exhibit some seasonality in nearly all localities. In the Malay peninsula this results from the alternation of north-east and south-west monsoon winds bringing rain from the South China Sea and the Indian Ocean respectively (ch. 1). Water stress is probably the main determinant of when leaves are shed and renewed and of the timing of flowering (Wycherley, 1973; Whitmore, 1975), although it has been suggested that sunshine plays a major role in the latter (Ng, 1977). Even Singapore, which lies almost on the equator, suffers periodical water deficits (Nieuwolt, 1965). Studies accumulating from the three main blocks of tropical rain-forest in South-east Asia, Africa and Central and South America confirm that seasonal plant activity is the rule (Montgomery, 1978). For example, in Malaya a sample of upper-canopy trees in hill forest on the slopes above Kuala Lumpur showed clear seasonality in leaf and flower cycles over a number of years (Medway, 1972), and an equally clear, though quite different, rhythm was found on the coastal plain below (Putz, pers. comm.).

Flowering and fruiting cycles are not determined by climate alone, however. The lack of synchrony between species, and to a lesser extent within them, may be due in part to a rather even spread of climatic cues, but it is also promoted by interactions with the animal community exerting selective pressures on its plant foods, as will be discussed in detail in the next section.

In summary, it may be said that Malayan forest primates inhabit a rich plant habitat and experience an equable climate; they must, however, share the forest with a wide variety of closely and distantly related animal competitors. In these circumstances it seems probable that most of the selective forces of significance are exerted by one organism on another, rather than by the elements on the organism.

THE ROLE OF ANIMALS IN FRUIT PRODUCTION

The best-studied interaction between plants and vertebrate animals concerns the effect of the latter on the morphology of fruit (Plate V) and the timing of their production. Because this particular interaction is of central interest to the studies reported in the following chapters, it is discussed in some detail below.

In evolutionary terms, the most significant components of the reproductive parts of plants - flowers and fruit - are the seeds. The dispersal of seeds is considered to be of special importance in the tropics, since concentrations of seeds are likely to attract, and perhaps favour the increase of, animals that will destroy them. If measures to reduce seed destruction do not evolve and continue to evolve as seed destroyers learn to overcome them, the plants will cease to produce offspring. Dispersal is seen, therefore, as a way of reducing seed destruction, which in the widest sense includes the actions of micro-organisms and of insects as well as of higher vertebrates. Seed predation is believed to be a major factor maintaining plant diversity in tropical rain-forest (Janzen, 1971).

The term "frugivore" can apply to those animals which eat or otherwise destroy the seed, and to those which eat the pulp or rind and excrete or spit out the seed, thereby acting as seed dispersers for trees and epiphytic plants. Seed dispersal by animals is particularly common in tropical rain-forest. In a wet forest of Costa Rica over 90% of species produce fleshy fruit (Frankie et al., 1974). At least 67% of tree species in lowland Malayan forest are dispersed by pulp-eaters (at least 73% at Kuala Lompat). The importance of the distinction between seed dispersers and seed destroyers becomes evident when we consider some examples of seed protection mechanisms displayed by Malayan rain-forest plants:

(1) Attractive pulp

Nutritive pulp around a seed attracts animals which disperse seeds that might otherwise be taken by seed destroyers. There are fruit which seem to have evolved in conjunction with birds: since birds have no teeth or prehensile digits, the fruit have a thin rind, or, if the rind is thick, split when they open (i.e. when the seed is mature). They are often brightly coloured to attract attention, externally if the rind is thin, internally if the fruit splits open.

Fruit which appear to have evolved in conjunction with mammals, such as primates, fruit bats, civets and large terrestrial species, have a tougher rind. They tend to be brightly coloured if taken mostly by diurnal mammals, but dull coloured with a distinct odour if taken largely by nocturnal species.

"Bird fruits" may be classified as:

(a) "Unspecialised" - mostly small, yielding little nutritive value from each fruit in return for dispersal. Larger ones such as *Ficus* spp., *Eugenia* spp., *Maranthes corymbosum*, bear a thick pulp and are taken by mammals in addition to birds.

(b) "Non-pulpy" - with small, abundant seeds, eaten by birds, which digest most but excrete some unharmed.

(c) "Specialised" - nutritious, rather hard or waxy pulp, taken by fruit specialists, notably hornbills, pigeons and barbets. They belong mostly to the families Lauraceae, Meliaceae, Myristicaceae and Burseraceae.

The best strategy for a plant bearing fruit attractive to pulp-eating seed dispersers is to produce them at a different time from other plants bearing similar fruit. This phenomenon has been documented for the genus *Miconia* in Trinidad (Snow, 1965) and for strangling figs, genus *Ficus*, in Malaya (Medway, 1972). Evidence from Kuala Lompat during 1974-77, in particular, suggests that a wide variety of tree taxa (at family as well as generic level) include species which have evolved so as to space out fruiting (see fig. 2.11 and table 2.2 below). Asynchrony tends to occur between more than within species, presumably because members of the same species respond to the same cues for flowering.

Figs are a special case among fruit evolved for and with rain-forest pulp-feeders. They are not true fruit (although usually referred to as such for convenience), but collections of flowers enclosed completely by a pulpy container. Sterile flowers grow among the reproductive ones expressly to provide food for the larvae of small 'fig-wasps', which are obligate pollinators of the female flowers (Corner, 1940; Janzen, 1979). It is perhaps the protection afforded by the pulpy container which allows flowers to be produced at any time of year (as often as thrice yearly in some species), even when other flowers are likely to fail from bad weather. Strangling figs thus collectively provide fruit more continuously than does any other genus, which has great import for forest animals.

(2) Caching hard fruit

There is as yet no evidence from South-east Asian forests of a mechanism involving rodents and plants which produce seeds in a very hard, unattractive case. Tree squirrels in temperate forests (e.g. K. MacKinnon, 1976), ground squirrels in Africa (Emmons, 1975) and caviomorphs in Central America (Smythe, 1970) cache seeds in the ground, when they are available in abundance. Some are later removed and eaten in times of food shortage, but a proportion are missed and germinate.

(3) Predator satiation

One way of reducing seed predation is for most individuals of one or more taxonomic groups to fruit rarely and in synchrony. This mechanism of satiating predators is incompatible with mechanisms of seed dispersal by animals, and most plants exhibiting it are wind-dispersed or else have no obvious means of dispersal. It appears to be successful, since it is shown by most of the abundant tall trees of Malayan forests: *Shorea* and *Dipterocarpus* spp. (Dipterocarpaceae), *Koompassia* spp. (Leguminosae) and several genera of the Sterculiaceae.

(4) Wind dispersal

Some tall trees and climbers which bear fruit in the upper canopy can afford to produce very light-weight seeds for wind dispersal, which are so small not to be worthwhile for mammals to harvest.

(5) Mechanical protection

This mechanism operates by making it more difficult for vertebrates to extract seeds from the fruit, or for invertebrates to penetrate them. It includes: (a) thick, tough rinds; (b) woody shells around the seed; (c) tiny irritant hairs on the fruit surface; (d) sticky resins in the rind or pulp; and (e) sharp spines externally. Mechanical protection does not wholly protect seeds, since large squirrels, for example, can open even the hardest fruit. Such animals may prefer, however, to select the less well-protected species, so as to minimise the time and energy expended to gain access.

(6) Toxins

There is an immense diversity of phenolics, alkaloids and other secondary compounds among rain-forest plants, substances of no obvious metabolic use to the plants (Ehrlich and Raven, 1964; Janzen, 1975; Feeny, 1975). Since it is impossible to alter leaf structure and production patterns to prevent destruction by insects, plants have evolved chemical defences. The pressures on plants to defend themselves chemically against insects, and on insects to find ways around those defences, are thought to have contributed to the massive diversification of plants and insects found in rain-forest today.

(a) WIND-DISPERSED

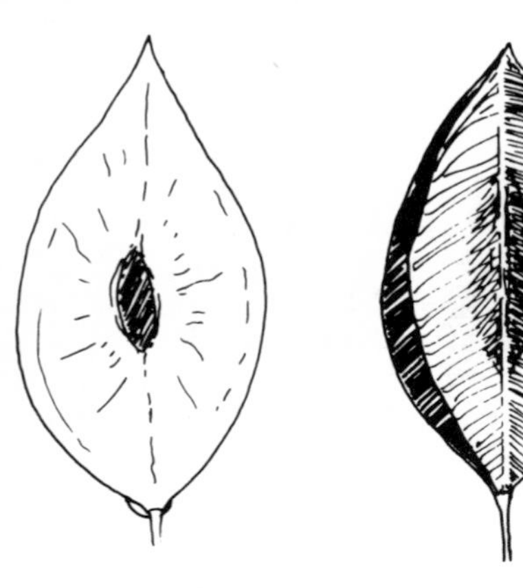

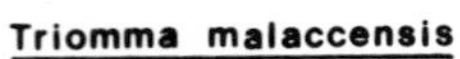

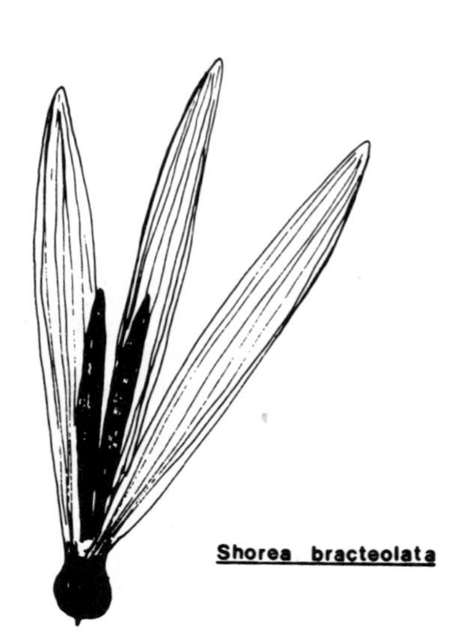

(b) MECHANICALLY PROTECTED

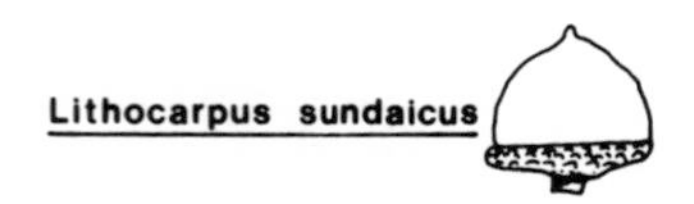

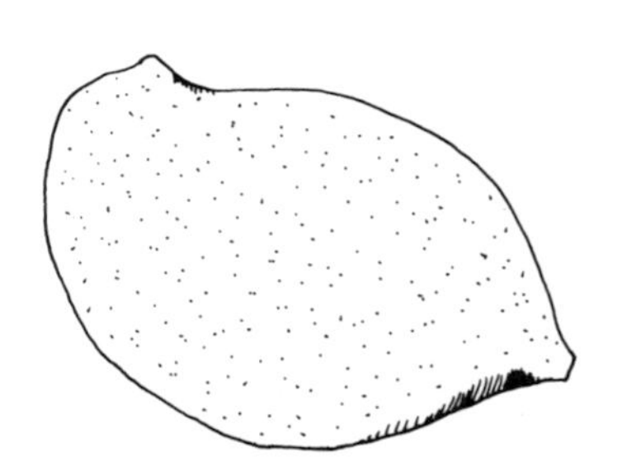

Cynometra malaccensis

(c) BIRD-DISPERSED

Dysoxylum (?) acutangulum

(d) MAMMAL-DISPERSED

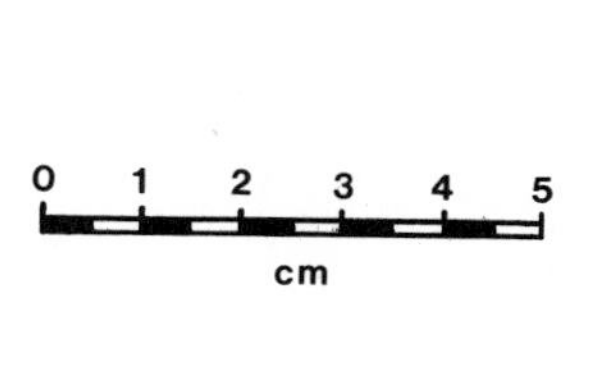

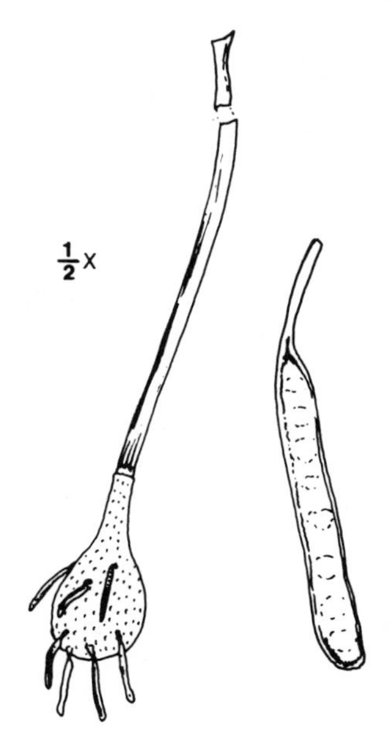

(e) PRIMATE-DISPERSED

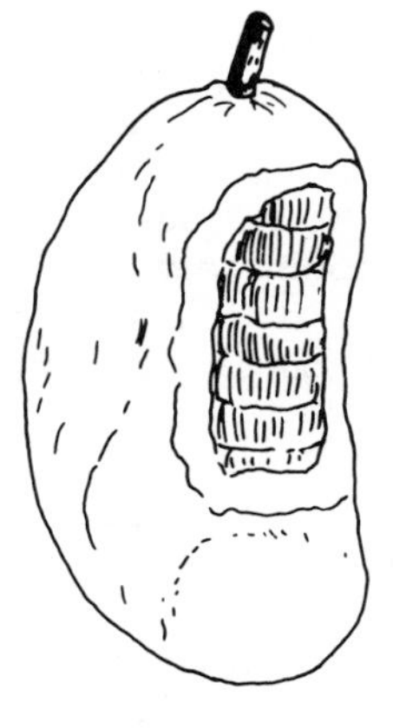

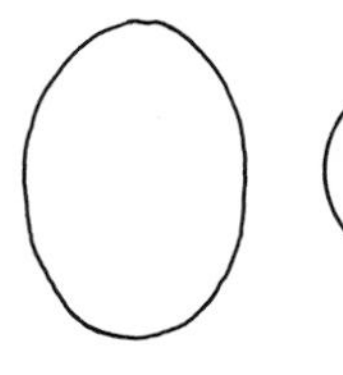

Plate V. Fruits of the tropical rain forest (JBP)

(a) dispersed by wind - dry, winged
(b) protected mechanically, perhaps dispersed by rodents - dry, woody
(c) dispersed by birds but not primates - dry, bitter aril/endocarp
(d) dispersed by some primates and terrestrial mammals - beans in pod
(e) dispersed by primate pulp-eaters - sweet juicy pulp

Secondary compounds are also expected to affect mammals, especially since they have more varied diets than host-specific insects which can evolve very specific means of defeating defences (Freeland and Janzen, 1974). They affect (a) the maximum quantities which can be consumed, (b) the time which must elapse before eating the same species again, and (c) the differential exploitation of plant parts. Many fruit are also protected by toxins, particularly prior to ripening. Nevertheless, there are few, if any, fruit that are not taken by at least one type of avian or mammalian frugivore. Leaf monkeys and certain birds appear able to detoxify some fruit never eaten by other frugivores, e.g. the strychnine-containing fruit of the climber *Strychnos* (eaten by pheasants and leaf monkeys), and the cyanide-containing seeds of a locally common lowland forest tree, *Elateriospermum tapos* (eaten unripe by leaf monkeys).

Leaf monkeys clearly obtain some protection against a large sector of toxins from the neutral environment and sheer bulk of their fermenting fore-stomachs (Bauchop and Martucci, 1968; Kuhn, 1964). Gibbons and macaques may rely for protection on more specific detoxifiers, such as microsomal enzymes and substances ingested along with the toxins, or else by restricting consumption of any one toxic food. Conclusive evidence of the limiting effect of plant toxins on primates is actually lacking, although there are some very suggestive associations in extreme environments (Gartlan et al., 1978; McKey, 1978; Glander, 1978; Whitten, 1980). Toxicity might explain such intriguing observations as that in which gibbons at Kuala Lompat were seen to eat the new leaves of *Elateriospermum tapos* (whose seeds contain cyanide) when they first appeared but not subsequently. There is no doubt, however, that primates select foods for their positive values such as protein and sugar contents, and most argue that this is a more important force in food selection by primates (C.M. Hladik, 1978).

Fruit-eating animals may thus exert two opposing pressures on the timing of fruiting. On the one hand, by destroying seeds they may cause plants to fruit synchronously so as to swamp predatory activity. On the other hand, by dispersing seeds they may encourage plants to fruit asynchronously so that each may increase its chances of being noticed and dispersed. Since the tropical climate allows such flexibility in the time of year at which plants flower, and in the interval between flowering and the ripening of fruit (Ng and Loh, 1974), there is plenty of opportunity for plants with resident dispersers to stagger their fruiting. Some plants may, of course, be subjected to both pressures by different animals, and the same animals can exert both pressures on the same plant, as do those which cache seeds for later eating but fail to find all of them again. We have still to examine the germination success of seeds removed from trees by primates, but it is likely that gibbons are important seed dispersers to certain trees and climbers, whereas leaf monkeys destroy many seeds and macaques destroy some seeds and disperse others.

If related species stagger their fruiting, but individuals within each species are synchronised, a disperser can lock into a search image and also benefit from accumulated knowledge of the location of such trees in its home range. Competition between individual plants, however, may lead to asynchrony within species, which will confound the predictive powers of the disperser. This to some extent applies to Malayan forests (see below), and it must place a premium on the primates' ability to plan economical routes between current food sources, since it increases the ratio of searching to feeding time in a habitat where all woody species occur at very low densities.

KUALA LOMPAT

This section presents a more detailed description of the principal site of the studies reported in subsequent chapters, Kuala Lompat, illustrating some of the generalities mentioned above. The account is drawn from Raemaekers (1977) and Payne (1979b) (see also Chivers, 1974; Curtin, 1976a).

The site

Kuala Lompat lies some 50 m (164 ft) above sea level on the eastern edge of the 510 m^2 Krau Game Reserve, at the junction of the rivers Lompat and Krau (Plate VI). The study area (fig. 2.2) can be divided into at least four habitats on the basis of water régime:

(1) a shelf alongside the rivers, which is rarely and briefly flooded with fast-flowing water;
(2) swamps which are seasonally flooded for days or weeks;
(3) parts well drained and never flooded;
(4) sloping ground, to the north side of the study site.

The shelf has been formed by flooding and in places extends as far as 220 m away from the Lompat river. Only twice within living memory has flooding reached beyond the shelf (1927, 1971). The habitats differ in plant species composition and in plant structure. Some trees occur only or mainly on the riverine shelf (e.g. *Saraca thaipingensis*), while others occur mainly in swampy areas (e.g. *Nauclea* spp.), and still others prefer permanently dry ground (e.g. *Elateriospermum tapos*). The swamps themselves differ floristically from one another.

The forest can be classified as lowland evergreen dipterocarp (Plate VII), but it is actually relatively poor in dipterocarps and unusually rich in large Leguminosae. Dipterocarps comprise

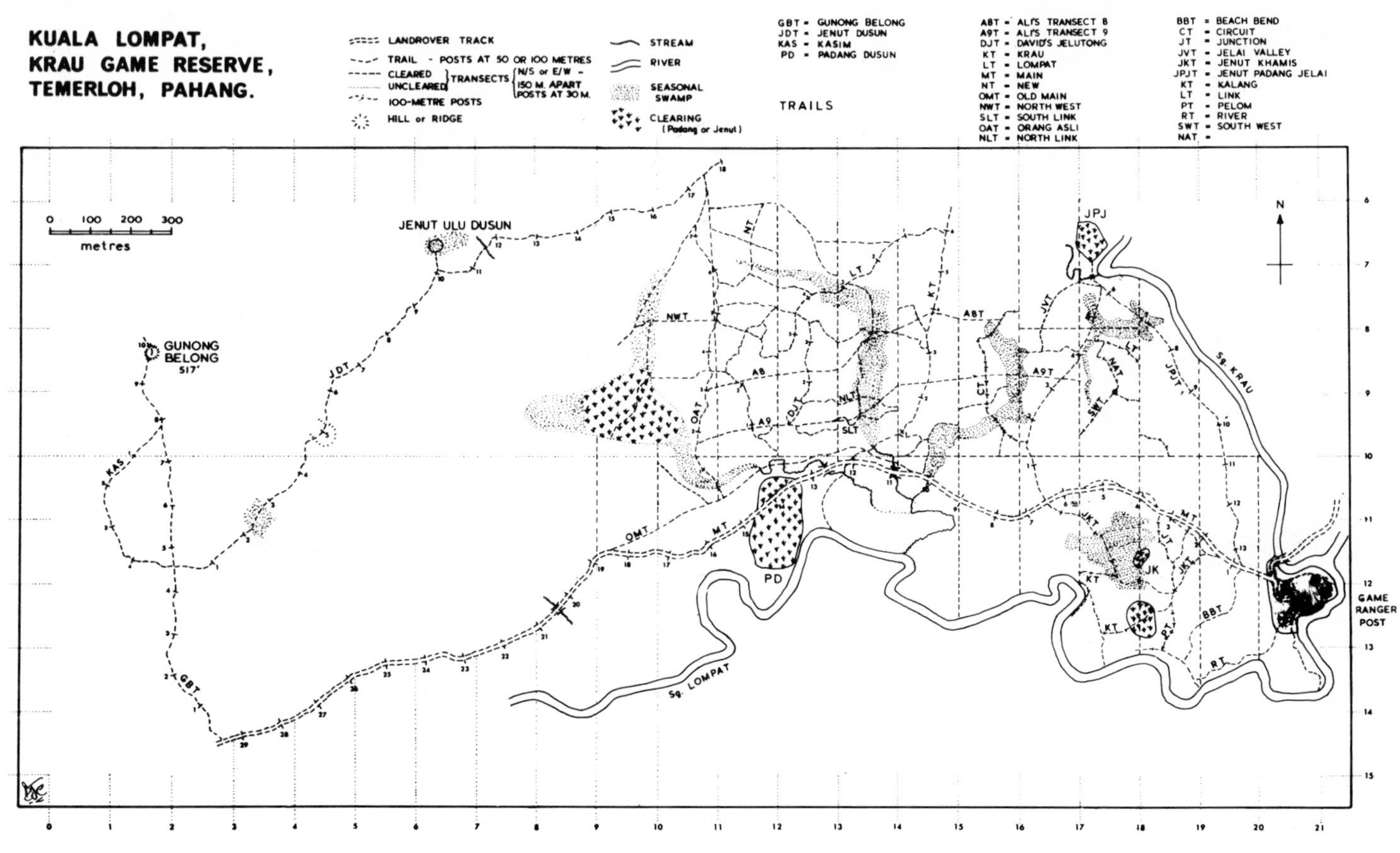

Fig. 2.2. Kuala Lompat Post, Krau Game Reserve, Peninsular Malaysia.

Plate VI. The Kuala Lompat Post of the Krau Game Reserve from the south-west (DJC)

(a) the clearing and study area started by Sheila Curtin
(b) the study area further west up to the old clearing started by David Chivers

(figures vary according to the distribution of the sample) some 8% of trees over 50 cm g.b.h. (girth at breast height) and 21% of trunk basal area. Since they are commoner further from the rivers, it may be that it is the proximity of the two rivers which is responsible for their low incidence.

The action of man is a further source of variation within the site. There are some clearings, beginning to grow over, into which the primates do not venture (fig. 2.2). There also appears to have been some felling of large trees in the eastern part of the site 30 or more years ago, resulting in a less-continuous upper canopy and denser lower growth than in the west. There has been no logging on a commercial scale, however.

Kuala Lompat lies in the rain shadow of Gunung Benom (2107 m, 6914 ft a.s.l.) and is, therefore, rather dry for the region, averaging 1982 mm (78 in) of rain annually (mean of 1969, 1975, 1976, 1977 and 12 consecutive months 1970-71). The rhythm of rainfall is rather variable, but there is a trend towards a wet season at the turn of the year, followed by a dry season early in the year (fig. 2.3a). Temperature is very stable (fig. 2.3b, c).

The botanical sample

The botanical sample from which the data presented below are drawn, except where otherwise stated, consists of 10 plots, 3 of 1 ha and 7 of 0.25 ha (fig. 2.4). The 29-month period for which phenological data are presented may be divided into two parts, during either of which only 9 of the 10 plots were monitored, amounting to 3.75 ha. During the first part, from April 1975 through March 1976, the plot P10 was omitted; during the second part, from April 1976 through August 1977, P10 was included but E7 was omitted. The first part will be referred to below as 1975/6.

The plots, excepting B8 and P10, are arrayed on a grid of 300 m interval over an area of some 60 ha. The sample consists of all free-standing trees over 50 cm g.b.h. rooted in the plots, together with the strangling fig trees which they support. About 900 stems are included in the sample at any one time. These were girthed, mapped and identified, over 90% of stems being identified at the specific level.

During the whole period 1975-77 each tree was scored once a month for young leaves (including shoots), flowers and fruit. Each monthly round of observations spanned 2 - 5 days. During 1975/6 alone, trees were also scored on a scale of 4 intervals for each plant part: 0, 1 - 10, 11 - 50 and 51+ % of crown surface area. For fruit of different sizes, correction was made subjectively by estimating what proportion of the crown would have been bearing

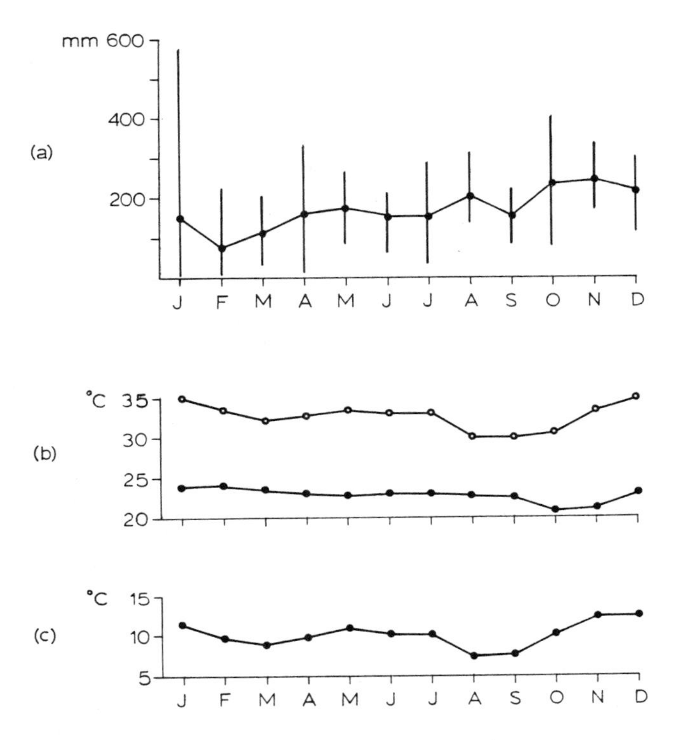

Fig. 2.3. Weather. (a) Rainfall: monthly means and ranges compiled from all records taken 1968-77. (b) Temperature: mean daily maximum and minimum in each month, 1975/6. (c) Temperature: mean daily range in each month, 1975/6.

fruit had they been of an arbitrary standard size, 2 cm diameter. To obtain a unit of relative abundance for each plant part, the mid-points of these percentile ranges, i.e. (0), 5, 30, 75 were multiplied by trunk basal area calculated from girth. Basal area is a rough estimate of crown mass and thus of surface area. Corrections were made for the tree's load of mechanically-dependent plants, since these lower the ratio of the host's crown mass to basal area (details in Raemaekers, 1977, in press). Since a tree might carry the same fruit for longer than one month, scores represent the standing crop each month rather than production. No distinction was drawn between immature and mature flowers or fruit during the scoring process.

Forest structure

There are some 200 - 300 trees over 50 cm g.b.h./hectare (mean = 243). As the girth limit is lowered, so the number of tree species rises steeply in a J-curve (fig. 2.5). In profile the canopy is

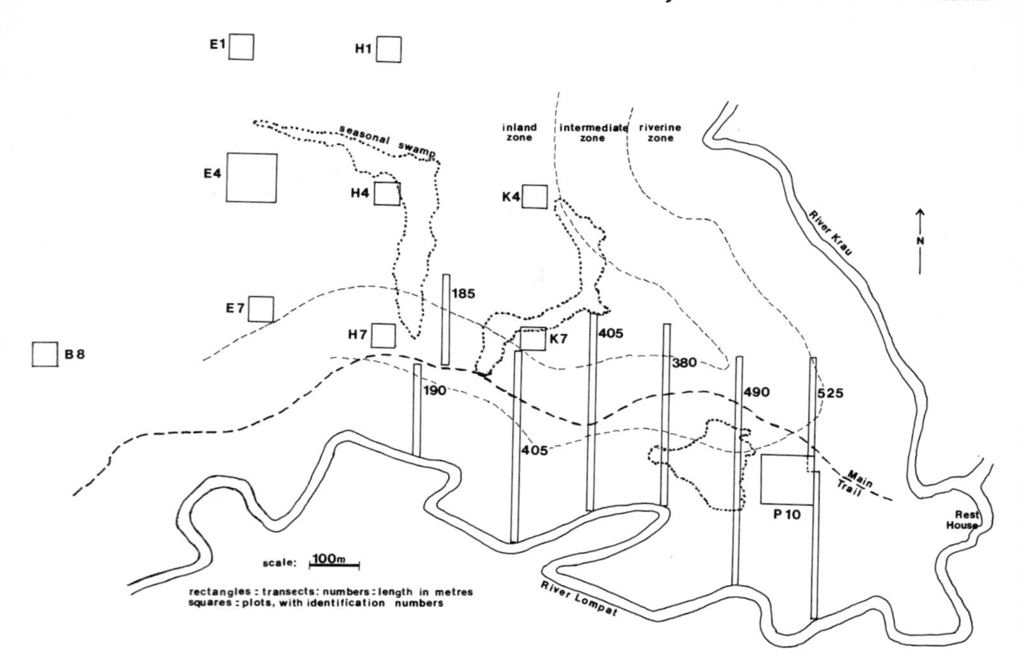

Fig. 2.4. Study site, showing location of botanical plots. Large plots are of 1 ha, small ones of 0.25 ha. Dotted areas are clearings. The line of oblique strokes represents the approximate dividing line between virtually virgin forest to the west and old regrowth of disturbed forest to the east.

tall but often interrupted (fig. 2.1). The network of climbers to some extent bridges gaps, at least lower down, and the profile diagram gives a false impression of openness by omitting small trees. It nonetheless shows some of the variation with height above ground of three structural features which affect arboreal animals: exposure to direct sunlight, canopy volume and horizontal continuity. Canopy volume and horizontal continuity, together with mean branch diameter, reach their maxima at middling heights. The top of the canopy consists of the discontinuous tips of the highest branches of tall trees. The lower part of the canopy is also composed of more

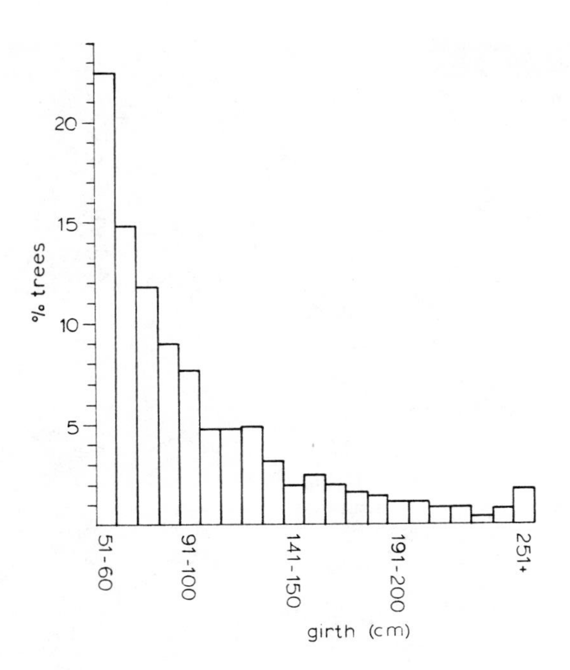

Fig. 2.5. Frequency distribution by girth of trees exceeding 50 cm girth at breast height. 911 trees in all plots except P10.

pliable and vertical elements than are the middle levels. The general implications for primates are that

(a) most food will grow in the middle levels where there is an optimum combination of vegetation mass and direct sunlight, and
(b) the most stable, continuous and horizontal substrate also occurs in the middle levels, favouring the kind of locomotion displayed by Malayan higher primates, especially the quadrupedal monkeys.

The climbers which partly bind together the tree canopy are not evenly distributed across the area: the percentage of trees bearing climbers varies between plots from 50 to 92, presumably according to the degree and recency of natural and man-made tree falls. There is, however, no association between the amount of climbers and distance from a river. Two independent assessments of the distribution of climbers among trees of different sizes were made. One, which

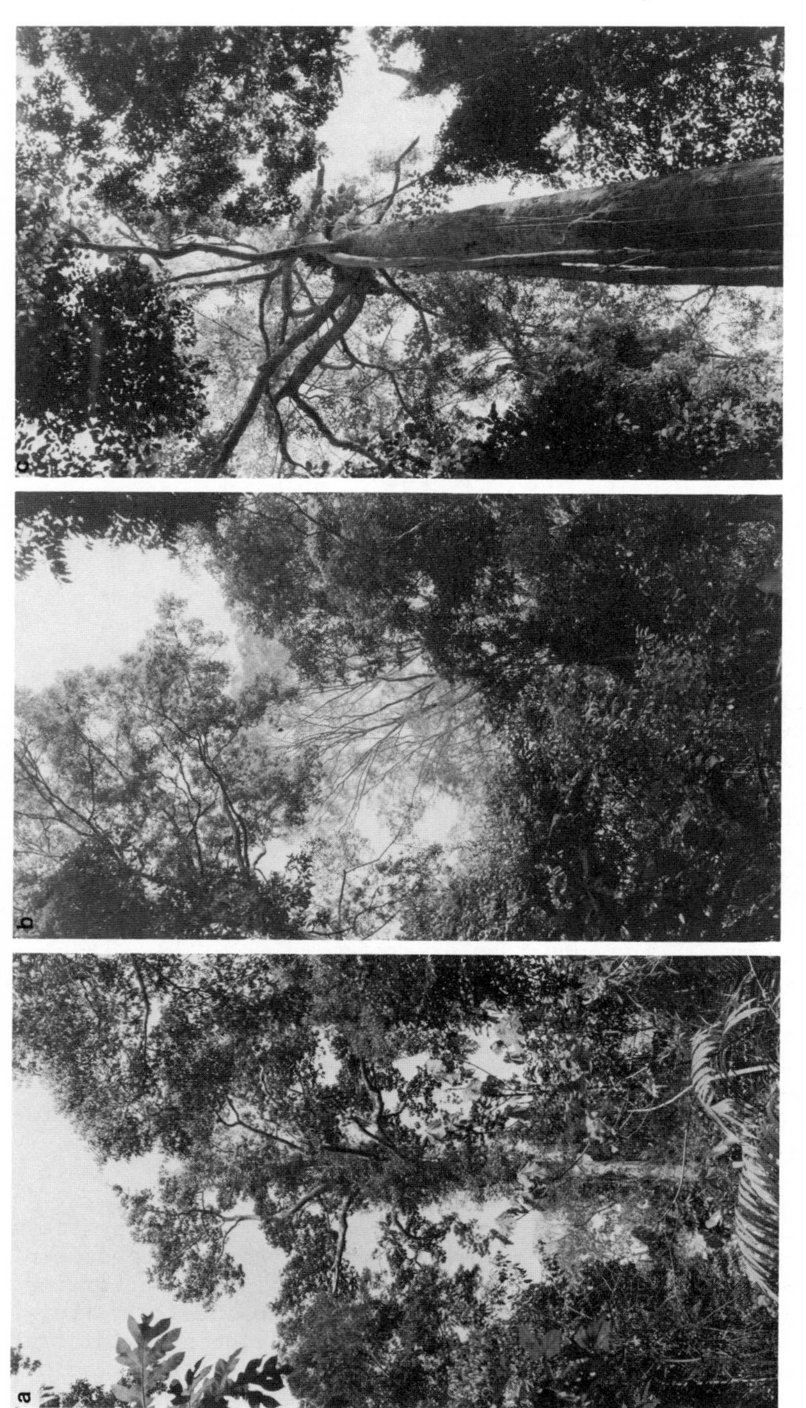

Plage VII. Trees in the lowland dipterocarp rain forest at Kuala Lompat (JJR)

(a) *Dipterocarpus cornutus*, the commonest tall tree, with a heavy load of vines
(b) *Koompassia excelsa*, the tallest tree in Asia
(c) *Ficus* sp., strangling its emergent host; note the thick old descending roots on the left and the fine young ones on the right

classed trees by their girths, showed a very slight decrease in climbers among the largest trees (fig. 2.6). The other, which classed trees by their heights, showed this trend more strongly (table 2.1). About 18% of the total tree and climber canopy is composed of climbers in the undisturbed forest (815 trees over 50 cm g.b.h.).

The strangler is one of the more striking structural elements of the tropical forest. Almost all stranglers at Kuala Lompat are figs and most figs are stranglers (Plate VII). The strangling fig trees at Kuala Lompat tend to occupy large host trees, presumably because these provide broad or hollow forks where the fig seed may lodge and begin epiphytic growth (Raemaekers, 1977). Assuming that the mass of the strangler is proportional to that of the host, fig trees will therefore tend to grow large. They also fruit copiously (see below) and, as a genus, are relatively common: 28 stranglers and 1 free-standing tree out of 915 trees, and 11 climbers. For the reasons mentioned in the previous section, the fig trees hold a unique position in the food chains of the forest.

Species diversity

The numbers of trees and climbers in a single plot of 1 ha which are accounted for by different taxa are shown in fig. 2.7. This plot is average for the study site in species diversity, numbers of trees and numbers of climbers, and includes a strip of swamp. Nearly 60% of tree species are represented by a single stem each; studies in similar forest indicate that the number of tree species would double if the lower limit of girth were halved (Wong, 1967). The plot also contains in excess of 65 climbing species, and

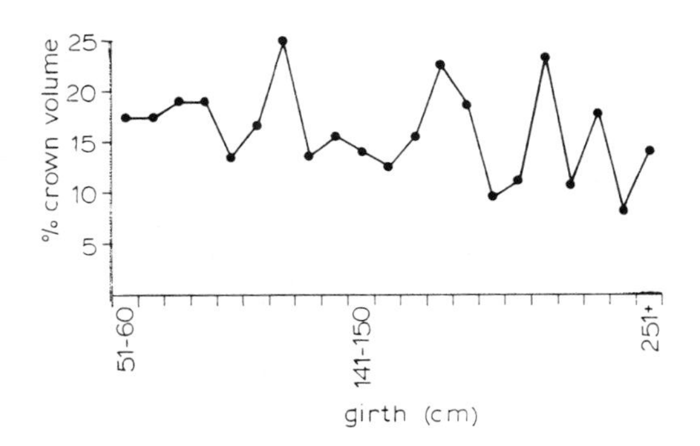

Fig. 2.6. Distribution of climbers between trees of different sizes: percentage of (tree + climber) crown volume which is comprised of climbers plotted against tree girth. The variance increases with tree girth because of decreasing sample size. 799 trees in all plots except K4, P10.

Table 2.1. Relative contribution of climbers to the crowns of trees in different height categories at Kuala Lompat.

Tree height (feet)	% of canopy with climbers present			
	0%	1-9%	10-50%	>50%
<40	29	21	13	37
41-60	22	20	20	38
61-80	24	17	20	39
81-100	47	13	14	26
>100	54	12	12	22

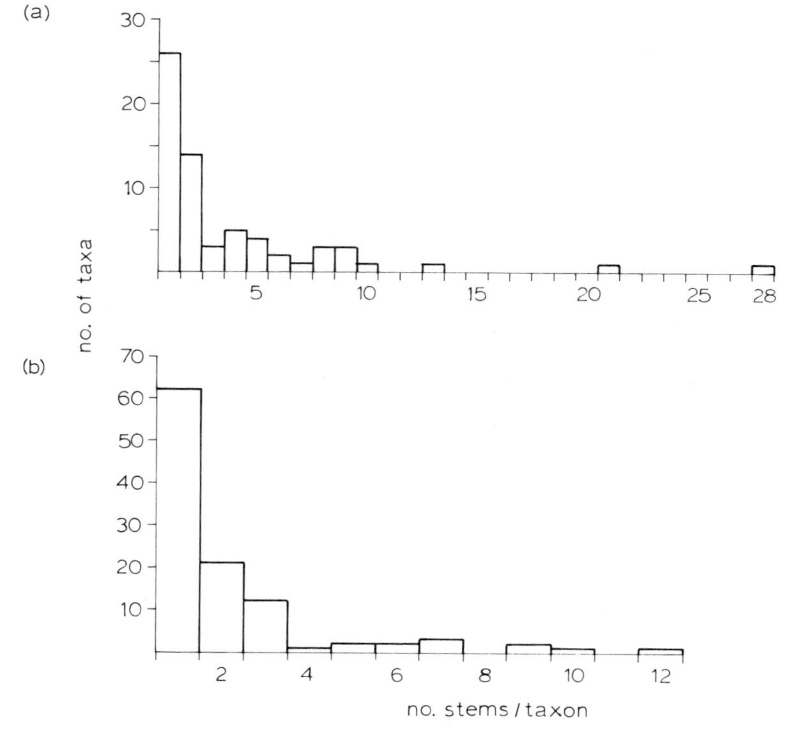

Fig. 2.7. Species diversity of a hectare plot (E7). (a) Climbers. One stem = presence of taxon on one tree over 50 cm g.b.h. True diversity considerably underestimated. 65 taxa, 245 stems. (b) Trees over 50 cm g.b.h. rooted in the plot. All 5 fig trees treated as a single species. Diversity only slightly underestimated. 107 taxa, 226 trees.

of those distinguished taxonomically 25 are represented by a single score each (one score = climber present on one tree).

The number of tree species tallied continues to rise steeply after all plots are pooled (fig. 2.8). By the end of 1977, 395 species of trees of all sizes had been tallied in the study area of about 150 ha. A primate social group, occupying a home range of 10 - 60 ha, has access to perhaps 400 species of trees over 50 cm g.b.h. and of climbers on them. Primates using the under-storey and covering larger home ranges (notably the pig-tail macaque) increase greatly the number of plant species theoretically available to them by including smaller plants and different habitats respectively.

It is difficult to compare the diversity at Kuala Lompat with that elsewhere, because there is no standardised format of description. The minimum girth used varies, as do the size and dispersion of sample plots. Poore (1968) identified 375 spp of trees contributing to the canopy in 23 ha of Jengka Forest Reserve 30 km from Kuala Lompat, which is comparable to the diversity at Lompat itself. As mentioned in the first section of the chapter, diversity is generally greater in the Malesian region than in Africa or America. For example, Rollet (1969) found 365 spp of trees and lianas over 31 cm girth (10 cm diameter) in 64 ha in Guiana, which is less than at Kuala Lompat or Jengka when the minimum girth and inclusion of lianas are taken into account. Some sites outside Malesia may, however, reach comparable diversities. A. Hladik (1978) recorded 62 spp of trees over 31 cm girth in 0.4 ha of Ipassa Forest,

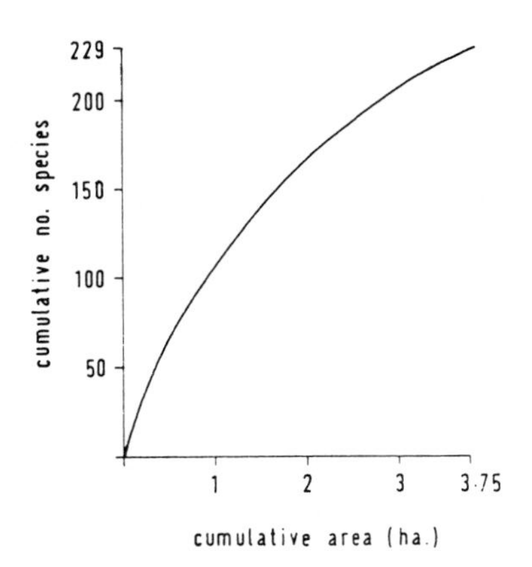

Fig. 2.8. Tree species/area curve: increase in the number of species of trees exceeding 50 cm girth tallied as the area of the botanical sample is increased. The curve shown is the average of three almost identical plots obtained from three random orderings of all plots except P10.

Gabon. This is very similar to Bukit Lagong and Sungei Menyala in Malaya, and about average for a series of sites in Brunei (Borneo), though appreciably less than at Kemasul, Malaya (data in figs. 1.3, 1.4 and 1.5 in Whitmore, 1975).

The high diversity means that all plant species occur at low densities and are therefore relatively hard to locate. Among species, such as primates, which eat many species of plants, this is expected to select for stable and perhaps small home ranges, along with good spatio-temporal memory, in order to increase the efficiency of food-finding. It also affects the study of animal ranging behaviour. To find out whether and how patterns of ranging are related to elements in the distribution of food, it is desirable to compare the plant composition and production of quadrats with the animals' use of those quadrats (e.g. Struhsaker, 1974; Clutton-Brock, 1975a). At Kuala Lompat this approach is not feasible, because it is impossible to predict the floristic composition of a quadrat from a sample of that quadrat, and because the alternative course of enumerating all plants in all quadrats entered by a social group is impracticable (there are, for example, 15,000 trees over 50 cm g.b.h. within the territory of one lar gibbon study group). Extrapolation of leaf and flower cycles from some individuals of a plant species to others, in order to estimate production by whole quadrats, is also open to error, as will become clear in the following paragraphs.

Phenology of production

The recurring natural phenomena, the cycles of production, of the total tree community in the sample are shown in figs. 2.9 and 2.10. Two measures are used: (1) the percentage of trees active (1975-77), and (2) the number of relative abundance units/hectare (1975/6 only). The latter is more sensitive to the volume of production, but the former has the advantage of simplicity and ease of comparison with the results of other studies. Besides, the number, that is the density, of trees producing food may be quite as important in determining animal ranging patterns as the actual quantity of food produced.

Abundance varies considerably between months - up to 70-fold in flower abundance units. Flowering peaks follow the dry season early in the year (compare figs. 2.9b and 2.9c). The subsequent fruiting peaks, insofar as one can deduce from a sample of one whole peak and two half peaks, are a little lower and longer. This is to be expected because not all trees which flower set fruit, and because of the variable period between flowering and the ripening of fruit. It could also be partly a methodological artefact of scoring unripe fruit as well as ripe, given that fruit take longer to develop than flowers.

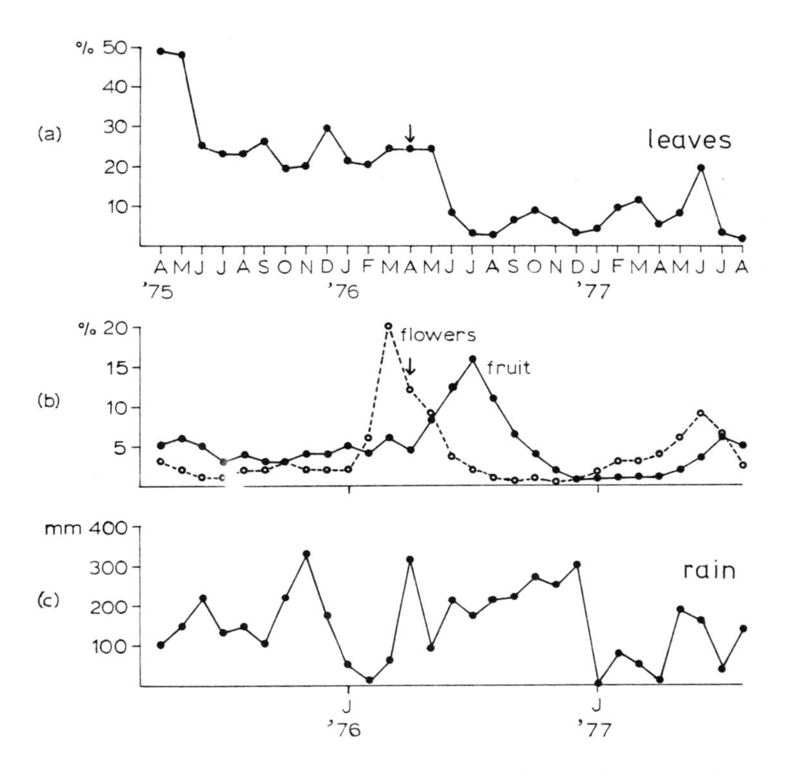

Figure 2.9. Phenology of the tree community. Percentages of trees exceeding 50 cm girth bearing (a) young leaves, and (b) flowers and fruit, in each month April 1975 - August 1977. Arrow represents change of observer, which could account in part for the lower percentages recorded in the second part of the period. c.900 trees in all plots except P10 up to March 1976 inclusive, and all except E7 from April 1976 on. (c) Rainfall during same period.

Since primates eat only some of the species in any quantity, the cycles of the whole community may be a poor guide to those of the species which they do eat. The possibility of seed dispersers exerting an influence on the timing of fruiting was discussed in the previous section; there is some evidence of such an effect at Kuala Lompat. The community of tree species which bear fruit attractive to gibbons and siamangs (i.e. eaten by them in 10+ feeding observations in a year) spreads its fruiting slightly more evenly across the year than does the community of unattractive species (fig. 2.11). Similarly, it is evident at Kuala Lompat that, within tree families, related species which produce similar fruit eaten by pulp-eaters tend to bear them at different times (table 2.2).

Table 2.2. Asynchronous fruit production by sets of tree species with morphologically similar fruits, Kuala Lompat 1976-77. N = number of trees observed with fruit. C = canopy levels in which fruit is borne: E, emergent; U, upper; M, middle; L, lower. L = length of fruit (mm). At Kuala Lompat, eaten by: gibbons (G), macaques (M), hornbills (HB), pigeons (P), barbets (B), other small birds, including bulbuls, flowerpeckers, leafbirds (SB) (all mainly seed dispersers); leaf monkeys (L), *Ratufa* (R), *Callosciurus* (C), *Sundasciurus* (S) (all mainly seed predators).

Family and genera	Species	Months with ripe fruit	N	C	L	Colour of ripe fruit externally	Eaten by
a. Euphorbiaceae							
Aporusa	sp.	Apr 1976	1	L	17	Red aril	B, P, SB
Baccaurea	*racemosa*	Jun, Jul 1976	2	L	15	Blue aril	
"panchau"		Sept 1976	1	M	15	Red aril	R, S
Sapium	*baccatum*	Oct 1976	1	M	10		
Austrobuxus	*nitidus*	Mar 1977	1	U	15	Red aril	
b. Burseraceae							
Dacryodes	*rugosa*	Jul, Aug 1976	4	L	17	purplish	3HB, P
Dacryodes	*rostrata*	Nov 1976	1	U	28	purplish	B, SB
Santiria	*laeviga*	Feb, Mar 1977	3	U	15	purplish	
Santiria	*rubiginosa*	Apr 1977	1	M	11	purplish	R, C, S
Santiria	*tomentosa*	May 1977	1	U	19	purplish	
c. Myristicaceae							
Horsfieldia	*superba*	May 1976	1	U	?	yellow	
Knema	sp. A	Jun 1976	1	M	45	yellow	HB, M
Horsfieldia	*sucosa*	Aug 1976	2	U	55	pink	
H.	(?) *macroma*	Nov 1976	2	U	?	yellow	L, C
Knema	sp. B	Dec 1976	1	M	40	yellow	
Myristica	sp.	Feb 1977	2	M	65	yellowish	
Knema	sp. A	Jun 1977	1	M	45	yellow	

d. Lauraceae							
(genera unknown)	A	Dec 1976	1	U	27	dark green	2HB
	B	Feb 1977	1	M	30	black	
	C	Jun 1977	2	U	28	dark green	L, R
	D	Aug 1977	1	U	30	pale green, white spots	
e. Myrtaceae and Guttiferae							
Eugenia	A	Jun 1976	1	U	?	green	G, M, B
Eugenia	B	Jul 1976	1	L	25	green	SB, P
Eugenia	C	Jul, Aug 1976	4	U	25	green	
Eugenia	D	Sept 1976	1	U	20	green	R, S
Eugenia	E	Nov, Dec 1976	1	M	47	green	
Eugenia	F	Jan 1977	1	U	14	black	
Calophyllum	*floribundum*	Apr 1977	1	M	?	green	
Eugenia	G	May 1977	1	M	?	green	
Calophyllum	*curtisii*	Jun 1977	4	U	10	greenish	
f. Leguminosae							
Dialium	*procerum*	Oct 1976	2	U/E	44	black	G, M
Dialium	*platysepalum*	Oct, Nov 1976	5	U/E	30	brown	
Dialium	*patens*	Dec 1976	2	U/E	25	brown	L, R, C
Dialium	*laurinum*	Feb 1977	1	U/E	?	brown	
g. Anacardiaceae							
Mangifera	*quadrifolia*	Apr 1976	1	U	100	dull green	G, M
Mangifera	"temuor"	Jul 1976	1	U	100	dark green	
Mangifera	"rawa"	Apr 1977	1	M	33	pale green	R, C
Mangifera	"temuor"	Aug 1977	1	U	100	dark green	

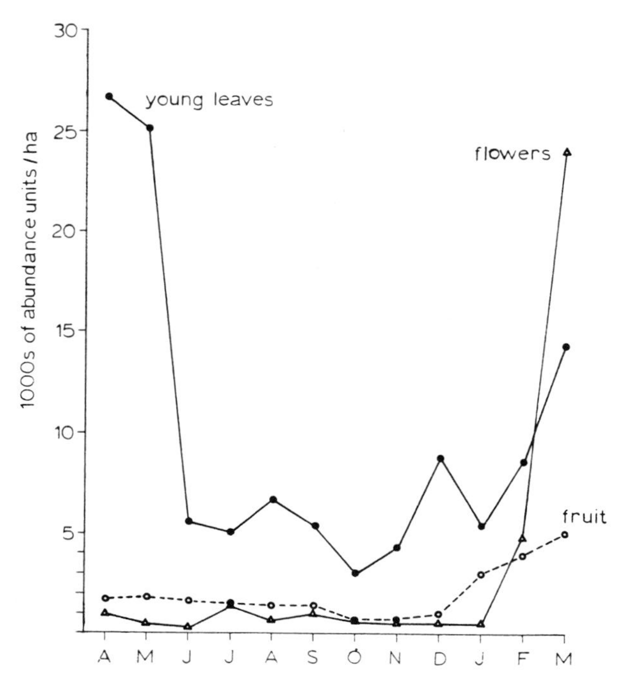

Fig. 2.10. Phenology of the tree community. Relative monthly abundance of flowers, fruit and young leaves, April 1975 - March 1976. See text for explanation of abundance units.

A test to find out whether two seed predators, the giant squirrels *Ratufa bicolor* and *R. affinis*, exert the reverse pressure causing their prey species to fruit more synchronously, revealed no such trend, however. This does not mean that no such trend exists, of course, for the analysis may simply have been too crude to eliminate the swamping effects of other pressures.

Within tree species, leaf production cycles range from fairly continuous output on the one hand, to complete deciduousness with almost complete synchrony within species on the other (fig. 2.12). Continuous leaf production is actually rather rare and is confined to small trees: less than 2% of trees over 50 cm g.b.h., belonging to 4% of species, bore young leaves in 9 or more months of 1975/6 (fig. 2.12a). These trees bear few new leaves at a time (fig. 2.13a) and for this reason tend to be unattractive to primates as sources of food.

Deciduousness, whether partial or complete (fig. 2.12b), is a feature mainly of tall trees whose exposure to direct sunlight presumably subjects them to strong water stress in the dry season. Nevertheless, only some of the tall trees, notably the Leguminosae,

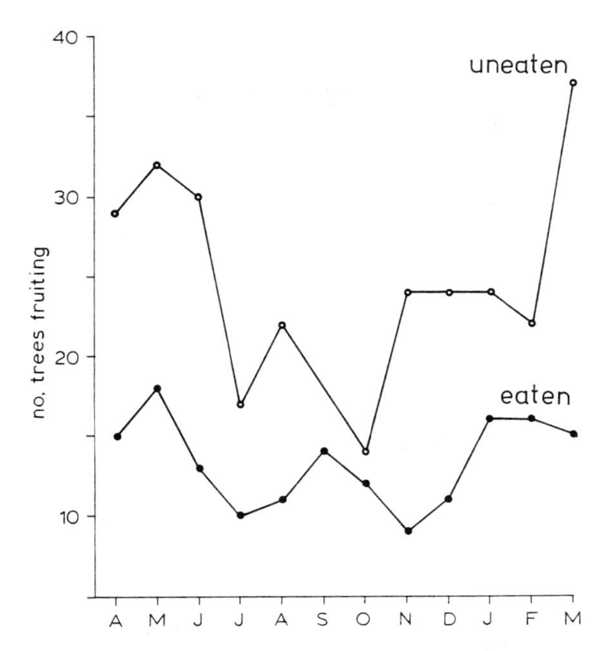

Fig. 2.11. Fruiting strategies. Comparison of how evenly spread through time is the fruiting of species whose fruit are eaten by siamangs and gibbons (dots) with that of species whose fruit are not (circles). All plots except P10; 155 trees of 21 taxa of eaten species which fruited in 1975/6, and 277 trees of 54 taxa of uneaten species which fruited in 1975/6. 'Uneaten' species include those which received less than 10 feeding observations from one or other animal study group, and 3 taxa which were more eaten but whose seeds were dropped *in situ* instead of swallowed and excreted whole, and were therefore not dispersed by these animals.

are deciduous; the dipterocarps are all evergreen. A. Hladik (1978) remarks that in a lowland rain-forest in Gabon there are always some large deciduous trees in leaf flush, providing scattered but large and conspicuous sources of food for folivores. This does not appear to hold for Kuala Lompat. Inspection of the activity of Leguminosae over 100 cm g.b.h. in the 3.75 ha sample shows that during the year 1975/6 there were only 4 months in which at least one tree bore young leaves over 50% or more of its crown (although there were such trees with just a few young leaves in all but one month).

The great majority of trees at Kuala Lompat are evergreen and fall between these extremes of continuous production and deciduousness. The medium-sized *Sloetia elongata* exemplifies this (fig.

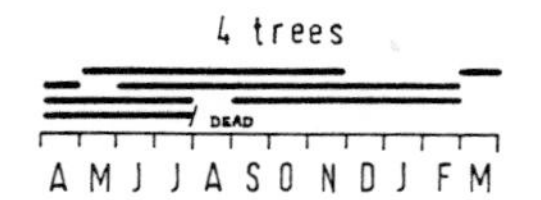

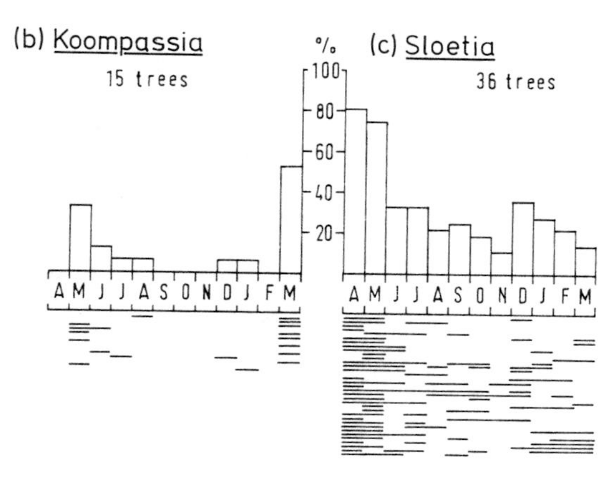

Fig. 2.12. Leaf production of trees. (a) *Randia scortechinii* (Rubiaceae), a continuous producer. (b) *Koompassia malaccensis* (Leguminosae), deciduous. (c) *Sloetia elongata* (Urticaceae), intermediate.

2.12c). Each tree produces intermittently and for periods of varying length, and there is much asynchrony between trees. Nevertheless, during 1975 this species as a whole followed a rhythm recognisably like that of the whole tree community. Trees following such intermediate rhythms bear many of their new leaves a few at a time (fig. 2.13c). The primates tend to select only the larger sources (e.g. fig. 3.13), but these are of short duration: a tree seldom bears new leaves over more than 10% of its crown in two consecutive months (fig. 2.14). The short duration of food sources will naturally tend to increase the difficulty of locating sources of species which are anyway rare.

The variety of leaf cycles is repeated among flower and fruit cycles. Deciduous trees tend to shed their leaves, flower, renew their leaves and then fruit. Most evergreen species show partial synchrony between trees (fig. 2.15a), at least within an area of a few hectares, though such areas may be out of phase. Note that only 5 of the 12 *Ixonanthes* trees in fig. 2.15b flowered in both of this species' 1975 peaks. Those trees which flowered during the first peak, but not the second, presumably failed to recoup enough reserves to do so again when the next set of climatic cues came round. One must be careful to draw a distinction, however, between maladaptive failure to reproduce on cue because of lack of reserves, and an adaptive strategy of reproducing less often than once a year,

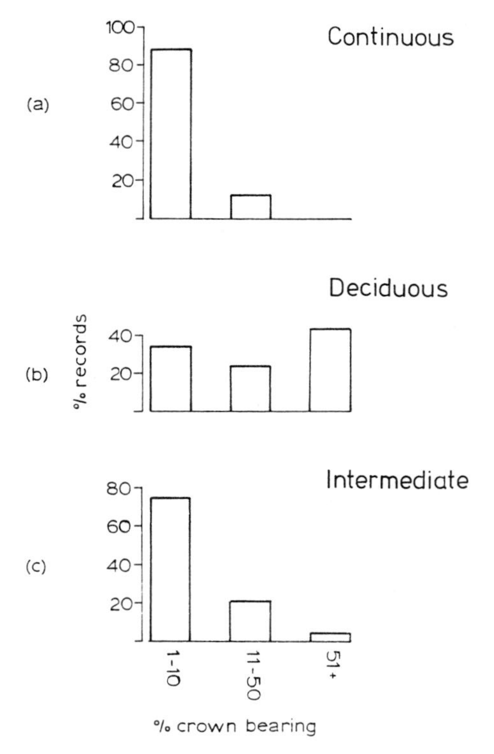

Fig. 2.13. Proportion of tree crown bearing young leaves at one time. (a) Continuous leaf producers, 124 records. (b) Deciduous trees, 35 records. (c) Intermediate trees, 333 records. One record refers to one tree in one month.

that is, to remain one jump ahead of seed predators whose numbers depend in part on abundance of the seeds in question and fluctuate accordingly. During the year 1975/6 only 56% of tree species and 31% of individuals flowered, and only 33% of species and 18% of individuals fruited. Similarly, few trees which bore fruit in 1976 did so in 1977. The failure to fruit reflected in these figures is doubtless in part an adaptive strategy against seed predation.

A few of the smaller evergreen species reproduce almost continuously, for example *Pternandra echinata* (fig. 2.15c). There appears to be a positive association between continuous flowering and continuous leaf production. Whereas primates tend to neglect the small sources of young leaves produced by such trees, they may find the small sources of flowers and fruit attractive enough to harvest them (e.g. gibbons and *Randia scortechinii* and *Pternandra echinata* fruit). On the whole, however, the larger the source, the more attractive it is (a rule which also appears to apply to squirrels,

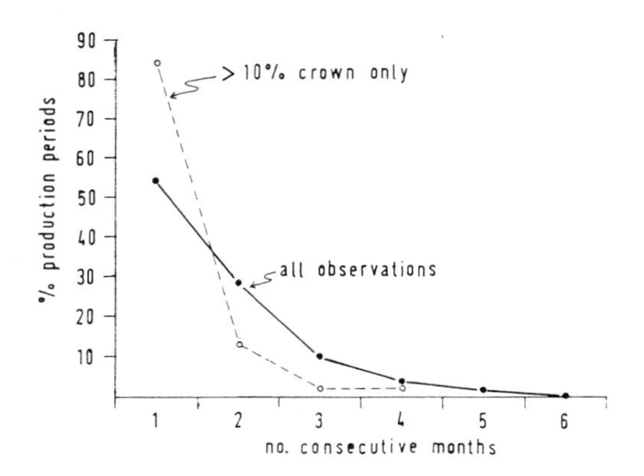

Fig. 2.14. Length of leaf production periods. One period is the number of consecutive months in which one tree bore young leaves. (a) All periods, (b) only periods during which more than 10% of the tree crown bore young leaves. 105 trees of 60 species, 200 periods of (a) and 62 of (b), 1975/6.

insofar as larger species use larger sources than do smaller species (Payne, 1979b: 358-360). A congener of *P. echinata, P. capitellata*, provides an example of a species of small tree which produces a large food source because the trees are clumped and fruit in synchrony.

Since figs are so much eaten by animals, their production warrants special mention. Fig trees generally fruit more often than do other species (Hladik and Hladik, 1969; Medway, 1972). At Kuala Lompat they do not appear to fruit particularly often (12 out of 28 trees fruited in the year 1975/6, of which 2 fruited twice and 1 thrice), but when they fruit they produce larger crops than do trees of other animal-dispersed species (fig. 2.16). Figs are continuously available because of the large number of species (over 20 at Lompat), the asynchrony between species, and the partial staggering within species attendant upon the method of pollination by fig-wasps. The constancy of their availability is reflected in Table 2.3, which shows the distribution of feeding across species of figs by a siamang group and a lar group during 1975/6, with at least one fruiting fig tree to every 6 ha in each month.

In conclusion, we may say that Kuala Lompat shows the typical features of lowland rain-forest in South-east Asia: great stature, very high plant species and structural diversity and complex cycles of production, nevertheless overlaid by a seasonal community rhythm. The relationship of community flowering peaks

(a) 36 trees Sloetia

flowering

A M J J A S O N D J F M

fruiting

(b) 19 trees Ixonanthes

40
30
20
10
% trees
40
30
20
10
A M J J A S O N D J F M

flowering

A M J J A S O N D J F M

fruiting

(c) 5 trees Pternandra

Fig. 2.15. Flowering and fruiting of trees. (a) *Sloetia elongata* (Urticaceae). (b) *Ixonanthes icosandra* (Erythroxylaceae). (c) *Pternandra echinata* (Melastomaceae).

Table 2.3. Availability of figs at Kuala Lompat, judged by feeding on them by a siamang and a lar group. April 1975 - March 1976.

(a) Months in which the groups ate different species of figs

species	month A	M	J	J	A	S	O	N	D	J	F	M	
a	x	x			x						x		
b	x	x	x	x	x				x		x	x	
c	x			x		x							
d	x	x	x	x									
e	x		x	x	x	x					x		
f		x	x					x	x		x	x	
g			x	x	x	x	x	x	x	x		x	
h			x		x		x						
i			x										
j				x	x		x			x	x	x	
k					x	x			x	x	x	x	
l					x	x							
m						x							
n							x			x	x	x	
o							x						
p							x						
q								x					
r									x	x			
s									x	x			
t												x	
total	5	4	7	6	8	6	6	3	6	6	7	7	mean 5.8

(b) Number of fruiting fig trees fed in each month

	month A	M	J	J	A	S	O	N	D	J	F	M	
siamang group (range 47 ha)	4	5	6	5	5	11	7	2	5	8	7	12	mean 6.5
lar group (range 57 ha)	9	9	9	5	8	5	6	6	5	4	13	9	mean 7.3
combined (range 67 ha)	11	10	11	8	11	14	12	6	8	10	15	14	mean 10.8

There are thus, on average, 5.8 species and 10.8 trees of figs available each month in an area of 67 ha, or one tree to 6.2 ha.

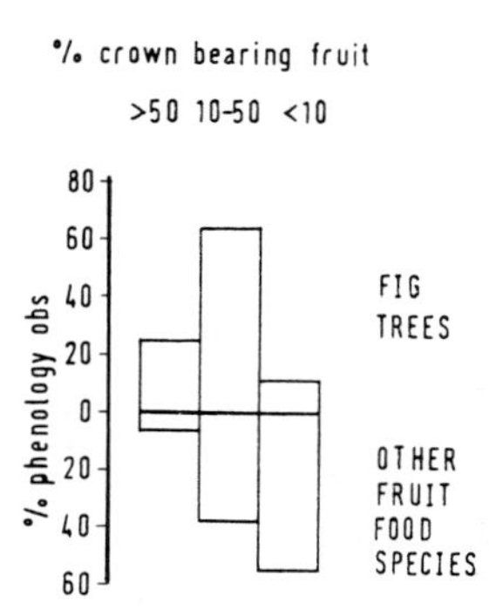

Fig. 2.16. Fruit crop sizes of fig trees compared with those of other tree fruit species eaten by gibbons and siamangs.

to weather is clearer than that of leaf production peaks. The fruiting cycles of individual trees and species will be discussed further in Chapter 8, drawing on feeding data collected over the 10 years of study.

J J RAEMAEKERS

SIAMANG, LAR AND AGILE GIBBONS

S.P. Gittins and J.J. Raemaekers

Sub-department of Veterinary Anatomy and Department of Physical Anthropology, University of Cambridge

INTRODUCTION

Gibbons are light-weight arboreal apes, remarkable for their long, strong arms, which are associated with the predominance of suspensory behaviour in locomotion and posture (see below, plates X and XI; chapter 7).

Three species of gibbon are found in the Malay Peninsula: siamang, lar and agile (Plate VIII; fig. 1.6). The siamang, *Hylobates syndactylus continentis* [THOMAS, 1908], the largest living gibbon, has an adult weight of 10 - 12 kg, males and females being of the same size (Schultz, 1933, 1974). The coat of long hairs is completely black in colour, except for a little whitening of the hair around the mouth. The siamang has a large throat pouch, which inflates to act as a resonator while vocalising. It lives sympatrically with the lar gibbon over much of the Malay Peninsula and also in the north of Sumatra. There may be no sympatry between the siamang and agile gibbon in the Malay Peninsula today (Gittins, 1979; cf. Chivers, 1974), although there is extensive sympatry between the two species in Sumatra, south of Lake Toba (Wilson and Wilson, 1977; Gittins, 1979). The siamang is commoner in hill forest than the smaller gibbons but is rarely found above an altitude of 1500 m (4920 ft) above sea level (Caldecott, 1980).

Like all the small gibbon species, the lar and agile gibbons are allopatric. The lar gibbon is distributed over the south of the Malay Peninsula, sub-species *H. lar lar* [LINNAEUS, 1771], and also northwards from the Thai border, sub-species *H. lar entelloides* [GEOFFROY ST. HILAIRE, 1842]. The agile gibbon, *H. agilis* [CUVIER, 1821], is distributed in a band across the north of the Malay

Plate VIII. Gibbons of Peninsular Malaysia

(a) siamang, *H. syndactylus*, female of GS16B in Ulu Gombak (DJC)
(b) lar gibbon, *H. lar*, male and female of TG2 at Kuala Lompat (DJC)
(c) agile gibbon, *H. agilis* (SPG)

Peninsula, separating the two sub-species of lar. Both species have an adult weight of 5 - 6 kg and differ very little anatomically, so little in fact that they have been regarded previously as sub-species of *H. lar* (Groves, 1972). In both species the coat colour is very variable, from almost black to pale blond. Individuals may show marked colour contrasts: for example, the back and chest may be one tone, and the arms and legs another. Coat colour does not provide a means for distinguishing the two species.

Perhaps the most easily recognisable difference is the colour of the hair covering the hands and feet. In the lar gibbon this is always white, giving the appearance of wearing white gloves and bootees' (Harrison, 1966); in the agile gibbon the hair on the cheiridia (hands and feet) is the same colour as the rest of the coat. Both species also have white marking around the face, which takes the form of a complete ring in the lar gibbon and of separate eyebrow and cheek markings in the agile gibbon. The cheek markings are not always present in the agile gibbon, usually being missing in adult females (Groves, 1972; Gittins, 1979).

The common name currently used for both lar and agile gibbons in the Malay Peninsula is 'wa-wa', derived onomatopoeically from their calls. In areas where both species are found living in close proximity, local residents distinguish between them by the colour of cheiridia and by calls. The name 'wak-wak' is retained for the lar gibbon, but the name 'ungka' is given to the agile gibbon, since it describes more closely the bisyllabic 'whoo-aa' call of this species (Gittins, 1979).

The first study of gibbons in the wild, in fact one of the first primate field studies, is the now classic work of C.R. Carpenter (1940) on the lar gibbon in North Thailand. Carpenter observed several groups of lar gibbons over a 3-month period and was able to describe the salient features of their behaviour and ecology. They lived in family units of an adult male, his mate and up to four offspring. Each family lived in a territory defended by loud territorial singing and by fights on the boundary with neighbouring groups. Group members moved around the territory together, and the most important component of their diet was succulent fruit. It was more than 20 years before gibbons were again studied in the wild. This second study was also of the lar gibbon, but this time the work was carried out in Malaysia over 20 months by John Ellefson (1968, 1974). This longer study confirmed Carpenter's findings and helped extend our knowledge of the behaviour and ecology of gibbons.

The second species of gibbon to be studied in detail was the larger siamang, also in the Malay Peninsula (Chivers, 1974). Before this two-year study only brief reports had been published on the behaviour of the wild siamang (McClure, 1964; Kawabe, 1970; Koyama, 1971). Chivers (1972) compared his findings on the siamang with those of Ellefson on the lar gibbon, and found their basic social organisation to be the same but that several aspects of their ecology differed. Siamangs travelled less each day than lar gibbons, lived in smaller home ranges and ate more leaves. These ecological differences were associated with small differences in social behaviour, such as the more cohesive nature of the siamang group and their less intense territorial behaviour.

These differences between siamang and lar gibbons have been studied in greater detail recently by simultaneous study of both species at Kuala Lompat for six months (MacKinnon, 1977; MacKinnon and MacKinnon, 1978) and for two years (Raemaekers, 1977, 1978a, 1978b, 1979). These studies were aimed at discovering how two closely-related species of ape with apparently similar feeding niches could coexist; in both cases the subjects included the siamang group studied by Chivers and the lar group who shared the same area of forest. Complementing these studies at Kuala Lompat is the only field study of the agile gibbon, in the north of the Peninsula (Gittins, 1978, 1979). Since lar and agile gibbons are so similar morphologically, a major aim of this study was to see if their allopatry could be explained in terms of similar ecological requirements, so that, following from the theory of competitive exclusion, they would be unable to coexist.

Apart from these long-term studies of gibbons in the Malay Peninsula, short studies of the Kloss gibbon, *H. klossii*, in the Mentawai Islands (Tenaza and Hamilton, 1971; Tenaza, 1975, 1976; Tenaza and Tilson, 1977), showing that this species is also monogamous and territorial, have been followed by a two-year study in another site on Siberut (Whitten, 1980). In most aspects of their behaviour they appear to be very similar to the lar and agile gibbons.

Hybrid zones do occur between populations of the smaller gibbons on mainland Asia, and an interesting account of the overlap between lar and pileated gibbons in the Khao Yai National Park has been given by Marshall et al. (1972); more detailed studies have continued (Brockelman, 1978; Marshall and Brockelman, in prep.). Gittins (1979) describes hybrids of lar and agile gibbons in the north-west of Malaya. Other long-term field studies include those of Dede Leighton (in prep.) on Müller's gibbon in East Kalimantan, Marcus Kappeler (in prep.) on the moloch gibbon in west Java and Srikosamatara, S. (in prep.) on the pileated gibbon in Thailand.

The only other accounts of free-ranging gibbons have been from brief sightings during surveys and long studies of other species or from short studies concerned mainly with social relations and song. A full bibliography of gibbon field studies is given in Baldwin and Teleki (1974).

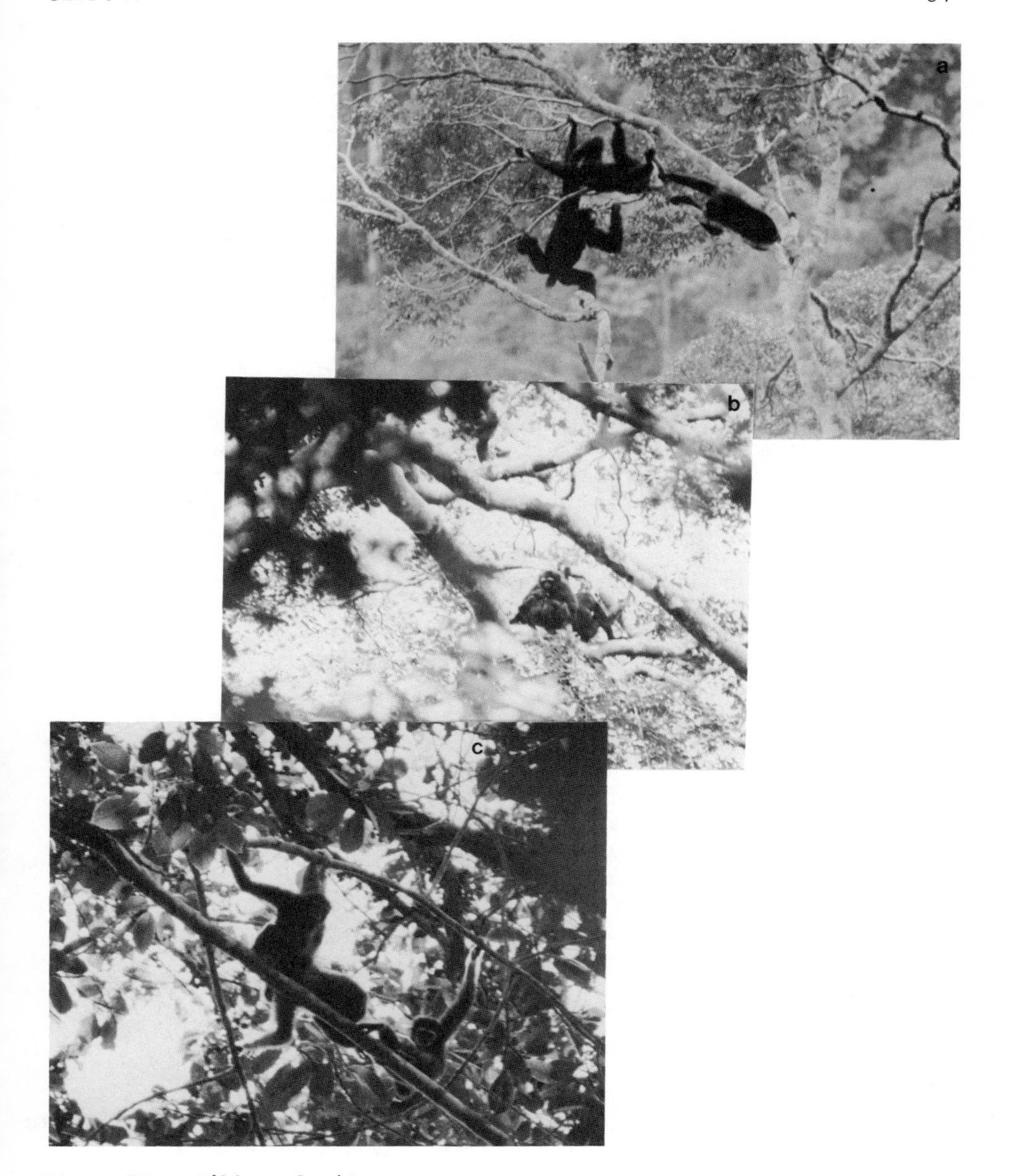

Plate IX. Gibbon family groups

(a) siamang in Ulu Gombak, GS19 - male, female, sub-adult female (DJC,1969)
(b) agile gibbons in Sungai Dal, DG1 - male, female, infant (SPG,1976)
(c) lar gibbon at Kuala Lompat, TG2 - male, female, juvenile female (DJC,1974)

Two captive colonies of gibbons have been established, one off the coast of Thailand and one in Bermuda. The Thai colony was set up in natural surroundings and many pertinent observations of group formation and social interactions have been made there (Berkson et al., 1971; Brockelman et al., 1973, 1974). The colony on Bermuda, although living in very unnatural conditions, yielded a very detailed ethogram of the gibbons' behaviour (Baldwin and Teleki, 1976).

The purpose in this chapter is to present a general picture of the way of life of family groups of siamang, lar and agile gibbons. Information on the siamang comes from 11 years of study of TS1 at Kuala Lompat, supplemented with observations of a neighbouring group TS1a and of groups in the hill forest of Ulu Sempam (Chivers et al., 1975; Chivers, 1974). To facilitate a comparison with the shorter studies of the other two gibbon species, the data presented here on the siamang come mainly from Raemaekers' study, in which equivalent data are available for the lar gibbon group TG2 (GA of MacKinnon and MacKinnon, 1977, 1978) and for the agile gibbon group DG1 at Sungai Dal. Where relevant, the results of Carpenter (1940) and Ellefson (1974) are compared with those presented here.

The main period of data collection for the siamang and lar gibbons at Kuala Lompat was April 1975 to March 1976 inclusive, and for the agile gibbon at Sungai Dal from April 1975 to February 1976. During this period a monthly schedule of observations proved most practical. The siamang and lar gibbons were followed for 10 days each month; almost all the 120 days for siamang are unbroken from dawn to dusk, but some of the 118 days for lar gibbons are incomplete. The agile gibbons proved more difficult to follow, and efforts were made to follow them on two 10-day blocks each month; this resulted in 55 complete days (out of 177) of dawn-to-dusk observation and 890 h of visual contact. The location of study groups on the study areas are shown in fig. 3.1.

SOCIAL ORGANISATION

Group composition

Gibbons live in small family groups of an adult male, his mate and up to four offspring (Plate IX). They live for about 20 - 30 years and appear to pair for life (see chapter 8). This high level of fidelity is important in a species in which the maturation rate is slow, and where the young are not fully independent until 7 or 8 years of age. The usual group size in the wild is about four individuals (table 3.1). Young are born singly at intervals of 2 - 3 years, with sexual behaviour tending to occur for only a few months at these intervals. The gestation period in captivity is given as 210 days for lar gibbons and 230 - 235 days for siamang (Napier and

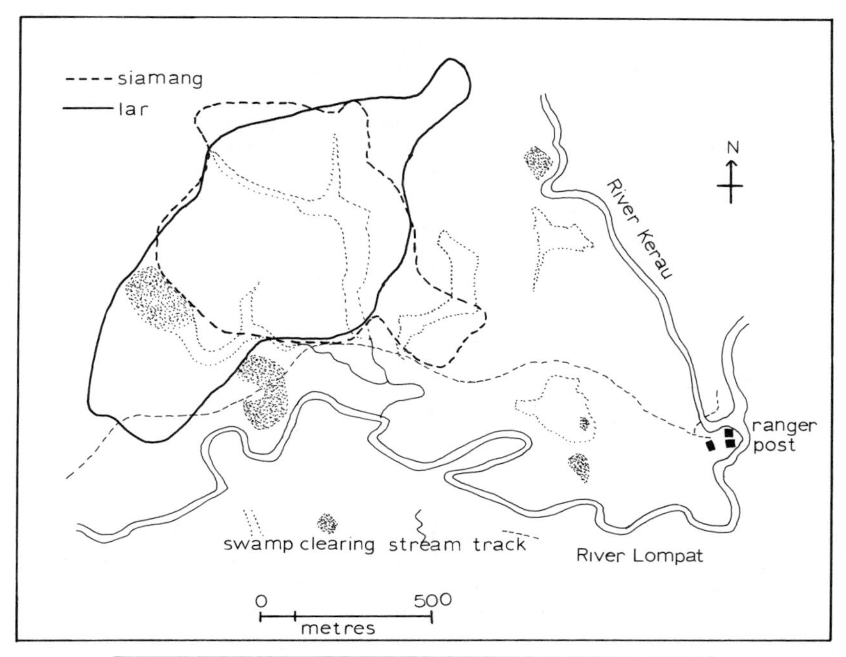

b)

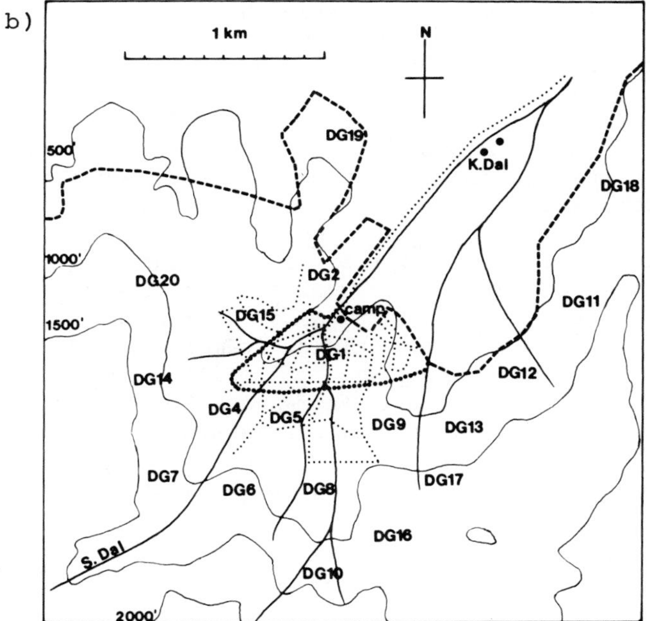

Fig. 3.1. Study sites and locations of study groups (a) Kuala Lompat, (b) Sungei Dal (boundary of Forest Reserve shown as dashed lines).

Table 3.1. Mean group size. (1) MacKinnon and MacKinnon (1978); (2) Ellefson (1974); (3) Carpenter (1940).

	Mean group size	n
siamang		
Kuala Lompat[1]	3.0	6
lar gibbon		
Kuala Lompat[1]	3.3	6
Tanjong Triang[2]	3.3	4
Doi Dao[3]	4.4	21
agile gibbon		
Sungai Dal	4.4	7

Napier, 1967).

Five age classes can be recognised according to the size of the animal and its behavioural development. The infant, from birth to 2 - 3 years, is very small and is carried by the adult female for the first year of life; during the second year, the infant lar or agile gibbon is still carried by the female during major travel bouts, but the infant siamang is carried by the adult male. This kind of paternal care is very uncommon among mammals, and is presumably related to the monogamous social organisation found in gibbons (see discussion).

On the birth of a new sibling, the young gibbon enters the juvenile stage where it gradually becomes more independent. If births occur frequently there will often be two juveniles in the group at the same time. The juvenile-1 stage extends from 2 - 4 years; the individual is small, travels independently, but still tends to orientate closely to the adults. The juvenile-2 stage, from 4 - 6 years, comprises a medium-sized gibbon which often travels and feeds alone. Since infancy is more extended in siamang, and the social group more cohesive, it is difficult to differentiate the juvenile stage, which extends from about 3 - 6 years. The sub-adult stage, from six years of age, consists of a full-sized gibbon which still lives with the family, or which is peripheralised, and is unmated. Adults are full-sized animals living in pairs, and usually with offspring.

The compositions of the study groups during the study period were:-

siamang (TS1) - adult male, adult female, sub-adult male, juvenile-1 female;
lar (TG2) - adult male, adult female, sub-adult male, juvenile-2 female;
agile (DG1) - adult male, adult female, juvenile-2 male, infant.

The process of young gibbons leaving the group has been observed in both siamang and lar gibbons, but so far only for sub-adult males. As he matures there is increasing antagonism with his father; he is forced to remain outside food trees until the rest of the group has fed, and he spends more and more time away from the group. Carpenter (1940) suggests that it is sexual jealousy of the father which motivates him to oust his growing son, which is supported by the coincidence of peaks of copulation between parents and aggression between father and son in siamang in 1970 (Chivers, 1974). Presumably the growing urge to find a mate, coupled with increasing aggression from its parents (mainly the male) results in the sub-adult (male or female) finally leaving the group to establish one of its own.

There are few observations on the actual process of group formation, but the most likely sequence of events is as follows. Unmated young males sing for long periods, either within the parental home range but at some distance from the rest of the group, or from areas of forest unoccupied by other groups (Ellefson, 1974; Aldrich-Blake and Chivers, 1973; Tenaza, 1976; MacKinnon and MacKinnon, 1977). These males eventually attract females, presumably by their unaccompanied song, and the pair court. Females rarely sing alone, and are not known to attract potential mates by singing.

A new pair of lar gibbons at Kuala Lompat courted for several months before mating was seen, and a young male siamang courted with several females without settling down before he disappeared. Such care in choosing a mate is understandable, since, if they are to have only one such partner, a bad choice would greatly reduce their reproductive success.

Behaviour within the group

Gibbons live in small family groups and are able to coordinate activity and ranging behaviour with the minimum of social interactions. Overt communication between members of the group is infrequent, and activity appears to be coordinated by social perception (Altmann, 1962); that is, the awareness of the activity of other members of the group, and awareness of changes in activity.

The adult pair are extremely tolerant of each other, and aggressive interactions are infrequent. They often feed close together and very rarely does one supplant the other. The lack of sexual dimorphism between male and female is probably a major factor in the lack of dominance of one over the other in group behaviour; this led Carpenter (1940) to coin the term 'co-dominance' to describe the relationship between the adult pair.

The male and female show considerable tolerance of the infant, who climbs over and jostles them with impunity. As juveniles mature, however, they become the object of brief attacks, mostly by the male. Juveniles and sub-adults, of the smaller gibbons in particular, are often chased from food trees, and they frequently rest apart from the adults. In lar and agile gibbons the adult male threatens the juvenile with an open-mouth stare (Ellefson, 1974), and the juvenile responds by moving away or making an appeasement gesture; this consists of a grimace and a whimper call (Ellefson, 1974). When juveniles move toward adults they sometimes make this call and expression with a hunched posture. Similar behaviours have been described for siamang (Chivers, 1976).

Grooming occurs during the longer rest bouts and is initiated by one individual moving to another sitting close and presenting, which involves leaning forward almost flat and displaying the back for grooming. Grooming may then occur, but after a few minutes the groomer presents and may be groomed in return; this alternation continues for about 10 minutes before the gibbons rest and/or move apart. During Raemaekers' study, members of the siamang group spent 1% (5 min) of the activity budget grooming, and those of the lar group 3% (18 min). During Chivers' study, members of the same siamang group spent 6% (41 min) of their alert period grooming; this figure includes, however, the small amount of grooming that may occur between adult and infant or small juvenile as they settle for the night, and, as such, is not recorded as part of Raemaekers' activity budget. Since members of lar and agile gibbon groups sleep spread out in different trees, no grooming takes place after they retire to night trees. Grooming was so rare in the agile gibbon group that it does not score in the calculations of the activity.

Thus it is evident that gibbons spend only a small part of their day grooming. There is some suggestion that siamang may groom more than the smaller gibbons, and this may be related to the more cohesive nature of the siamang group. The commonest grooming association in the siamang group was between adult and sub-adult males, and in the agile gibbon group between adult and juvenile males. It could be that this is a result of the tenser relationship between adult males and older offspring.

Young gibbons have no peers in their group with which to play. If play occurs, it must be either solitary or with an individual at least two years older or younger. Solitary play involves various forms of acrobatics, from hanging and swinging to chasing around in circles; twigs and leaves are also played with, being manipulated, bitten and chewed. Play involving two individuals is started by one gibbon moving to another and grappling with it. Sometimes there is biting, and soft grunts and growls. Play tends to occur during rest bouts while older animals groom; it took up 1% (8 min) of the siamangs' activity period, and 0.2% (1 min) of the lar gibbons'. Play was observed several times between juvenile and infant in the agile gibbon group, but it was so infrequent that it does not score in calculations of the activity budget.

The siamang group is remarkable for its cohesion and synchrony of activity. During the day, the five members of the siamang group spent more than half their time less than 10 m apart, and they were all engaged in the same activity in 74% of scans (Chivers, 1974). By contrast, lar and agile gibbon groups are less 'tightly knit'. The male, female and juvenile siamangs were all visible in 96% of scans, but the equivalent members of the lar gibbon group were all visible in only 57% of scans. The male and juvenile agile gibbons were more than 10 m from the female for more than 50% of the time, and more than 30 m for 12%. The siamang are also more cohesive in their sleeping habits; they sleep in one or two adjacent trees, whereas the smaller species scatter, sometimes over hundreds of metres.

The order in which a group of animals travels is often stable, with certain classes of individuals habitually taking up positions in the front, middle or rear of the group. In the siamang, travel from one part of the home range to another occurs in discrete bouts; the group moves in single file, with the female usually leading (65% of bouts). It is difficult to establish if this is true leadership, however, since the male may be directing movement from his central position (Chivers, 1974). In the smaller gibbons, movement around the home range is of two kinds: either the group moves in single file, or individuals disperse to follow separate paths along a broad front. During the former, male and female agile gibbons showed an equal tendency to be first, with the juvenile following behind, whereas any member might lead in the lar gibbon group.

Behaviour between groups

Each group occupies a territory, which it advertises by singing and which it defends during conflicts involving chases and calls. The song is delivered in bouts lasting about 15 min, given mostly in the morning (fig. 3.2) from anywhere within the territory. The singing of one group stimulates others to sing, and the songs pass around the population, like a chorus in the smaller gibbons but more sequentially in siamang. In addition to these song bouts, paired agile gibbon males frequently sing long solos, mostly starting before dawn; similar dawn bouts occur occasionally in lar gibbons. Unmated males sing solo in all species.

The group song is divided clearly into male and female parts. Either or both give a series of preliminary *whoops*, then the female utters her *great call* (a protracted series of notes rising to a climax), after which the male joins in with a *coda*. Each sequence of calls lasts about 20 - 30 sec, and is given at 2 - 3 min intervals, between which various whoops are given. The song of the two smaller

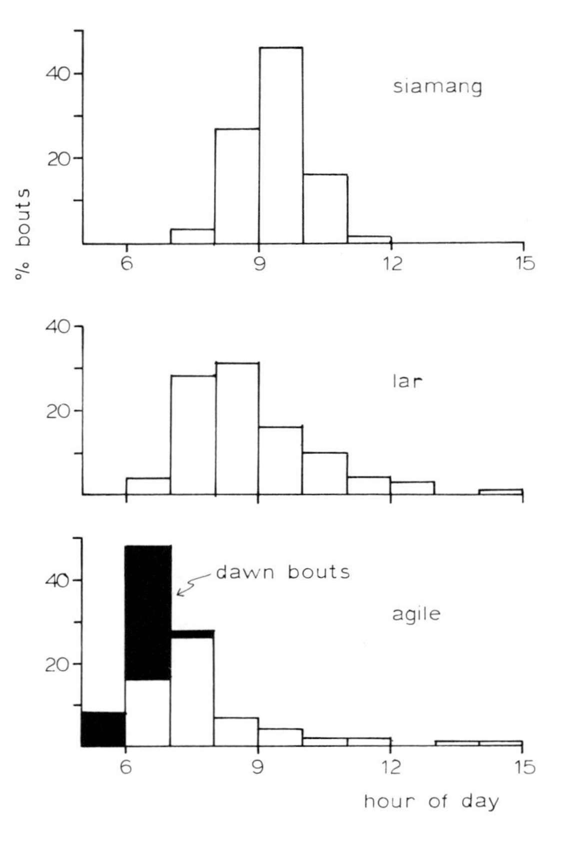

Fig. 3.2. Distribution of singing through the day, as percentages of song bouts starting in each hour (study groups only: n = 37 bouts by siamang, 107 bouts by lar gibbons and 330 bouts by agile gibbons).

gibbons are similar in form; that of the siamang diverges from them both in form and voice quality. Siamangs can be heard over about $1\frac{1}{2}$ km in flat terrain, and the smaller gibbons up to 1 km.

Conflicts occur when two groups come close together near the boundary separating their two territories, and last from a few minutes to over two hours. During this time the males give soft calls, sitting for much of the time between swift chases. Females usually remain in the background, but may give great calls and groom the males. Siamang calls in such situations are noisy but more discordant than in song bouts. Sometimes a food tree is the object of dispute, when one group arrives to find another feeding in a 'favourite' tree. Mostly, however, the groups seem to dispute nothing more than an imaginary boundary fence.

The frequency of singing and of conflicts varies widely between populations and over time. Weather, food abundance, sexual activity, and population density and the appearance of new neighbours all affect singing rate. During one period at Kuala Lompat the siamang group sang 3 times in 59 days, yet during another they sang on most days, and twice on many of them. As a rough guide, siamang sing on about one day in three, lar gibbons about once a day and agile gibbons twice a day (once before dawn and once during the morning).

The rate of conflicts is also much affected by the number of neighbouring groups, and especially by the appearance of new neighbours. At Kuala Lompat, where the study groups have few neighbours, conflicts can be as rare as 7 in 59 days for the lar gibbons and 0 in 59 days for siamang. By contrast, at Tanjong Triang conflicts occurred among lar gibbons on 50% of days (Ellefson, 1974), which is of the same order as 38 disputes in 55 complete days for the agile gibbon group at Sungai Dal. The low level of territorial behaviour exhibited by TS1 is a consequence of having few immediate neighbours; the siamangs in Ulu Sempam had more conflicts (on one day in three), but at a level still considerably less than that of the smaller gibbons.

Singing and conflicts between groups are very important features of gibbon life, consuming much time and energy and occurring mostly in the morning when the gibbons are also seeking their preferred foods. Ellefson (1974) estimates that a male lar gibbon spends 6% of his waking hours in territorial behaviour; the comparable score for the agile gibbon male at Sungai Dal was 13%. No other Old World primate appears to place such emphasis on territoriality, and few primates of any kind maintain so effectively exclusive access to a permanent territory.

RANGING

Home range size

From the location of territorial disputes between neighbouring groups of gibbons, it is possible to delimit very precisely the boundary of the territory; this boundary remains fairly static over long periods (two years at least in the case of the agile gibbons at Sungai Dal, fig. 3.3). These boundaries seem to have some psychological significance for the gibbons (Ellefson, 1974), as they rarely wander over them, even when neighbouring groups are absent at the time. The home range, defined as the total area used by the group in a given period, is larger than the territory. The relative unimportance of the area outside the territory is best illustrated by the observations that the territory of the agile gibbon group at Sungai Dal was 76% (22 ha) of the home range (29 ha), and yet 92% of the distance travelled in the 55 complete days was

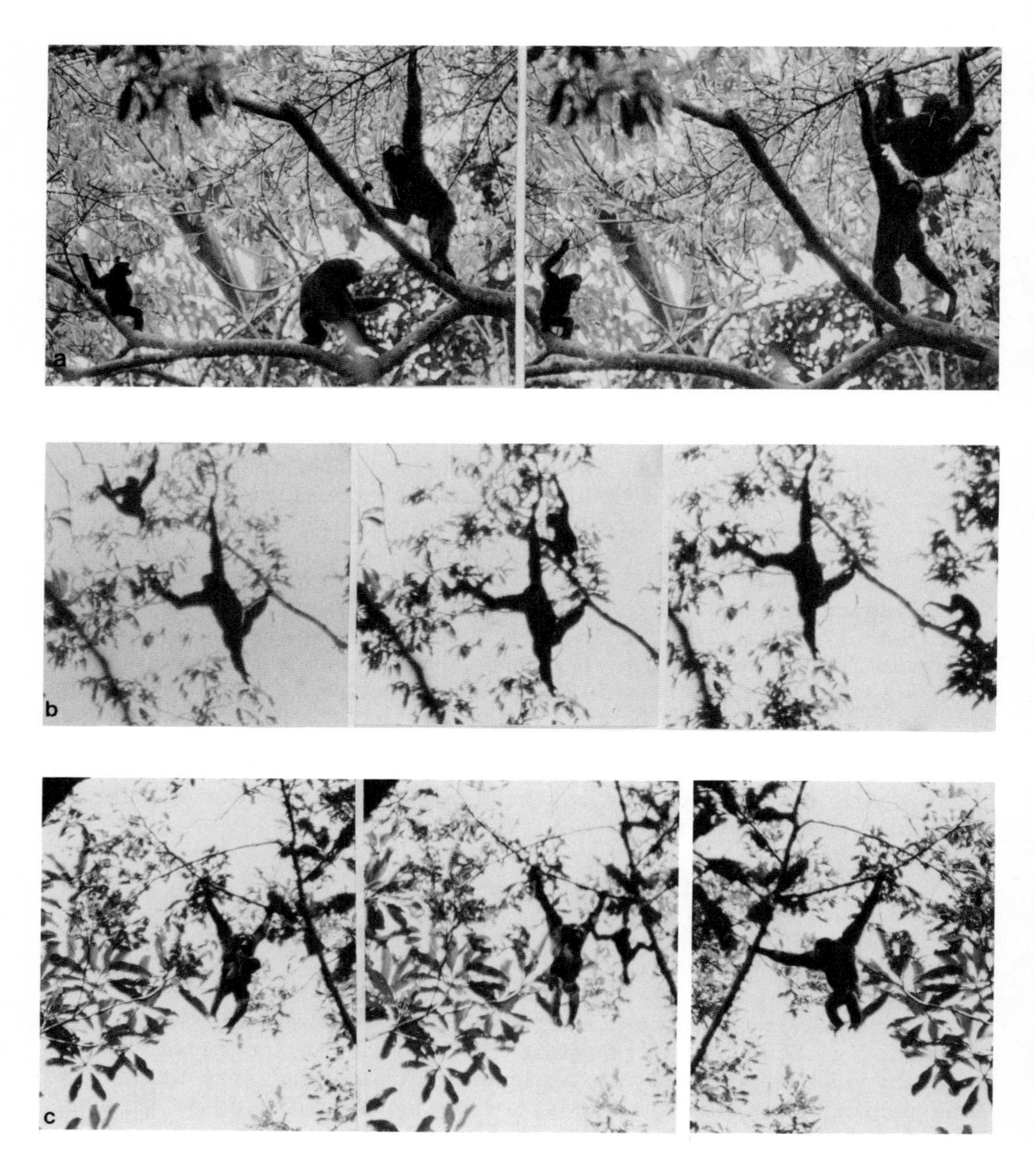

Plate X. Gibbons travelling

(a) siamang in Ulu Sempam, RS2 - female, male, infant male (DJC,1970)
(b) siamang at Kuala Lompat, TS1 - male helping infant between trees (DJC,1970)
(c) agile gibbons in Sungai Dal, DG1 - female and infant, male (SPG,1976)

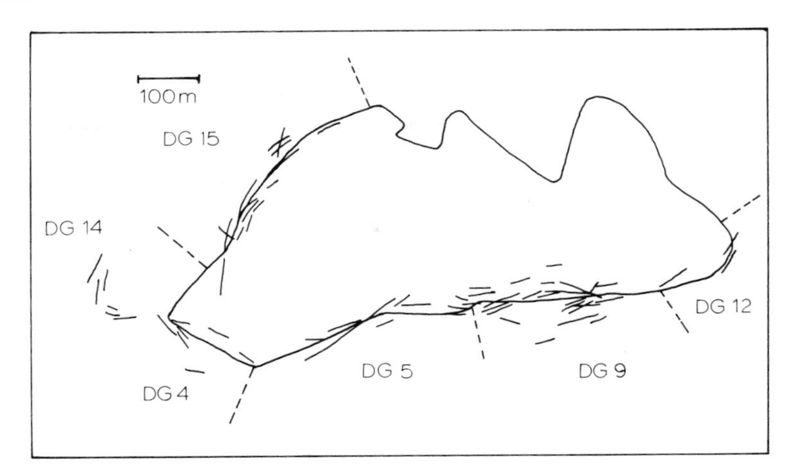

Fig. 3.3. Locations of territorial disputes involving agile gibbon group DG1 at Sungai Dal (territory marked by solid line and neighbouring groups labelled, n = 76 disputes over 177 days).

within the territory.

In dense populations such as Sungai Dal and Ulu Sempam, territories abut onto each other with little space between groups. In sparser populations such as Kuala Lompat, groups may be so widely spaced that territories cannot be defined and even home ranges may not abut. From present evidence it seems that lar and agile gibbons occupy territories and home ranges of similar size, whereas those of siamang are smaller (Table 3.2). Considering the small social groups, all the gibbons live at a much lower density than the other primates living in the same forest (see chapter 10).

Home range use

The two major factors affecting a group's coverage of its home range are (1) the distance travelled each day by each group member and (2) the dispersion of the group. Lar and agile gibbons have relatively long day ranges for arboreal primates, and group members are widely dispersed while feeding, which results in extensive and frequent coverage of all parts of the home range (fig. 3.4). The siamang, on the other hand, has a shorter day range and group members keep closer together. Lar and agile gibbons travel about $1\frac{1}{2}$ km each day (no significant difference), and siamangs travel significantly shorter distances, about 1 km or less (table 3.3; fig. 3.5). The siamang's shorter routes are also associated with fewer, longer visits to food sources each day. Such differences between the daily ranging behaviour of siamang and lar gibbon are displayed in fig. 3.6.

Table 3.2. Home range size. (1) Chivers (1974); (2) Ellefson (1974); (3) Carpenter (1940).

	Mean ha	Range	n (groups)
siamang			
Kuala Lompat 1968-70[1]	34		1
1975-76	48		1
Ulu Sempam[1]	15		1
lar gibbon			
Kuala Lompat	54	50-58	2
Tanjong Triang[2]	59	20-46	4
Doi Dao[3]	25	16-32	3
agile gibbon			
Sungai Dal	29		1
Ulu Muda	25		1

Since the members of a gibbon group spread out while moving, the group covers more of the forest than any one of its members, whereas members of the more cohesive siamang group tend the follow the same route. For example, at Kuala Lompat members of the siamang group followed partly different routes on only 10% of days, whereas members of the lar group did so on 78% of days over the same period. Thus, if the total behaviour of the two groups were compared in fig. 3.6, such contrasts in dispersion and forest coverage would increase the distinction.

A simple method of investigating the way these gibbons use their home range is to divide the area into quarters about its centre. The siamang group at Kuala Lompat took much longer to enter all four quarters of its home range than did the lar group (table 3.3). The use of the home range by the agile gibbons at Sungai Dal was investigated by dividing the area into thirds from east to west- they covered their range at about the same rate as the lar group (table 3.4).

This difference in home range use is also reflected in the number of visits made to each hectare in the home range over a year.

Table 3.3. Day range length.

	Mean	Median	Range	n
siamang	738		200-1700	119
lar	1490		450-2900	92
agile	1217	1260	650-2200	55

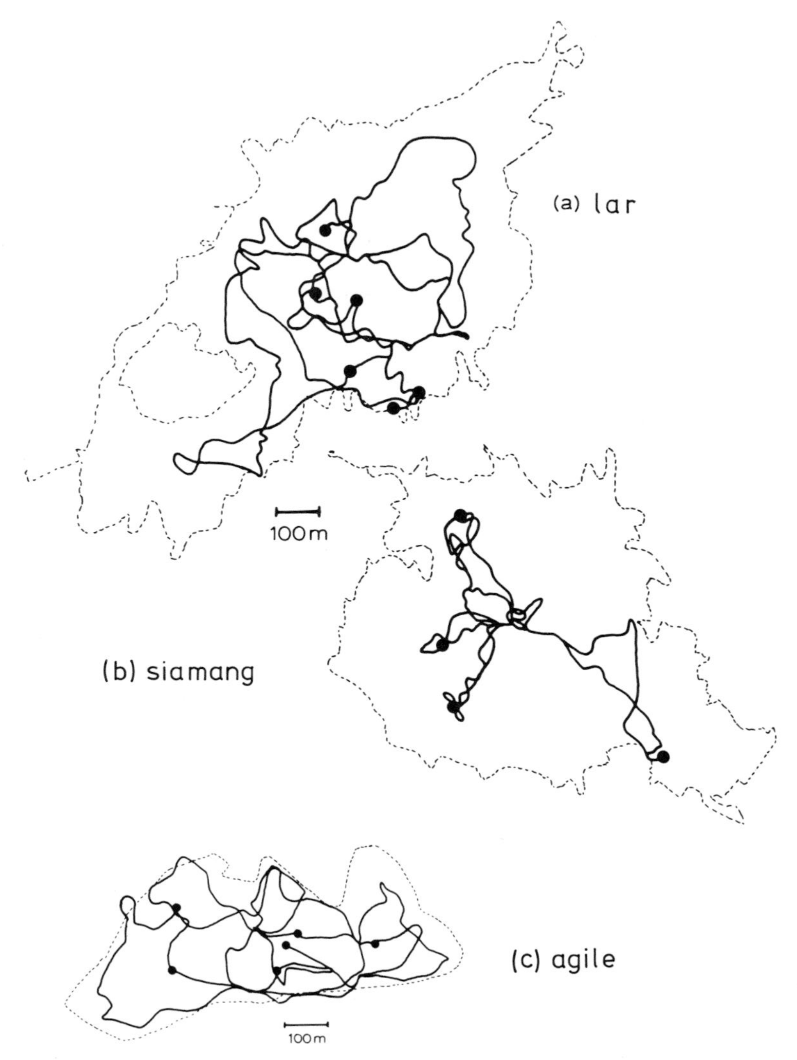

Fig. 3.4. Patterns of ranging over 5-day periods. (a) lar gibbon, (b) siamang and (c) agile gibbon (home range boundary marked by broken line, sleeping sites by black dots; siamang and lar gibbon routes recorded over the same 5-day period.

A lar gibbon visited each hectare of its home range more often than a siamang visited each hectare of its home range (fig. 3.7a). In the long term, however, siamangs and lar gibbons distribute their time over the home range in a similar way, as is shown by expressing the data as percentages (fig. 3.7b), and so do agile gibbons.

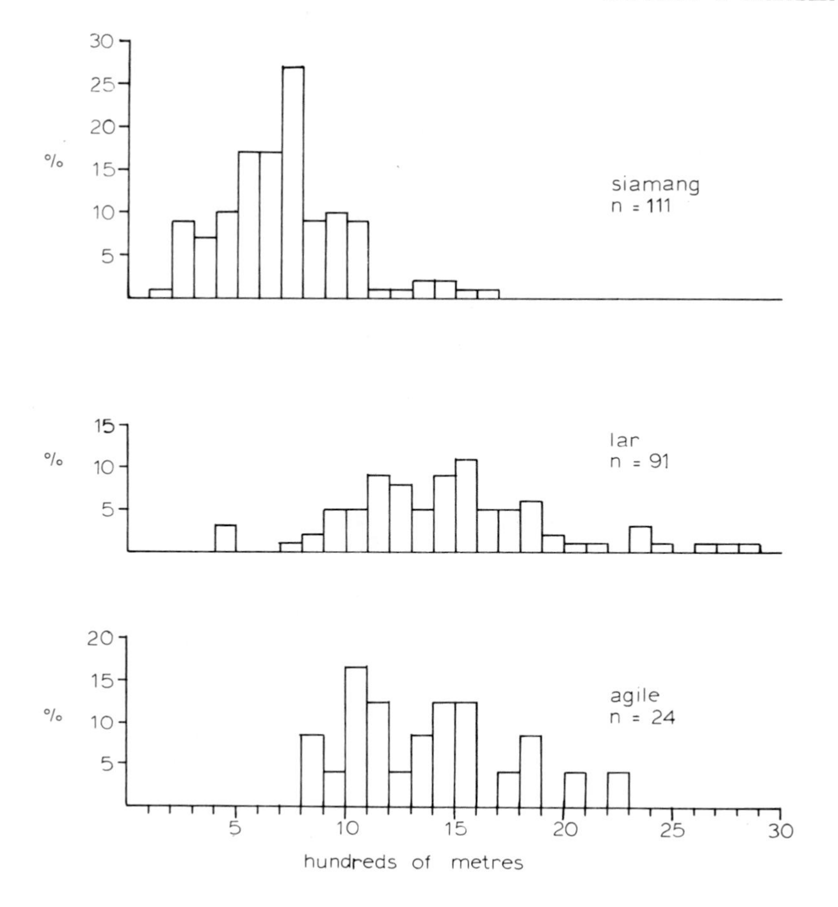

Fig. 3.5. Frequency distributions of day range lengths, as percentages of routes falling in each 100-m class (to represent the seasons equally for agile gibbons, 8 routes selected randomly from each season).

Despite this long-term similarity, in the short term the siamang has fewer opportunities to monitor any part of its home range. Since most food sources last for only a few days or, at most, weeks, this difference is relevant to the ability of each species of harvest food.

Canopy use

In order to analyse the vertical distribution of activities in the forest canopy, individuals of the siamang and lar gibbon groups at Kuala Lompat were scored during scans every 10th min as being low (below 20 m above the ground), medium (between 20 and 40 m, 65 - 131 ft) and high (above 40 m). Heights of agile gibbons were estimated

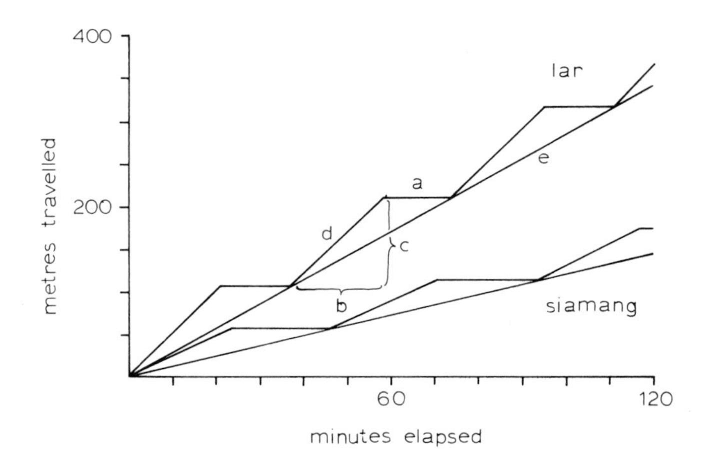

Fig. 3.6. Comparison of daily ranging behaviour of siamang and lar gibbon: the slope drawn for each species is the average for one adult male and one adult female over 6 months (n = equivalent of 119 animal-days for siamang and 92 for lar; a = duration of feeding visit, b = interval between feeding visits, c = rate of progression between feeding visits, and e = overall rate of progression during activity period).

to the nearest metre and classified subsequently into lower canopy (5 - 15 m, 16 - 49 ft, above ground), middle canopy (15 - 25 m), upper canopy (25 - 35 m, 82 - 115 ft) and emergents (above 35 m).

The distributions of the heights for different activities are shown in fig. 3.8. Singing occurs from the highest tree tops, which increases the distance that the song carries and which could allow the acrobatics accompanying the song to be seen by surrounding groups. Sleeping also occurs high in the canopy, probably as a precaution against predators. The gibbons travel in the middle and upper canopy, rather than in the discontinuous emergents and lower canopy; agile gibbons travelled mainly in the upper canopy, where there were many spreading trees with open networks of branches that

Table 3.4. Number of days to cover home range. The difference between the siamang and lar is significant (median test: $\chi^2 = 6.24$, df = 1, $p < 0.02$).

	Mean	Range	n	Home range
siamang	6.0	1 - 21	125 (days)	47
lar	2.5	1 - 10	121 (days)	57
agile	2.2	1 - 5	24	29

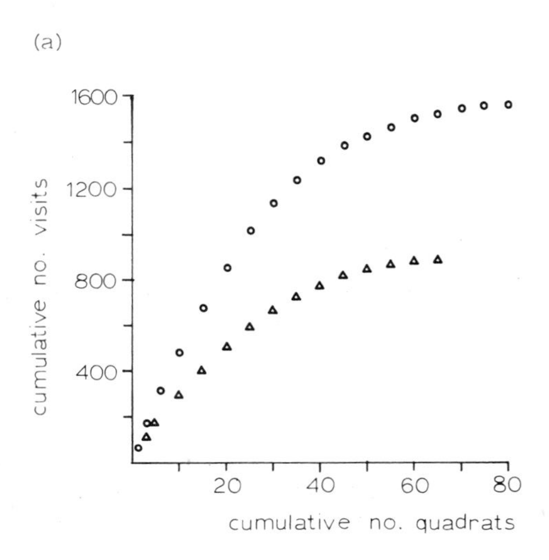

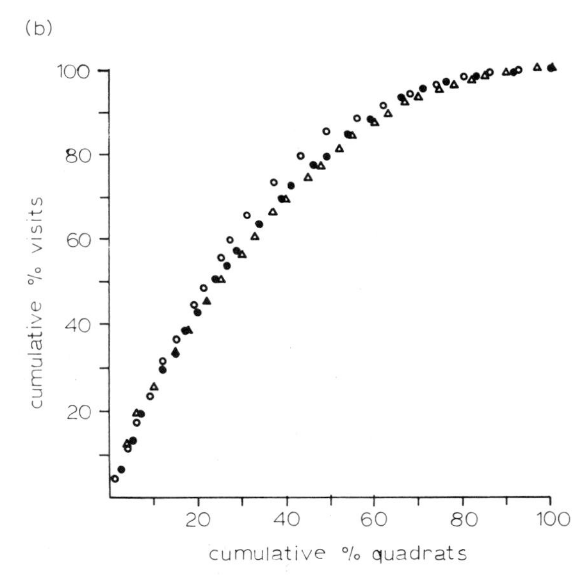

Fig. 3.7. Distribution of time about home range: (a) cumulative number of visits to hectare quadrats plotted against cumulative number of quadrats ranked by the number of visits each received (siamang and lar gibbon only, 12 months); (b) the same data converted to percentages, with data for agile gibbon also.

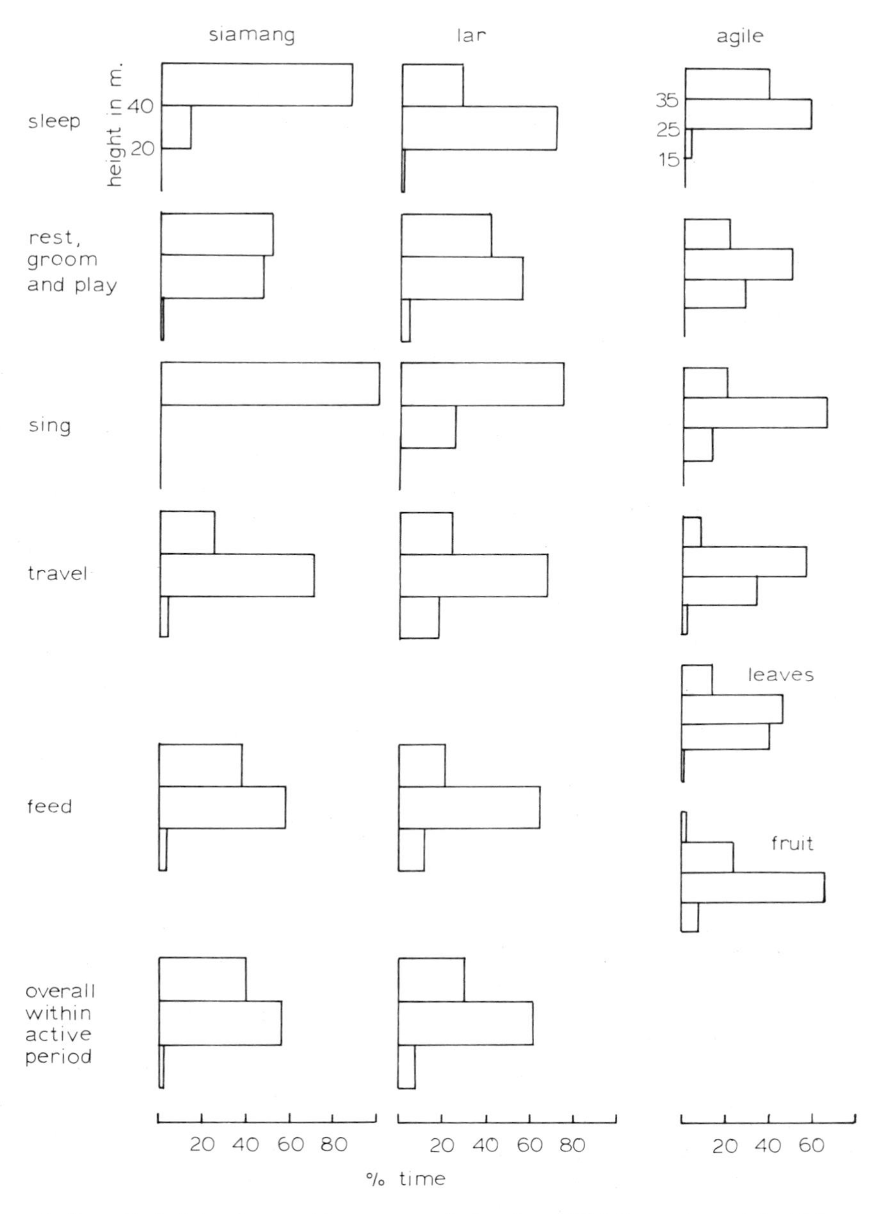

Fig. 3.8. Height above the ground during different activities. For lar gibbon and siamang the canopy is divided into three layers, for agile gibbon into four. Note also that heights while feeding on fruit and leaves are calculated separately for agile gibbon. (n = 5464 observations for siamang, 5984 for lar gibbon and 2852 observations for agile gibbon).

allowed them to brachiate freely.

Feeding (Plate XI) also tended to occur at medium heights, probably because the emergents are mainly dipterocarps, which produce little food for gibbons. Many of the small trees in the lower level will be young trees which have not yet reached reproductive age (Ng, 1966), so they will not attract gibbons with fruit. The division of the main canopy into upper and middle levels at Sungai Dal reveals an interesting relationship between the height at which gibbons feed and the type of food eaten: fruit-eating occurred mainly in the middle canopy of smaller trees, whereas leaf-eating occurred significantly higher.

Vertical stratification has been found to contribute to the ecological segregation of sympatric bird species (Lack, 1971; Cody, 1974); it also appears to be important among forest primates, since congeneric leaf monkeys and macaques at Kuala Lompat are also segregated by height (chapter 6). Although the differences between siamang and lar gibbons are not large, the siamang's preference for higher levels for all activities lies in the direction predicted from several considerations.

Siamangs favour larger food sources than the smaller gibbons; since larger food sources tend to be taller trees, siamangs will feed higher. As a function of their more folivorous diet, siamangs also spend more of their feeding time eating the leaves of tall trees, such as *Durio singaporensis* and *Koompassia excelsa*. It has already been shown for agile gibbons that leaf-eating tends to occur higher than fruit-eating. Thus, differences in diet could be the main cause of the siamang feeding higher than the lar gibbons.

A second reason relates to the structure of the forest at different heights. The smaller size of lar and agile gibbons allows them to move more easily on flexible supports, especially in smaller trees. The agile gibbon tends to brachiate on smaller supports than the siamang, and to feed more often from smaller supports (Gittins, 1979). Although the siamang is able to travel and feed on thin branches, by using more limbs to spread its greater weight over more supports (Fleagle, 1976c), it is reasonable to suppose that it is clumsier and less efficient at travelling and feeding on such supports than the smaller gibbons. Thus the latter may be better able to exploit the smaller trees with finer branches in the lower levels of the forest, and this increased mobility may represent some degree of niche separation.

Plate XI. Gibbons feeding

(a) siamang at Kuala Lompat, TS1 - female (1969) and juvenile female (1977) (DJC)
(b) lar gibbon at Kuala Lompat, TG2 - male (JJR,1975)
(c) agile gibbon on Penang, juvenile male (SPG,1975)

Plate XII. Fruits eaten by gibbons

(a) *Antidesma coriacea* (Euphorbiaceae)
(b) *Sarcotheca griffithi* (Oxalidaceae)
(c) *Bouea oppositifolia* (Anacardiaceae)
(d) *Knema laurina* (Myristicaceae)
(e) *Baccaurea motleyana* (Euphorbiaceae)
(f) *Vitis* sp. (Vitaceae)

(h) *Ficus bracteata* (Moraceae)
(i) *Ficus annulata*
(j) *Ficus auriantacea*
(k) *Ficus heteropleura*
(l) *Ficus sumatrana*
(m) *Ficus stupenda*

FEEDING

Food types

Gibbon foods can be divided into three classes: (1) reproductive plant parts, flowers and fruit, (2) vegetative plant parts, leaves and shoots, and (3) animal matter.

From a gibbon's viewpoint, a fruit is a parcel of more or less succulent pulp sandwiched between unwanted rind and seed (Plate XII). Gibbons eat few unripe fruit, and probably never chew up seeds, confining themselves to the ripe pulp of fairly juicy species. They swallow the seeds of most species, and excrete them whole; those which they do not swallow they drop whole *in situ*. Since they are likely to be hundreds of metres from the fruit source when they excrete its seeds, gibbons transport seeds well away from the parent source. Since they revisit fruit sources often during their productive periods, they are also reliable seed transporters. Provided that they do not damage the seeds in their guts, they should be ideal seed dispersers for trees. Hladik and Hladik (1969) have shown that in Panama those monkey species which eat most ripe fruit pulp often improve the proportion of seeds germinating; passage through the primate gut also seems to hasten the onset of germination. It seems very likely that gibbons have a similar effect. In aiming for juicy pulp gibbons neglect the hard, dry, wind-dispersed seeds of dipterocarps, legume pods and seeds and the hard fruit of the Burseraceae, which between them make up a high proportion of the forest's fruit production - they comprise 23% of trees over 50 cm g.b.h. at Kuala Lompat.

The choice of flowers is not so obviously related to their gross structure. At Kuala Lompat both siamang and lar gibbon have consistently shown an exceptionally strong liking for the catkins of the nettle family tree *Sloetia elongata*.

Most of the green plant parts eaten are young leaves, which have opened but not yet reached mature colour, size or texture. Unopened leaf shoots are also eaten, and, more rarely, mature leaves, petioles and pithy vine stems. Young leaves of all colours from white to deep purple are taken.

Animals, which are with few exceptions insects, are eaten in two distinct fashions according to their distributions. Massed social insects (ants, termite columns on trunks) and caterpillar infestations are eaten fast and usually for only a short time on each occasion. Scattered spiders, galls and the like are found much more slowly during foraging through mature or dead leaves. Perhaps only one item is caught each minute, which, in contrast to other activity, may last for up to two hours at a time.

Table 3.5. Features of major food classes. (Number of plusses indicates relative degree)

	Relative abundance	Ease of harvesting	Available energy	Protein content
fruit pulp	+	++	+++	+(+)[1]
flower	+	+	+++	+(+)
young leaf	+++	+++	++	+++
mature leaf	infinite	++++	+	++
insect	+	+	+	++++

[1] Note that figs often contain animal protein in the form of fig wasps and their parasites

The probable abundance of these food classes, and their content of available energy and protein, are sketched crudely in table 3.5. The term 'available' is used advisedly, for though it would obviously pay to eat mostly mature leaves, in a forest consisting mostly of mature leaves all the time, to extract much energy from them requires fermentation by symbiotic gut organisms, which can break down fibre to usable lengths of carbohydrate (McCance and Lawrence, 1929). Unlike the sympatric leaf monkeys, gibbons are apparently unable to do this. The high fibre content alone may be enough to render leaves unattractive to gibbons, but it is also likely that they contain a large variety of toxins which repel them. These toxins may account for the strong selection between species of young leaf shown by gibbons (see below). Toxins are also present in unripe fruit to protect the young seed (Janzen, 1971, 1975), and one may surmise that gibbons pass over unripe fruit for this same reason. Ripe fruit, in contrast, are rich in free sugars, which are very easily and quickly assimilated (Hladik et al., 1971). The ability to cope with toxic substances in plants may be a major distinction between apes and monkeys, with major implications for their respective social organisation (Wrangham, 1979).

Dietary proportions

The percentages of total feeding times over one year devoted to different food classes by a social group of each of the three gibbon species are shown in fig. 3.9. Over shorter periods the dietary proportions vary, of course; the most divergent proportions recorded in monthly proportions are shown in table 3.6. Figs, the fruit of *Ficus* species, are classed separately, because of their unique life history, and also their unique importance in terms of bulk in the gibbons' diet.

It is at once apparent that the smaller gibbon species eat relatively more fruit and fewer young leaves than does the siamang (fig. 3.9). Otherwise, the siamang and lar gibbons share the great

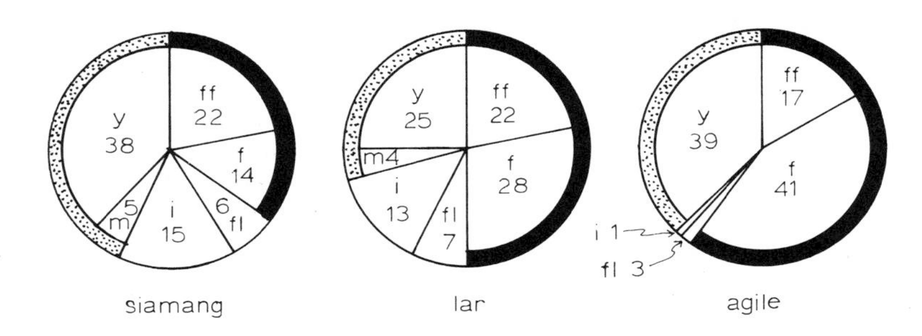

Fig. 3.9. Dietary proportions, as percentages of feeding time spent on different food classes (ff = figs; f = other fruit; fl = flowers; y = young leaves and shoots; m = mature leaves; i = insects and any other small animals; n = 13142 observations for siamang, 6852 for lar gibbon, and 1724 observations of feeding and foraging by agile gibbon adjusted for daily and seasonal variations in sample size, see Gittins, 1979).

majority of their foods, defined by plant part and species. During the year of study, however, the two groups at Kuala Lompat overlapped quantitatively in diet by only 53%; monthly values varied from 22% to 64%, which was only 10% less than the overlap between different members of the same group.

There is an interesting variation in the proportion of different foods taken at different times of day (fig. 3.10). The siamang, lar gibbon and, to a lesser extent, the agile gibbon all eat mostly fruit in the first part of the morning, and figs are strongly selected during the first and last feeding bouts of the day. Possible reasons for this pattern of feeding, discussed in detail by Raemaekers (1978b), concern the best ways of meeting energy demands when they are greatest. Thus the early selection of fruit, especially figs (larger sources), with their more readily available and abundant nutrients, provides most easily the urgent needs after the long, cool overnight fast. It follows that the peak of leaf-eating would have to be later in the day, but leaves are likely to be more nutritious after several hours of daylight (because they only

Table 3.6. Range of dietary proportions observed in different monthly samples. Ratio of [all fruit and leaf] to [all leaf].

siamang	0.23 - 0.67
lar	0.42 - 0.75
agile	0.39 - 1.00

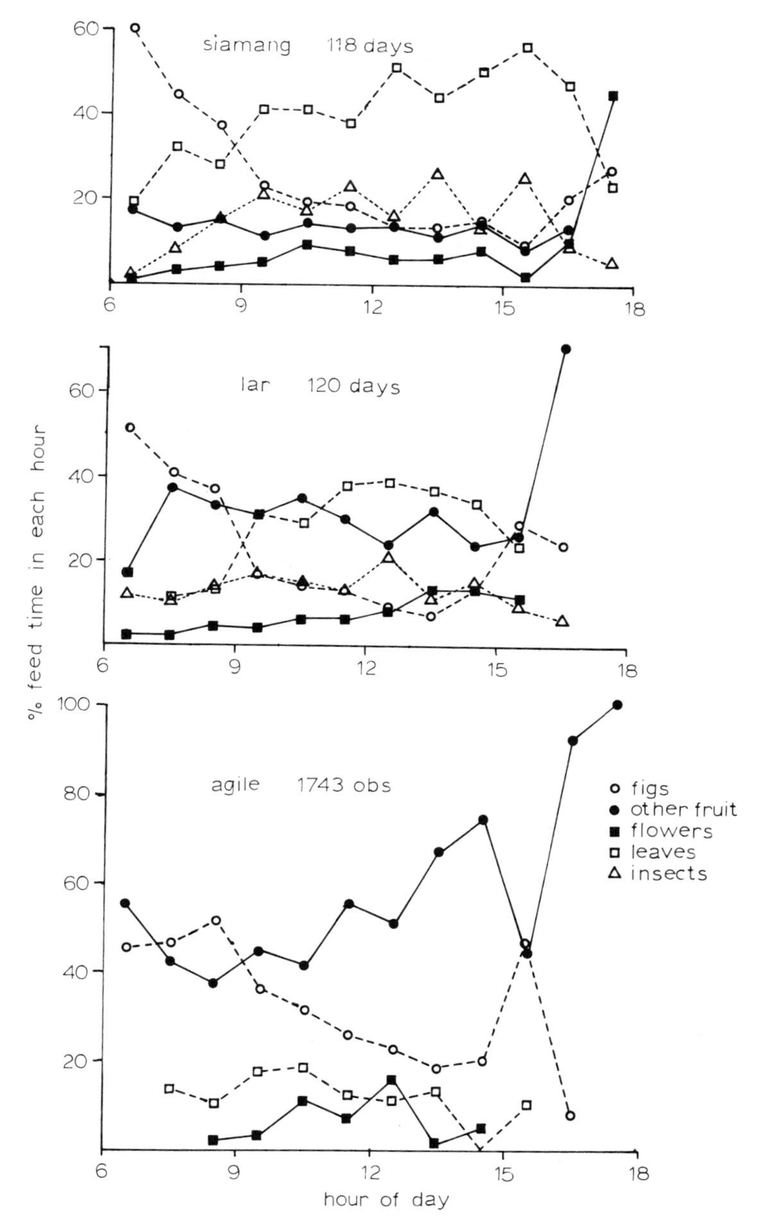

Fig. 3.10. Changes through the day in dietary proportions, by percentages of feeding time in each hour spent on each food class (foraging excluded for agile gibbon).

photosynthesize by day), and their slower digestion will help the animal through the night.

Other aspects of food selection

All three species eat a very large number of different foods. For example, during the year covered by Raemaekers' study the lar gibbon group ate at least 261 foods, defined by plant part and taxon, and the siamang group ate at least 284, in addition to animal foods. Most of these foods were only nibbled, however, and the superficial diversity of the diet belies a concentration on a very much smaller number of foods (fig. 3.11).

The gibbons are, in fact, extremely selective feeders. Reference has already been made to their selection of ripe fruit pulp and young leaves, and they are equally selective about which species they eat. Comparing the proportion of time which siamang and lar gibbon spent eating foods growing on free-standing trees with the frequency of those trees in the botanical plots (fig. 3.12), most feeding concerned tree species occurring at densities of one tree or less each hectare. To these rare food species may be added the figs, of which there were over 20 species and each of which was rare, and those food species which were too rare to occur in those plots at all. Thus, the bias toward feeding on rare species is much stronger than is shown in the figure. The anomalous column at the

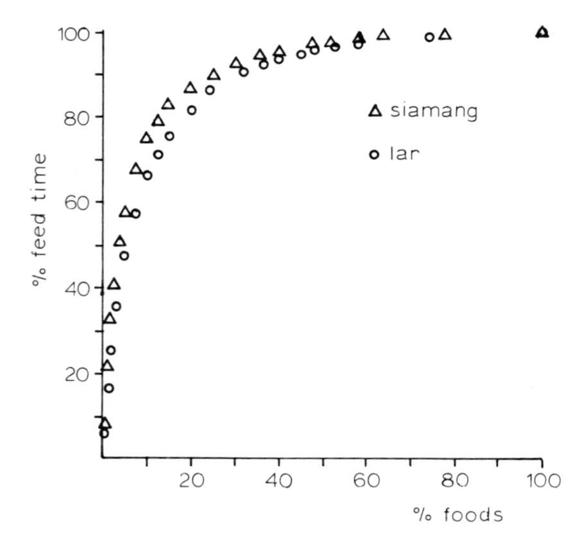

Fig. 3.11. Distribution of feeding time across plant foods for siamang and lar gibbon. Cumulative percentage of feeding observations plotted against cumulative percentage of foods ranked by the number of feeding observations each food received (a food is defined by plant part and taxon; n = 13142 observations for siamang, 6852 for lar).

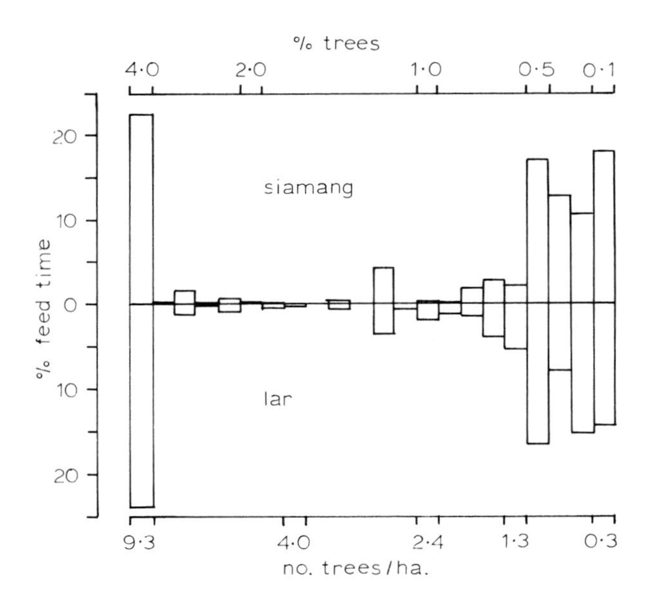

Fig. 3.12. Selectedness of siamang and lar gibbon diets over 12 months, by percentage of feeding observations on free-standing tree species occurring in the botanical sample plotted against the frequencies of those species in that sample (expressed as no. of trees/ha and % of trees).

left end of the histogram is composed entirely of one species, *Sloetia elongata*, which had the distinction of being both the commonest tree and the one most eaten by gibbons. This emphasis on a relatively common species appears exceptional, however, since no comparable records come from other sites.

Apart from selection between plant species, the gibbons clearly select between plant parts. This could be tested in a crude way at Kuala Lompat, by investigating the use of known foods in the botanical plots by siamang and lar gibbons. Trees bearing figs, other fruit, flowers and young leaves were counted by species if eaten for more than 10 observations during the year's study. Each tree was scored once for each month it bore food - an opportunity provided - and once for each month in which it was eaten. When opportunities taken are expressed as percentages of those provided (table 3.7), it is clear that both groups selected figs much more strongly than other fruit and flowers, and these much more than young leaves. Fruit and flower availability is over-estimated, since immature fruit and flowers were included in the botanical records but were not eaten. Hence, although one may wish to label siamang and lar gibbons as leaf- and fruit-eaters, respectively, both are selecting more strongly for fruit, especially figs.

Table 3.7. Relative abundance of and selection for different food classes. Siamang and lar gibbon, 12 months. See text for details.

food category	group	feeding opportunities provided	feeding opportunities taken	taken as % provided
fig fruit	siamang	26	10	39
	lar	25	10	40
other fruit	siamang	71	7	10
	lar	97	9	9
flowers	siamang	19	2	10
	lar	15	2	13
young leaves	siamang	308	6	2
	lar	273	7	3

A third aspect of food selection concerns the size of food source. The species which gibbons eat do not differ in size from those which they neglect, but, given the choice of two sources of the same food, gibbons differentially visit the larger one (fig. 3.13). The size of source is estimated by the proportion of its crown area, but over a large sample of species this should not matter. Functionally, large food sources are attractive to gibbons, because they spend less time and energy in travel, finding and revisiting one large source, than they would over several smaller sources for the same amount of food. Although difficult to demonstrate statistically, it seems likely that lar gibbons do not select as strongly for large sources as do siamang. This follows from (1) the heavier siamang's greater food requirements, (2) the greater cohesion of the siamang group while feeding, and (3) the smaller number of food sources visited by the siamang each day.

DAILY ACTIVITY

Activity patterns

Gibbons become active at dawn, remain active for about nine hours (ten hours for siamang), and retire to their night trees several hours before dusk (table 3.8). They differ from sympatric monkey species by taking a less marked mid-day siesta, and by retiring to their night trees very early (fig. 3.14). Why gibbons should follow an activity pattern different from other primates and many other diurnal animals inhabiting the same environment is discussed in chapter 10.

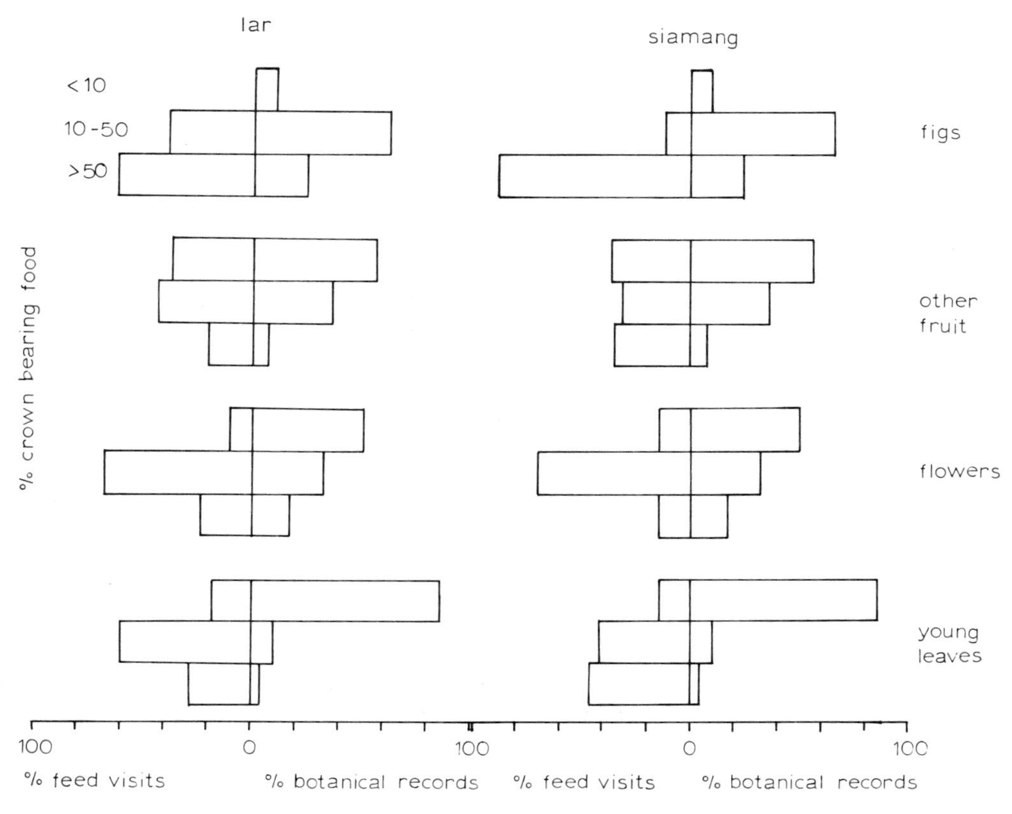

Fig. 3.13. Selection by siamang and lar gibbon of food sources of different sizes, in terms of percentage of feeding visits to trees bearing food over 1 - 10, 11 - 50 and over 50% of crown surface areas, contrasted with records of food production, independent of feeding observations, on all trees of those species in the botanical sample.

Activity budget

The activity budgets for siamang, lar and agile gibbons are shown in table 3.9. The two small gibbons are basically alike, although the agile gibbon at Sungai Dal remained active slightly longer each day, having rested more during the day and fed slightly less. The siamang's longer activity period is doubtless related to its longer feeding time; being twice the size of the smaller gibbons, it should require $1\frac{1}{2}$x more energy. This follows from the general mammalian formula relating basal metabolic rate to body weight ($BMR = W^{0.75}$; Kleiber, 1961). The siamang eats leaves faster than the smaller gibbons, and fruit only a little slower, if at all (Raemaekers, 1979). Even so, the siamang must devote

Table 3.8. Daily activity periods of the three species.

	group	mean	range	number of days
start time	siamang	0642	0610 - 0800	111
	lar gibbon	0643	0619 - 0735	98
	agile gibbon	0642	0615 - 0731	106
stop time	siamang	1632	1400 - 1831	114
	lar gibbon	1518	1310 - 1650	112
	agile gibbon	1548	1310 - 1740	55
duration (hours)	siamang	10.3	6.5 - 11.9	109
	lar gibbon	8.6	6.2 - 10.2	92
	agile gibbon	9.1	6.5 - 11.0	34

Table 3.9. Daily activity budgets. Figures represent activities of average group members. (a) Number of minutes daily, (b) percentage of daily activity period. 'Forage' observations on agile group divided equally between 'feed' and 'travel'.

	(a)			(b)		
	siamang	lar	agile	siamang	lar	agile
rest	155	103	158	25	20	29
groom	5	18	-	1	3	-
feed	310	217	196	50	42	36
travel	136	166	163	22	32	30
(sing)	5	14	27	1	3	5
play	8	1	-	1	.2	-
sleep	821	923	896			

more of its day to feeding to obtain sufficient energy.

DISCUSSION

Feeding niche and ranging strategy

This section outlines the feeding niche and ranging strategy of the small gibbons. Those of the siamang are similar but less extreme; differences between it and the small gibbons are discussed in a later section.

The results of the Malayan studies indicate that gibbons lack specialisations of the gastro-intestinal tract either for dealing with bulky, fibrous food or for dealing with toxic secondary plant compounds. They are in consequence selective feeders by comparison

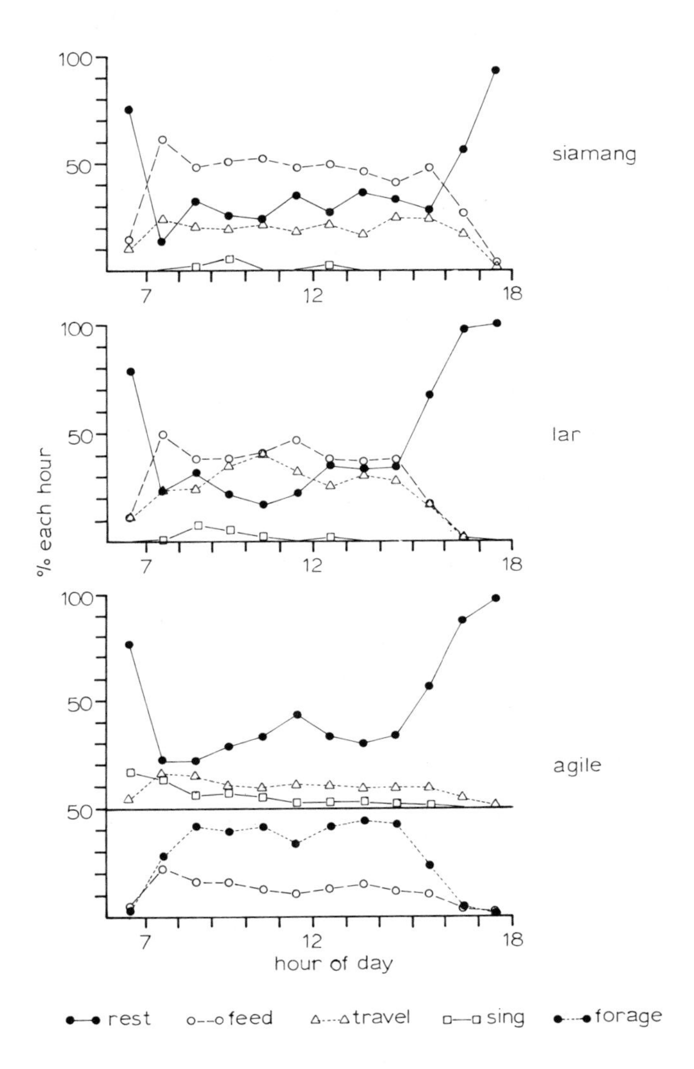

Fig. 3.14. Daily activity patterns, by percentages of each hour of the day spent in different activities (n = 5488 observations for siamang, 3242 for lar gibbon - both 2 days/month for 12 months - and 10871 observations, from 2 x 10 days/month for 10 months, for agile gibbon).

with other primates, restricted to a diet of succulent fruit pulp, certain young leaves, flowers and insects. Inability to use coarse foliage high in fibre may be the factor that restricts the gibbons to relatively aseasonal forests throughout their geographical range from Assam to Borneo and Java. Poor detoxification may be the factor that restricts them to the most diverse forests within this region by compelling them to adopt a strategy of eating only a little at a time of many foods, thereby avoiding the overloading of detoxification systems. In these respects gibbons contrast with nearly all sympatric monkeys, which can occupy seasonally drier and floristically simpler forests, and which persist longer in badly degraded habitats.

The most conspicuous element of gibbon diets everywhere is the fig. The attractions of figs are evident: they are soft and succulent, individual tree crops of them are large, and they seem to be more or less continuously available in most forests inhabited by gibbons. Yet figs are also the gibbon food most eaten by other frugivores, and are therefore not the kind of food which best segregates gibbons from their potential competitors. The gibbons' feeding niche appears to be the succulent fruit occurring in smaller, scarcer sources, which are less conspicuous and less attractive, at least to species which move in large groups or flocks. Prominent among such fruit in the diets of Malayan gibbons are members of the Meliaceae, Sapindaceae, Guttiferae, Myristicaceae, Annonaceae, Anacardiaceae and Melastomataceae.

At least some of these species appear to have evolved with specialist frugivores. For example, the fruit of *Knema cinerea* (Myristicaceae) splits open to reveal an irresistible bright red torpedo-shaped aril which comes away very easily when picked. This is almost certainly a specialised "bird fruit" (chapter 2). Although the small gibbons are not true fruit specialists, in that they do not depend exclusively on fruit, still less on certain species of fruit, they are well suited to such a fruit. They can afford to make frequent visits to the few trees bearing at any time and pick the few fruit which ripen each day.

The hard-shelled, many-seeded fruit of *Randia scortechinii* (Rubiaceae), another species much relished by Lompat gibbons, could be an equivalent mammal-orientated specialist. The tree is quite rare, rather small, and fruits more continuously than almost all other species at Kuala Lompat (chapter 2), all features which promote faithful visitation by specialist frugivores with narrow search images seeking to escape competition with opportunists.

Several consequences of feeding on small, scattered sources can be identified in features of the gibbon ranging strategy:

(a) The need to locate such inconspicuous sources and to monitor their fruiting periods necessitates the restriction of ranging to a limited area that is well-known. Gibbons live in moderate-sized home ranges, which they evidently know very well and for which they show marked affinity, remaining within the range even when surrounding forest is unoccupied.

(b) Sources may have to be visited daily, or even more than

once a day, to cull the ripe fruit. Thus, gibbons have to travel relatively long distances each day to obtain their food. Their wide and regular coverage of the home range enables them also to monitor the condition of imminent food sources. The gibbon's ability to hang while feeding, allowing it an easy reach in more directions than a sitting quadrupedal monkey (Grand, 1972), and its speed and mobility in flexible terminal branches, enable it to exploit small sources with low densities of food items relatively efficiently and rapidly.

(c) Small sources may necessitate the fission of even small social groups. The groups of small gibbons often split up to forage. The siamangs, with a lower proportion of non-fig fruit in their diets, i.e. fewer small fruit sources, are markedly cohesive by contrast.

(d) A diet of rare fruit results in low density and biomass relative to sympatric primates with less restricted diets (chapter 6).

Comparison of lar and agile gibbons

Lar and agile gibbons resemble each other very closely in their ecological adaptations. Their overall dietary proportions are alike, and they eat many of the same food species, using similar ranging strategies to exploit this food supply. The size of the home range for agile gibbons falls within the range of variation found in lar gibbons, the usual distance travelled each day is identical and the way the home range is used corresponds closely. The ways the two species travel and feed in the forest canopy are also alike, and the activity budgets and patterns are very similar; such differences as occur are very much less than the differences between either of these two species and the siamang.

Such evidence suggests that the requirements of the two species are so alike that they are ecological equivalents (Odum, 1971), which could not coexist for any length of time. Support for this comes from the discovery of mixed species groups of lar and agile gibbons in Ulu Muda, Kedah (Gittins, 1979). Individual lar and agile gibbons fed, ranged and bred together; the males of one species were able to defend their territories by song against males of the other species, as predicted by Wilson (1975) for ecologically equivalent species. He argues that the circumstances most favourable for this 'inter-specific territoriality' are the first contacts between two species recently evolved from a single parent species, which is almost certainly what has happened in lar and agile gibbons.

Ecological segregation of large and small gibbons

The small differences that have been described between the

siamang and very similar lar and agile gibbons, allow the siamang to live sympatrically with the smaller species, which are themselves incompatible and allopatric. The difference in body weight is the obvious clue to their ecological segregation. If we dare assume that the food supply limits both small and large species, then their dietary differences may be taken as the means of their segregation. This dietary difference may be related to body size in the following way.

The lar gibbon, like the agile gibbon, travels about twice as far each day as the siamang. Travel costs should be less per unit distance than for the siamang; it weighs only half as much and its stride is about as long, since (a) its armspan is not very much shorter than the siamang's, and (b) its armstride is longer than its armspan, because it often throws itself from hold to hold, unlike the siamang (Fleagle, 1976c). In general, the fewer the strides taken to cover a given distance, the cheaper the movement (e.g. Dawson, 1977).

The lar gibbon's cheaper travel enables it to reach the scattered sources of high-energy food embodied in fruit pulp. It needs this compact form of energy because of the presumably higher basal metabolic rate which follows from its more linear body form. Conversely, the siamang can make do with a higher proportion of the less compact energy source of young leaves, which is much commoner and thus emancipates the siamang from the relatively costly travel in search of fruit. For example, the siamang eats as many figs as the lar gibbon, but fewer other fruit (fig. 3.9); figs generally appear in larger masses than do other fruit (fig. 3.13; Raemaekers, 1978b), yielding more energy for the cost of reaching the source.

The interest of this comparison is its contrast with the prevailing condition whereby larger animals travel further than smaller ones of the same form, because the cost of travel does not increase as fast as body weight (Schmidt-Nielsen, 1972). This is presumably because stride scales in proportion to body weight, whereas in gibbons it does not. The anomalous position of the siamang is reflected in its home range area, which is smaller than that of the lar gibbon, although its body weight predicts that it will be larger (McNab, 1963; Milton and May, 1976; Clutton-Brock and Harvey, 1976). Its smaller home range, on the other hand, accords with its more folivorous diet, since, among the primates, folivory correlates with small home ranges (Clutton-Brock and Harvey, 1976), as is to be expected from the commonness of leaves relative to fruit.

Other physical differences between the siamang and smaller gibbons may likewise reflect their differences in diet. Though no gibbon shows special adaptations to leaf-eating (Chivers and Hladik, in press), the siamang has a proportionately longer colon

(Kohlbrugge, 1891), which may be associated with increased bacterial fermentation of fibre. The siamang also has a relatively larger palate, and its molars correspond with those of other folivorous primates in their efficiency of crushing, grinding and shearing, whereas the lar gibbon's molars correspond with those of other frugivorous primates (Kay, 1975, 1977). The siamang's larger palate and more efficient molars should allow it to chew faster; the lar gibbon's greater mobility about the tree should allow it to pick scattered items faster. Young leaves tend to grow in dense clumps and they can be picked faster than they can be chewed, whereas fruit are often dispersed about the source, so that they can be chewed faster than they can be picked or processed.

Thus, the siamang's faster chewing rate should predispose it to eat leaves faster, and the lar gibbon's mobility allow it to eat fruit faster, in accordance with their dietary proportions. The siamang does indeed eat a given species of leaf faster than the lar gibbon, while the latter generally eats fruit at about the same rate as siamang despite its smaller mouth (Raemaekers, 1979).

Territoriality and group size

There have been several attempts to explain how the social organisation and territoriality of gibbons could have evolved (Carpenter, 1940; Ellefson, 1974; Chivers, 1974). Such explanations have to overcome the major difficulty of separating cause from effect. Traits such as territoriality and group size are inter-related functionally and evolve together, so that it is difficult to establish the sequence in which they arose.

Some traits are less plastic than others, and it seems reasonable to suppose that these constrain evolutionary change in other variables (Wilson, 1975; Clutton-Brock and Harvey, 1977). The feeding niche of a species is constrained by many factors external to the species gene pool, such as geographical features and interspecific competition (Lack, 1971; Wilson, 1975). Thus, 'the most plausible assumption is that the feeding niche occupied by a species (and the morphological, physiological and behavioural adaptations which affect food exploitation directly) is likely to constrain evolutionary change in other variables' (Clutton-Brock and Harvey, 1977: 575).

This assumption is made here, and applied to try and explain group size and territoriality in the smaller gibbons as consequences of their feeding niche. Ellefson (1974) laid the basis of this argument, but he tended to argue from territorial behaviour to the food-getting mode and subsequent anatomical changes. The argument also owes much to a recent discussion by Wrangham (1979) on the evolution of social systems in apes; it is intended to apply in a milder form to the siamang, which exhibits a slightly less extreme form of the gibbons' system.

In a review of the data available for birds, Crook (1965) found that those which feed on small dispersed food sources tend to maintain year-round feeding territories, whereas birds which feed on foods that are locally abundant in sporadic patches tend to be gregarious. Jolly (1972) applied these findings to primates and concluded that a primate whose food is found in small dispersed sources will tend to be solitary and maintain a year-round feeding territory, whereas species whose food is either abundant or uniformly distributed, or found in very large patches, will tend to be social feeders.

An individual of a species that concentrates on small food sources will be subject to some degree of feeding competition if it shares those food sources with other members of its species. Furthermore, as group size increases so does the distance that has to be travelled, visiting more food sources, before the whole group is satisfied. Thus, other factors being equal, individuals of a species that concentrates on small, rich food sources should be solitary or, at most, live in small groups. So as to ensure a constant supply of small, scattered food sources, it is often advantageous to defend them against members of the same species, and territoriality develops (Wilson, 1975). The advantage of being territorial is that, by devoting a certain amount of time to defending an area, an animal can be sure of access to a certain minimum amount of food. An important consideration here, however, is that a resource must be 'economically' defendable before territoriality can evolve (Brown, 1964).

A large proportion of the gibbons' food supply consists of small, scattered food sources, and it is in the exploitation of these that they appear to differ from sympatric primates. Consequently, small group size would be expected in gibbons. For territoriality to evolve, their food supply must be defendable, and this appears to be the case. It hardly profits a neighbouring group to cross the territorial boundary looking for small, inconspicuous food sources, because, without a knowledge of the area, such food sources are difficult to locate. In addition, as a result of the regular coverage of all parts of the territory by the resident group, there is a high probability that an intrusion will be detected; the probability of detection increases with the amount of time the intruding group spends in the territory of the resident group. It will only really benefit a group to stray across the territorial boundary to feed when there is a large conspicuous food source close to the boundary. In this case no time is lost searching for food, and a large amount can be gathered rapidly, with less likelihood of being discovered. Such is the situation in the great majority of observations of territorial intrusion. Thus it seems that gibbons have an economically defendable food supply. The most cost-effective strategy available to gibbons would seem to be for a group to concentrate on obtaining a detailed knowledge of its own territory, rather than wasting time straying into a neighbour's territory.

Accepting that the gibbons' food supply gives a strong incentive for individuals to restrict movements to a limited area and to defend this area against conspecifics, the next problem is to explain the social organisation found in gibbons. Monogamy tends to evolve either when more than a single individual is needed to rear the young, or when the carrying capacity of the habitat is insufficient to permit more than one female to raise a litter in the same home range at the same time (Kleiman, 1977). In gibbons the male does not take a direct role in feeding the infant, and it is unlikely that his presence is essential, in the short term, for the development of the infant. It would appear much more likely that the second factor, concerning carrying capacity, has been more important in leading to monogamy in gibbons.

In most situations the optimal foraging strategy for a gibbon would be to feed alone, because competition over food will occur whenever the food sources are too small to satisfy all members of the group. Food sources are often of this sort, as is shown by the gibbons being spread out for much of their foraging time. If no other pressures were acting, females would be expected to live alone and to defend territories against other females and also males (Wrangham, 1979). Females are forced to accept males, however, because they would always be handicapped in territorial encounters against them. This is because males do not have the additional physiological cost of rearing infants, and so can devote more time to contending boundaries (Wrangham, 1979).

The next problem is to explain why males do not seek access to more than one female. In situations where the best interests of both males and females are served by maintaining exclusive territories against their own sex, three options are open to males (Wrangham, in press). They can (1) maintain a territory large enough to include the separate territories of two or more females, (2) defend a territory large enough to include the territory of at least one female and parts of the territories of other females, or (3) confine themselves to maintain territories, the boundaries of which coincide with that of a single female.

The first is the most attractive option for both males and females. The females have parts of their boundaries defended by the male, and feeding competition with the male is reduced as he shares his time over a larger area; the male benefits by gaining access to as many females as possible (Trivers, 1972). A single male gibbon, however, appears unable to defend an area large enough to support two or more females. This deduction is based on the observation that no gibbon group has ever been seen with more than one breeding female. To confirm the argument, and remove its circularity, the actual amount of food in a gibbon territory would have to be measured and compared with the requirements of the gibbons. This is a formidable task that would have to cover a period of food shortage, since it will be at such times that territory size will presumably be adjusted. Such a study has been carried out, however, on squirrel territories (Smith, 1968), and there was found to be sufficient food for only the one animal defending the territory.

In the second option, some females have territories shared by two or more males; the female loses by not having her boundary defended by a male, and the male loses by defending an area of forest containing a female with whom he is not necessarily breeding. Both males and females should prefer the boundaries of their territories to coincide, and so this option seems unlikely.

As it appears impossible for a male to defend a territory large enough to contain the territories of two or more females, his best alternative seems to be to confine his attention to one female. Once a male is committed to a single female, he can increase his own reproductive potential by increasing that of his mate - by assuming the major role in territorial defence. Not only is the male gibbon the main participant in territorial disputes, but he does the most singing for territorial advertisement. The female loses to some extent by having to share her food supply with the male, but his presence is unavoidable; freed from the burden of territorial defence, she can devote more time and energy to feeding, thereby increasing her reproductive potential.

The male should theoretically take on an increasing amount of the burden of rearing the offspring, and this does occur in the siamang where the male carries the infant during its second year of life. This behaviour may not have arisen in the smaller gibbons because of the greater burden of territorial defence. Other ways in which the male can help rear the young are feeding, defending and 'socialising' them (Kleiman, 1977). As mentioned earlier, the male gibbon does not directly feed the young, although he does tolerate their presence close to him in what are presumed to be the better parts of the tree (Chivers, 1974). It is doubtful that his knowledge of the food supply is any better than the female's. By defending an area of forest, however, the male is ensuring a constant food supply for his offspring.

The male agile gibbon of DG1 was tolerant of the infant, but was never seen to play or initiate interactions with it. This contrasts with the male siamang of TS1 and RS2, who did play with their infants. Thus, this seems to be a result of the closer relationships between male siamang and their infants, rather than just a difference in individual temperament.

In summary, the monogamous mating system in gibbons can be viewed as a response of males and females to maximise their own reproductive success. Their food supply seems to provide a strong incentive to feed alone and to defend a territory against conspecifics. The female is forced to accept the presence of a male, however, because she cannot defend economically a territory against males. The male appears to be limited in the area of forest he can defend, and so can only support the territory of one female. Having had to commit himself fully to one female, the male then helps her as much as possible by assuming the major role in defence of the territory and against predators. In the siamang this has proceeded a stage further, where the male takes on the burden of carrying the larger infant.

J.J. RAEMAEKERS

DUSKY AND BANDED LEAF MONKEYS

Sheila Hunt Curtin

Department of Anthropology
University of California
Berkeley

INTRODUCTION

The Asian colobines are the most abundant and taxonomically diverse of all non-human primates in Asia. Their adaptive success undoubtedly lies in their possession of a sacculated stomach, which permits digestion of cellulose by bacterial fermentation (Bauchop, 1971), and thereby allows exploitation of the vegetative part of the primary production, which is less available to most other primates and arboreal mammals. Of the three colobine species in Peninsular Malaysia, the behaviour and ecology of the two widely-occurring sympatric species are the subject of this chapter.

The dusky leaf monkey or spectacled langur, *Presbytis obscura* [REID, 1837], is a uniformly grey monkey (almost black in the north of the peninsula), with a long tail, dark hands and feet, and a dark grey face distinguished by white rings around the eyes and a white patch on the mouth (Plate XIII). Adults weigh about 7 kg (males 7.4, females 6.5 kg; Burton, in prep.), and infants are bright orange at birth, attaining adult colour by about six months. There is no sexual dichromatism.

The banded leaf monkey, *Presbytis melalophos* [RAFFLES, 1821], is similar in size to *P. obscura*, but shows less sexual dimorphism in body weight (males 6.3, females 6.4 kg; Napier and Napier, 1967; Chivers and Hladik, in press). They have a dark grey or brown back with light grey to white underparts, dark extremities and a dark grey face framed by white cheek ruffs (Plate XIII). In contrast to *P. obscura*, there is considerable geographic variation in coat colour (Medway, 1970). Neonates have a pale face with red-brown crown, white hands and feet and a white body marked by a red-brown stripe from nape to tail and across the shoulders (the 'crucifer' pattern, Pocock, 1934).

These two species occur together over most of the Malay Peninsula, from Johore to Tenasserim. Most habitats that support one species also have the other, although there is a tendency for *P. melalophos* to prefer slightly disturbed habitats and for *P. obscura* to prefer mature inland rain-forest (*P. melalophos* is the more abundant species in all habitats).

This chapter is based on a 12-month field study conducted between September 1970 and September 1971 in the Krau Game Reserve - Kuala Lompat Post. *P. obscura* was observed for a total of 356 hours and *P. melalophos* for 485 hours (243 work-days in all).

The majority of observations were made on two groups, o-1 (*P. obscura*) and m-2 (*P. melalophos*), which ranged over the same area of forest and were of very similar size and composition. Irregular observations were also made on one other group of *P. obscura* and four other groups of *P. melalophos* (fig. 4.1).

Groups were located by walking the trail system and listening for calls and sounds of movement. In addition to the network of survey lines, special trails were cut beneath much used arboreal pathways; searches along these trails, and in the core areas (see below) were usually productive. Once a main study group had been located, it was followed for the rest of that day and for 2-3 days thereafter, and then efforts were made to locate the corresponding group of the other species. Ideally groups o-1 and m-2 were followed from dawn to dusk for alternating periods of three days each. On days when the target group could not be located, another group of the same species was followed instead.

Continuous notes were taken on the number of animals in view, their age and sex, individual identification, activities (including social interactions, feed, rest and travel), location in the study area, height in the forest, and direction of movement. Animal location and travel routes were recorded in terms of tagged trees, which were in turn mapped against the network of survey lines and trails. Every 15 minutes the following data were entered on check sheets: number of animals visible, their age and sex, number of animals in partial view or heard nearby, and height in forest and activity.

SOCIAL ORGANISATION

Group composition

The two groups of *P. obscura* had an average size of 17

Plate XIII. Leaf monkeys of Peninsular Malaysia

(a) Banded leaf monkey, *Presbytis melalophos*, in Ulu Gombak and at Kuala Lompat (DJC)
(b) Dusky leaf monkey, *Presbytis obscura*, at Kuala Lompat (SPG) and in Ulu Gombak, party of group with orange infant (DJC)
(c) Silver leaf monkeys at Kuala Selangor (DJC)

a)

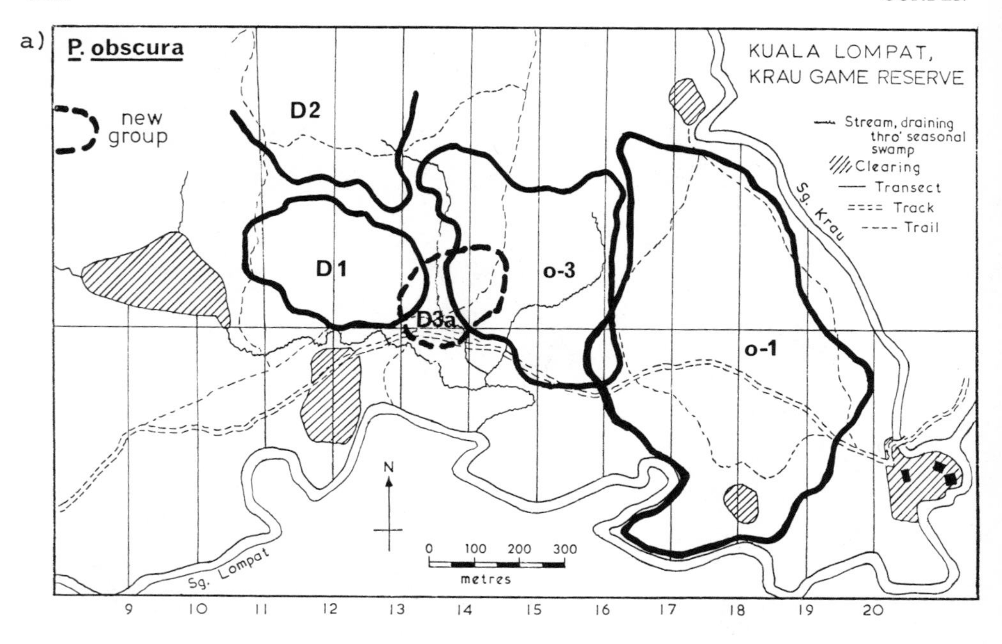

b)

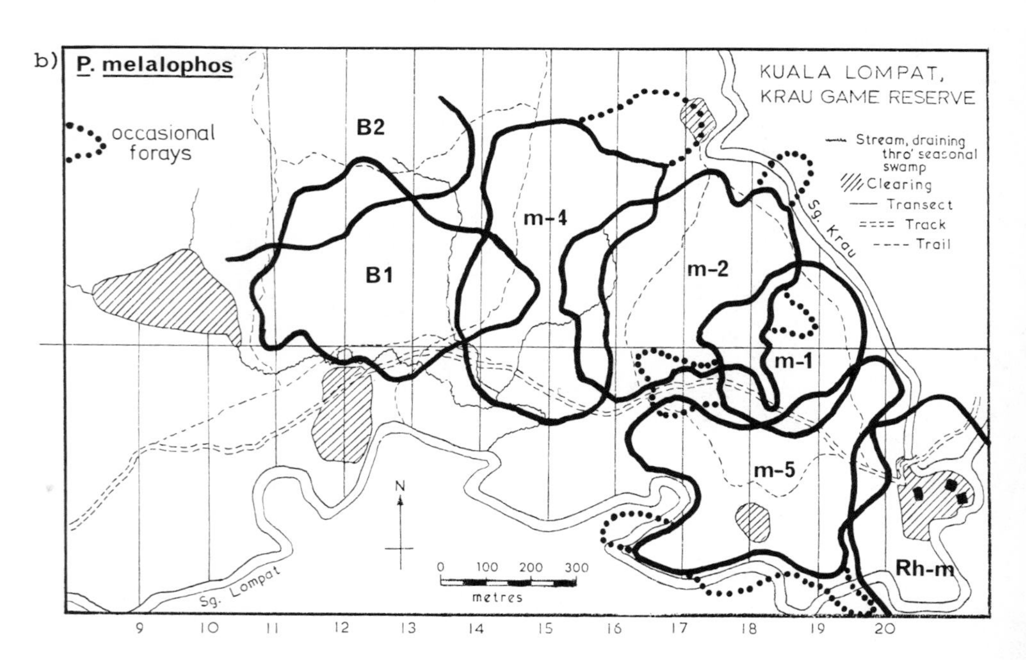

Fig. 4.1. Territories of dusky and banded leaf monkeys at Kuala Lompat, 1969-1970 and 1970-1971 (adapted from Curtin and Chivers, 1978).

individuals (table 4.1), with several adult males and many adult females and immature animals. Smaller groups, with a single adult male, have been reported from this Reserve (MacKinnon and MacKinnon, this volume), and from other parts of Malaysia (Bernstein, 1968a). All-male groups have never been reported for this species. Solitary adult males were observed twice (Curtin, 1976a).

The five groups of *P. melalophos* were smaller on average (13 animals, table 4.1), including large groups with several adult males, smaller groups with a single adult male and one group of four males. Bernstein (1967a) reports average group size as 15, and states that all groups contained only one adult male (Bernstein, 1968a). Harrisson (1962) observed one group of 7 banded monkeys that had only one adult male.

The two main study groups (o-1 and m-2) were very similar in both size and composition (table 4.1). Their home ranges overlapped almost entirely, with about 90% of m-2's range within the range of o-1.

On only one occasion during the 12 months were all 17 members of o-1 in view at the same time. Initially this was attributed to the difficult conditions of observations, but it became increasingly clear that other factors were involved. In contrast to their congeners, social groups of banded leaf monkeys appeared highly cohesive. The group fed, rested and travelled as a unit, and no consistent patterns of sub-grouping could be discerned (but see MacKinnon and MacKinnon, this volume, on splitting in their groups of *P. melalophos*).

The following analysis is concerned with the composition, diurnal variation and dynamics of sub-groups in *P. obscura*. A sub-group is defined as that portion of the group visible to the observer at one time, when no other animals are known to be within about 30 metres. Typically, the members of a sub-group were all in the same tree, more rarely in two or more adjacent trees.

The compositions of the more common of 300 separate sub-groups observed are shown in table 4.2. The most frequent sighting was of an adult female and a juvenile, accounting for 15% of all sightings of more than one animal, which was almost twice as common as the next most frequent category - two adults. Solitary individuals were seen 158 times. Adults were recorded alone significantly more often than their representation in the group would lead one to expect, while immature animals were recorded alone less often than expected (table 4.3).

Of 821 sub-groups sampled, those encountered in the early morning (0600-0800 hours) were larger ($\overline{x}$ = 3.9) than those

Table 4.1. Size and composition of social groups of *P. obscura* and *P. melalophos* at Kuala Lompat (September 1971).

Species	Group	Size	Composition										
			ADULT		SUB-ADULT			JUVENILE			INFANT		
			♂♂	♀♀	♂♂	♀♀	??	♂♂	♀♀	??	2	1	2 or 1
P. obscura	o-1	17	2	7	0	2		1	2	1		2	
	o-3	17+	3	5+	1+	1+				2+	1♀	1,1♀	2
P. melalophos	Rh-m	4	3	0	1	0		0	0		0	0	
	m-1	11-12	1	5-6	0	1		1	1	1	1♀	0	
	m-2	18	2	8	1	1				4	1	1	
	m-4	20-25	4	6+			1+			4+			3
	m-5	11-12	2	3-4	1		1	1	1	1	1		

Table 4.2. *P. obscura* sub-grouping patterns.

Sub-group	n	% total sub-groups
Af, J	45	15.0
A, A	26	8.6
A, J	19	6.3
Af, Af	17	5.6
A, 1-2 Af, 1-2 J	16	5.3
Af, I	12	4.0
Af, 2-4 J	12	4.0
Af, Af, 1-2 J	10	3.3
A, A, 1-2 J	10	3.3
Am, Af	8	2.7
Am, 1-3 Af, 1-4 J	8	2.7
A, A, 1-2 Af, J	7	2.3
3-5 A	7	2.3
Am, 1-3 Af, SAf, 1-2 J	6	2.0
Af, SAf	6	2.0
Am, A	5	1.7
Af, A	5	1.7
2-3 J	5	1.7
Total	224	74.7
Other sub-groups	76	25.3
TOTAL	300	100.0

A = adult. Am = adult male. Af = adult female. SAf = subadult female. J = juvenile. I = infant.

encountered during the rest of the day (0800-1900 hours, $\bar{x} = 2.7$); the difference between the means is significant at the 99.5% level. This pattern of diurnal variation in sub-group size suggests a relationship with daily activity. An analysis of 611 records of feeding, resting and travel according to sub-group size (fig. 4.2) reveals feeding to be about equally frequent for different sizes of sub-group whereas resting peaks for a sub-group of two; this accords with the observer's impression that resting sub-groups tended to be small, usually an adult female and juvenile or two adults. Travel

Table 4.3. Sightings of solitary* *P. obscura* (group o-1).

(a)

(N = 158)	AGE CLASSES Adult	Sub-adult	Juvenile	Infant
Ratio of class sightings to total sightings	$\frac{880}{1304}$	$\frac{51}{1304}$	$\frac{337}{1304}$	$\frac{36}{1304}$
O	138	7	13	0
E	107	6	41	4

$\chi^2 = 29.13$. Significant at 0.1% level.

(b)

(N = 67)	AGE-SEX CLASSES Adult male	Adult female
Representation in group o-1	$\frac{2}{9}$	$\frac{7}{9}$
O	21	46
E	14.9	52.1

$\chi^2 = 3.2$. Not significant at 5% level.

appears to increase with larger sub-groups.

Coordination among different sub-groups is not achieved by contact calls, which dusky leaf monkeys apparently lack. Instead, sub-groups regain contact with each other partly by visual means (surveying the surrounding forest from vantage points of large emergent trees) and partly through shared knowledge of traditional routes and arboreal pathways. Members of a group appear able to predict, presumably on the basis of experience, the movements of animals in other sub-groups. Such prediction is aided by the small size of the group range (see below), and the conservative nature of ranging patterns.

Behaviour within the group

In *P. obscura* the most frequent diadic interactors were adult females and juveniles (table 4.4a), a result which might be anticipated from their frequent association in the same sub-group (table 4.3). The next most frequently interacting pair was two adult

---

* Solitary animals were defined in the same manner as sub-groups.

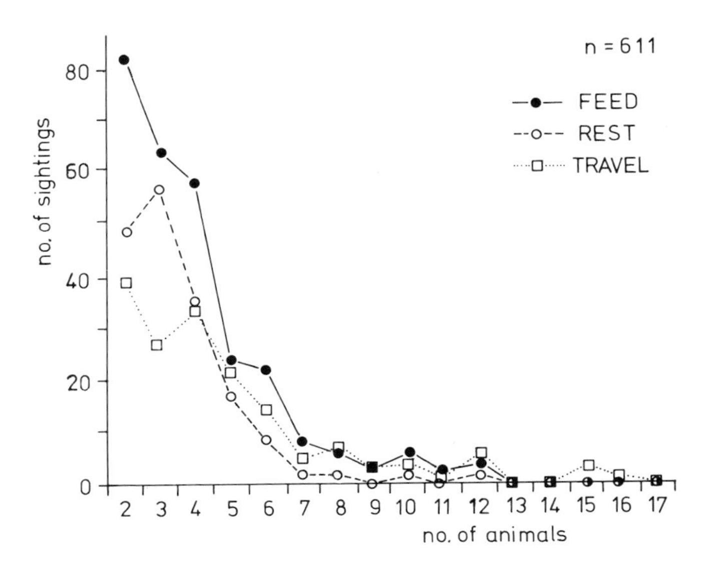

Fig. 4.2. *P. obscura* sub-group size and activity.

females, which were either senders or receivers in 71% of all social interactions. Juveniles were involved in more than 50% of all interactions, whereas adult males were participants in only 13%. Of 95 social interactions observed between adult females and juveniles, 26 consisted of the juvenile approaching and joining the adult female, and 23 were instances of passive body contact during resting ("huddling", Curtin, 1976a). Only 2 out of the total of 252 social interactions were overtly aggressive.

In *P. melalophos*, the commonest interactors were also adult females and juveniles. Of these 38 interactions out of a total of 166 (table 4.4b), 25% were approaches by the juvenile to the female, 25% were instances of huddling and the rest included such behaviour as grooming, maternal retrieval before travel, embracing and play. The next most frequently interacting pair were adult male and juvenile, and most of these 21 records were produced by two pairs of individuals:

(1) The adult male and a juvenile male from the one-male group m-1; other than a few instances of huddling, most of their interactions were approaches by the juvenile to the adult male or occurred in the context of cooperative displays directed at the observer.

(2) An adult male and juvenile/sub-adult male from the all-male group Rh-m; their interactions were more varied, including approaches, huddling, grooming and play. Many behaviours directed by the older to younger animal were of a maternal type; for example,

Table 4.4. Social interactions between age/sex classes

(a) *P. obscura*

			% interactions observed						total	expected	
		n	A	M	F	SA	J	I	observed	o-1	o-1, o-3
Adult,	A	52	7.0	2.0	6.0	0.4	4.0	0	20.2		
Adult ♂,	M	33		2.0	5.2	2.0	2.0	0	13.2	11.8	13.2
Adult ♀,	F	179			10.7	6.7	37.7	4.8	71.1	41.2	34.2
Subadult,	SA	23				0	0	0	9.1	11.8	15.8
Juvenile,	J	133					7.0	2.0	53.5	23.5	18.4
Infant,	I	16						0	6.8	11.8	18.4
		436 (218 interactions)									

(b) *P. melalophos*

			% interactions observed						total	expected	m-1,m-2,
		n	A	M	F	SA	J	I	observed	m-2	rh-m
Adult,	A	34	5.4	1.8	3.6	3.0	4.8	1.8	20.4		
Adult ♂,	M	53		9.0	1.2	6.0	12.7	1.2	31.9	11.1	17.6
Adult ♀,	F	70			1.8	3.6	22.9	9.0	42.1	44.4	41.2
Subadult,	SA	23				1.2	0	0	13.8	11.1	11.8
Juvenile,	J	84					4.8	5.4	50.6	22.2	20.6
Infant,	I	30						0.6	18.0	11.1	8.8
		294 (147 interactions)									

the adult male often carried the juvenile during routine travel between feeding trees, and always did so during tense situations, as when the other two males in the group displayed to neighbouring groups or to the observer. The most frequent participants in banded leaf monkey social interactions were juveniles, followed closely by adult females, with sub-adults last.

Because *P. melalophos* groups fed, rested and travelled as spatially cohesive units, while *P. obscura* groups were more typically fragmented into small sub-groups, an individual *P. melalophos* had twice the number of potential cointeractors available to a *P. obscura* (Curtin, 1976a). Although one might expect, therefore, rates of social interaction to be significantly higher in *P. melalophos*, they were in fact much lower (table 4.5). Social interactions were twice as frequent between dusky leaf monkeys as between banded leaf monkeys.

The two species also differed in details of the socialisation process, specifically in infant transfer, juvenile "monitoring", and the intensity of the bond between adult female and juvenile (Curtin, 1976a). The transfer of infants less than six months old to adult females other than the mother was observed only in *P. obscura*. Juvenile "monitoring", a behaviour pattern in which an adult female remains close and attentive to a playing group of juveniles while other adult females in the vicinity rest or feed, was also characteristic of *P. obscura*, but apparently absent from the repertoire of *P. melalophos*. Finally, *P. obscura* juveniles continue to be closely associated with an adult female (presumably the mother), usually resting in body contact with her and often being carried by her over difficult passages in travel, whereas *P. melalophos* juveniles spend more time in the company of their peers.

The roles of adult males merit special attention.

In *P. obscura*, adult males have three principal roles in the daily life of the group: detecting and warning of the presence of potential predators, increasing group cohesion and maintaining territorial boundaries. Both the adult males of group o-1 performed all three of these roles at various times during the study; neither animal seemed more prominent in the daily round of group activities.

Table 4.5. Rates of social interaction for *P. obscura* and *P. melalophos*.

Species	No.of interactions	No.of hours observation	Social interaction / Observation time
P. obscura	294	356	0.83
P. melalophos	198	485	0.41

(1) Predator detection. Adult males were to be seen frequently very high in the crowns of large emergent trees, looking out and down in the forest canopy - a behaviour pattern called "scanning". Even during feeding there was a tendency for males to be positioned higher in the feeding tree than adult females. When a disturbing stimulus was detected, one or both adult males gave a soft warning call, "whoo" (also given by other age/sex classes), followed by the two-phase honking call that gives this species its Malay name of "chengkong". In extreme disturbance the honk became three- or even four-phased ("cheng-cheng-cheng-kong"), and alternated with coughing calls and "whoos". Excluding conspecifics, adult male dusky leaf monkeys honked at (a) the observer, (b) passing Jah-Hut aborigines, (c) a hunting Malay serpent eagle, *Spilornis cheela*, and (d) an unidentified animal on the forest floor. Adult females never uttered the full resonant honk of adult males, although they were capable of a muffled honk, similar in structure but not in tone; perhaps this results from differences between the sexes in the anatomy of the larynx.

(2) Increasing group cohesion. On certain occasions the honk appeared to be used as a means of increasing spatial cohesion. It was directed at group members, rather than at disturbing external stimuli, and its effect was to cluster animals together that had previously been separated by 30 metres or more. This type of honking only occurred when the group was in relatively unfamiliar forest at the extreme limits of its home range.

(3) Maintenance of boundaries. The honk, combined with an arboreal display, was also given in areas of overlap with adjacent groups. Calling was spontaneous, elicited by no apparent stimulus other than the group's general location.

In *P. melalophos* males perform two roles that are central to the functioning of the social group. One is the diversion of potential predators by calls and arboreal display, while females and young retreat and hide in dense vine-covered trees. The second is the maintenance of territorial boundaries through (a) daily calling and display in the evening, through the night and in the early morning, (b) calling and display during the day when the group is entering an overlap zone, and (c) confrontation with males of neighbouring groups.

(1) Predator detection and diversion. The diversionary role of an adult male when faced with a potential threat to the group is dramatic. The initial warning may be a soft "whuh" (mild disturbance) or a ringing "churr-r-r-r" (severe disturbance) delivered by the first group member to detect the disturbing stimulus. Immediately one or more adult males approach the calling animal, follow its gaze, locate the animal or object and begin to call and display. Other group members retreat out of view, into densely-foliaged trees

hung with vines if available. In the course of his display, the male may move as far as 75 metres from where the other group members remain hidden and silent. This display was directed at the observer, at local aborigines, and once at a Malay serpent eagle, which had been hunting persistently in the area for several hours. The elements of the diversionary display are largely identical with those of the territorial display, and the call (long call, a strident "churr-r-r, churr-r-r, ka-ka-ka", which gives this species its Malay name of "cheneka") is also very similar to the territorial call. Only in the later stages of this study was the observer able consistently to distinguish territorial from alarm calls.

(2) Maintenance of boundaries. Territorial behaviour, including calls, display and confrontations between groups, will be discussed. It should be noted here, however, that the long call was given only by adult and sub-adult males, never by any other age/sex class.

The males from four groups were known as individuals. In groups with more than one adult male (three of the four groups), some males were more active than others in territorial calling and display, and in confronting potential threats to the group. For example, one adult male from group Rh-m displayed to the observer only once and did not join in either the rounds of evening calls or displays directed at neighbouring groups. In the first few months of the study, Rh-m consisted of the adult males White Eyes, Turtle and Stumptail and the juvenile male Junior. Of these, only Stumptail called and displayed at the males of the neighbouring groups, especially m-5; he also frequently led travel, although severely handicapped by his mutilated tail. He disappeared during the period December 1970 - January 1971, and was replaced by the adult male Snow, who was subsequently the most active in territorial displays. Snow and White Eyes associated closely, moving together and often resting within a metre of each other. White Eyes cooperated frequently with Snow in displays directed at the observer.

This unequal division of roles was also evident in m-5, where, of two apparently fully adult males, Lips was more active than Blackbelly in diversionary displays. Lips was also the probable aggressor in several brief scuffles involving the two males. Blackbelly, in contrast, was relatively passive in these situations; he presented to Lips once, and on another occasion was slapped by an adult female. These two males did not differ much in body size or canine development, and neither animal appeared old. There was, however, one physical difference between them: Lips' scrotum was black, while Blackbelly's was a pale flesh colour. The scrota of both the adult males of m-2 (Strider and Grizzle-chin) were black, while that of the sub-adult Hitam was pale. Each of the adult males performed territorial and diversionary displays on occasion, although generally one one did so at a time. Unlike Snow and White Eyes of the all-male group, and like the two males of m-5, they did not usually cooperate. Thus the male role tended to be occupied by only one individual at a time.

In summary, *P. obscura* at Kuala Lompat live in groups composed of sub-groups of variable size, which split and rejoined on a daily basis, being larger in the early morning. The commonest association, between adult female and juvenile, was so constant as to suggest they were mother-offspring pairs. The majority of social interactions were affiliative in nature, and bonds between adult females and immature animals formed the basis for daily patterns of social interaction, so that adult females constituted the cohesive core of group social life. Maternal behaviour included infant transference and juvenile monitoring, not seen in *P. melalophos*. Adult male roles included predator detection, maintenance of group cohesion during travel and the defence of territorial boundaries. In contrast, groups of *P. melalophos* were highly cohesive, with no sub-groups discernible, but with a surprisingly low rate of interaction, again involving adult females and juveniles mostly. Adult male roles included diversionary displays in response to disturbance, and daily territorial calling and display to neighbouring groups.

<u>Behaviour between groups</u>

A. Inter-group spacing

During the initial months of the study of *P. obscura*, honking calls were often heard in the early morning, and occasionally during mid-day and the afternoon. The day-time honks seemed to emanate from the narrow overlap zone shared by o-1 and o-3 (fig. 4.1a), rather than occurring randomly throughout the home ranges. The first observation of an adult male honking in this context was on December 17 1970:

1150 o-1 is feeding intensively on the blossoms of tree #530, on the western edge of its territory.

1203 Abruptly, an adult male (probably Little Joe) emits 20-30 two-phase honks in rapid succession. The other group members continue feeding in apparent unconcern.

1210 One more honk, and then another 30 seconds later. Feeding continues

1219 The adult male honks, grunts and runs east out of #530 followed by three other animals. Those remaining in #530 stop feeding and fall quiet. (The group rested without further incident until 1318 hours.)

Territorial honks could be distinguished from alarm honks by the following criteria: (1) They appeared to be given spontaneously, without any definite stimulus other than the general location of the group within its territory. (2) They were usually produced many times, while alarm honks seldom numbered more than three or four on any occasion. (3) Other group members did not react with flight or concealment, as they did to alarm honks, but tended to continue their normal activities. The observer was unable to detect any differences in tone or structure between the two types of call, but it appears from (3) that the monkeys did. Alternatively, visual clues, such as body posture and orientation of the calling male, provided crucial information. Occasionally, but not always, the honks were accompanied or followed by the male abruptly shifting position. Perhaps, however, it was that all adults and sub-adults were aware of the zone of overlap and could distinguish between the two types of honk on those grounds.

The distribution of honking through the day is shown in fig. 4.3. Of the 139 occasions on which honks were recorded, eight were prompted by the observer's presence; the remaining 131 calls average at 0.5 calls/work day.

Inter-group spacing mechanisms appear to have been highly efficient, since o-1 and o-3 occupied closely adjacent tracts of

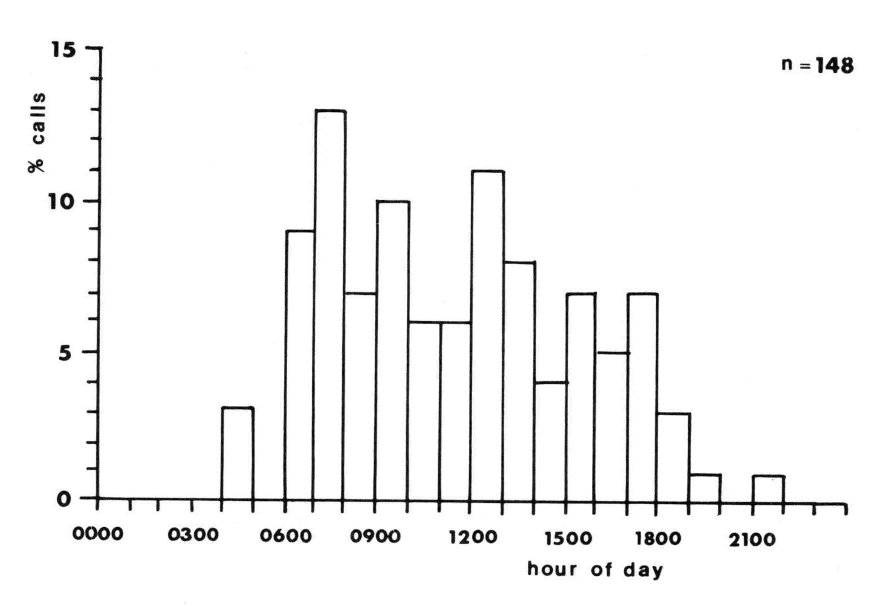

Fig. 4.3. Diurnal variation in calling by *P. obscura*.

forest, and yet were never observed to encounter each other directly Honking is probably too infrequent to be the sole mechanism; an additional factor may have been visual scanning by adult males, especially from night trees (John MacKinnon, pers. comm.).

The spacing of groups of *P. melalophos* was maintained by two principal mechanisms: (1) long calls given by one adult male of each group in the evening, night and early morning; and (2) long calls during the day. When these mechanisms failed and two or more groups met, adult males called, displayed, chased each other and occasionally fought. Such encounters, however, were the exception rather than the rule.

(1) Territorial "rounds". The evening calls were termed "rounds", because of their contagious character. They appeared also to be given by a single adult male, who called three or four times over about five minutes, the series beginning some 5-10 min after the group entered sleeping trees for the night. Calling by one male triggered calling by males of adjacent groups, so that a round of calls echoed through the forest. Such rounds were heard at about hourly intervals during the early evening, and then at longer intervals as the night progressed; they were seldom heard in the early morning, between 0000 and 0300 hours. The pre-dawn and early morning were, like the evening, times of intense calling (fig. 4.4). The early evening peak (1900-2000 hours) coincided with entry into sleeping trees and the dawn peak (0500-0600 hours) with exit from them.

Two sleeping trees bordered the Rest House clearing: #583 on the north (used by m-5) and #517 on the south (used by Rh-m). About five minutes after entering #517 for the night, one male (Stumptail in the first few months and Snow thereafter) generally moved to the top of the tree and sat looking intently north towards #583. Occasionally, on first entering the tree, the male took up a spread-eagle position, with all four limbs grasping supports and his black scrotum showing clearly against the white ventrum and inner thighs: he might maintain this position for several minutes while staring towards #583. After scanning the male called once and then leapt heavily downward in the tree, sometimes breaking a branch which crashed noisily to the ground. An m-5 male would call immediately, followed in a few seconds by a second call from the Rh-m male, who had regained the upper canopy in the interim, and was again looking intently towards #583. A third exchange of calls usually signalled the end of the round, although distant groups might still be heard calling for several minutes. At least some of the calls during the night were also accompanied by displays.

(2) Long calls during the day. Long calls were also given during the day by one or more males when the group was about to

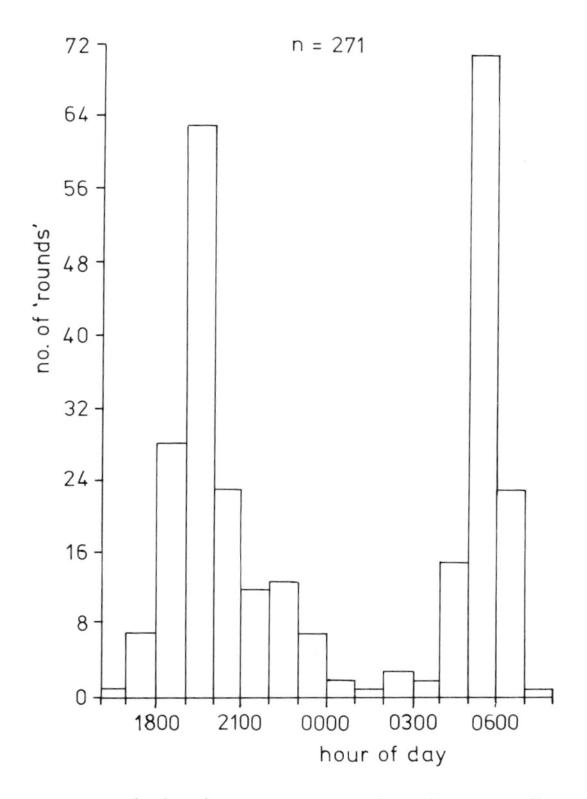

Fig. 4.4. *P. melalophos* territorial "rounds" by night.

enter an overlap zone. These calls did not usually elicit calling from neighbours, except on the rare occasions when two groups met. Calling by the male Strider of m-2 was described on January 27 1971:

1135 The group has been feeding on the fruit of #550 since 1110 hours. Now churrs (mild alarm calls) are heard, and several animals jump south out of #550. A long call from #550, then Strider moves deliberately to its very highest branches (65 m, 213 ft) and calls again. He races to the east side of the crown, calls a third time, then back to the first spot where he calls again. All his movements are strikingly noisy.

1145 Silence falls and the group rests until 1235.

The choice of tree #550 (a *Madhuca kingii*) as a calling point was apparently not a casual one, since it was used on several subsequent occasions. This was probably because of its location on the border of an overlap zone, and the tree's great size. Such a lofty tree provides both a superior vantage point and enough vertical space for the impressive locomotor display.

The 149 long calls heard occurred at all hours of day, but were more frequent in the morning than in the afternoon (fig. 4.5).

B. Inter-group encounters

Only one possible meeting between groups of *P. obscura* (o-3 and the group to its north) was observed in the course of the study, and this might have been only an interaction between a solitary male and the males of o-3. This incident took place on April 19 1971:

1135 o-3 resting. The sub-adult male suddenly runs north out of #645, no call.

1140 One adult in partial view in #646, resting.

1145 An adult rushes north out of #646, in same direction as sub-adult, no call.

1151 "Moan" call and a second adult runs north from #646.

1156 A third adult dashes 50 m north out of #646, "moan" call when it comes to rest.

1216 Two "whoos" from south, then two sharp honks to NNW, beyond the area in which the others came to rest.

1217 Third honk, then a fourth - very hoarse.

1221 Honks all over, from S of #646, about 65 m SW, from far NNW again, and to N where sub-adult male ran.

1225 3-4 animals have moved north in the direction taken by the #646 animals, towards the source of the honks to the far NNW.

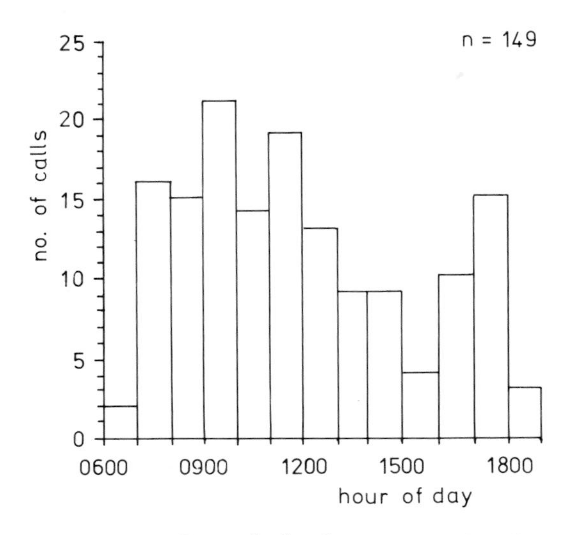

Fig. 4.5. *P. melalophos* calls by day.

1227 One male is still honking to the north. The two that were calling here around #646 have moved north towards that spot, as noted at 1225 hours. A brief glimpse is caught of 3-4 animals sitting close (body contact) in a large fork ... all gazing intently north. Then silence.

This interaction was apparently visual and vocal only, and no physical contact was involved. Judging from the calls, the active participants from o-3 were three adult males and one sub-adult male. Only one extra-group male appeared to be involved in the confrontation, presumably a member of o-2, although the presence of a solitary individual in this area could not be ruled out.

The group of *P. melalophos* whose territory overlapped least with those of neighbouring groups (fig. 4.1b) was m-5, who shared a relatively small part of the eastern edge of its territory with the all-male group Rh-m. Although only a small part of m-5's whole territory, this overlap zone was significant since m-5 frequently used sleeping trees within it and interacted regularly with Rh-m in evening call rounds. During these rounds the calling males of the two groups were in visual contact, orientating their displays only to each other, although at least one, and often two, other groups could be heard calling in the distance. No physical contact was seen between the males of the two groups. Once, however, an m-5 male crossed Sg. Lompat, and called and displayed in trees used almost daily by Rh-m; 5 min after the male had returned north across the river, Rh-m was seen retreating silently south. When Rh-m entered the overlap zone shared with m-5, they were notably tense and alert.

Group m-2 also directed territorial calls and displays mainly to one neighbouring group, m-5, although m-2's home range overlapped just as extensively with m-1 on the east. The relative lack of interaction between m-2 and m-1 was striking; on several occasions the two groups were within 130 metres of each other, in audible, and probably also visual, contact. On one of these occasions Hero, the single m-1 adult male, called, and on another an m-2 male called, but in neither case was the call answered, nor did any further interaction ensue. That m-1 contained only one adult male may explain the large overlap of m-1's range with those of adjacent groups (fig. 4.1b).

Aggressive confrontations between banded leaf monkey adult males were observed twice. Both encounters occurred to the west of the main study area, and involved animals unfamiliar to the observer. In neither case was it clear what precipitated the confrontation. November 25 1970:

1541 What appears to be an unusually large *P. melalophos* group (subsequently thought to have been m-4) is on the N/S transect 15 just north of the Main Trail. Several long calls and general movement north, including caller.

1542 Again, many continuous long calls. Now a second male (sounds like a sub-adult) also calls. Several "see-saw" (aggressive) calls, followed by what sounds like a fight.
Suddenly long calls ring out about 50 m WSW, and a second adult male approaches calling. Agitated long calls from the first adult male. The second approaches rapidly, calling all the way. The first male has almost reached the group.
More long calls now begin from a third direction, WNW. A pandemonium of calls from all three places. The first and second adult males meet, and what sounds like a scuffle ensues. "See-saws" and screeches intermingle with frantic calling.

1550 The two males seem to have separated. General quiet, broken by occasional "churrs" (alarm calls) to the N and WNW.

In the second confrontation actual physical contact was not confirmed, but the display by the one visible male was dramatic.
March 19 1971:

1045 Strident long calls near tree L277 (in Chivers' siamang study area, immediately west of Curtin's area). One adult male is running vigorously back and forth through the middle canopy. Another adult male starts long calls about 65 m to the north. Both males call continuously.
The first male races downward in a long diagonal, calling violently, to about 5 m above the ground, in the under-storey. He runs east extremely rapidly, still 3-7 m above the ground, leaping upright between small saplings, appearing hardly to use his forelimbs. After 70 m he halts, but still calls; the other male responds from the north. From the group out of which the first male came come sounds of agitated movements, then a scuffle and a panting screech. Silence falls.

1050 The first male to the east has fallen silent and is out of view. The male to the north continues calling for about 10 min.

In both encounters described above only one social group was definitely known to be in the area. Possibly an additional group (or two, in the case of the first encounter) was nearby, but hidden and silent except for the displaying adult males. Both encounters, however, appeared to take place between at least one solitary male and one group male. The possibility exists that these were attempts by a solitary male to penetrate or take over a group, rather than true territorial encounters between males of neighbouring groups.

C. Discussion of territoriality

Neither species of leaf monkey conforms to the classic type of territoriality, exemplified by the lar gibbon (Ellefson, 1974), which defends an exclusive area by vocal (calling) and physical (ritualised boundary confrontations) behaviours. Nevertheless, it is clear that *P. melalophos* is territorial by the criteria of exclusive use and defence (Burt, 1943). The criterion for exclusive use in this study was the failure to observe a non-resident group in a territory - the non-overlapping parts of adjacent home ranges. Physical confrontations between groups are rare, but vocal interactions occur many times each day through the calling "rounds", which thereby serve the function of territorial defence. For the 5 groups of *P. melalophos*, overlap zones amounted to 19% of the total area of all group ranges (Curtin, 1976a). For the 2 groups of *P. obscura*, calling between neighbouring groups was less frequent, and ritualised confrontations were absent; overlap zones, however, amounted to only 3% of total range area.

In any event, the patterns of group interaction and territorial maintenance are quite different in the two species. Adjacent groups of *P. melalophos* interacted with each other daily by means of territorial calls. From dusk to dawn calls were highly contagious: calling by one male almost always elicited calls from the males of as many as four neighbouring groups. In contrast, day-time long calls, even in overlap zones, were seldom answered. The energy devoted to the calls and the accompanying displays was significant: the same male has been observed participating in as many as six discrete calling bouts in one night (from 1900-0600 hours). Possibly there is monthly variation in the frequency of call rounds, and 6 rounds/night represents a peak frequency. Chivers (1973) has suggested that calls are more frequent during the main fruiting season (June-August), which may also coincide with a mating peak (Curtin, 1976a); no clear peaks of calling, however, were evident in this study.

Long calls of *P. obscura*, by contrast, were not contagious and were given less often. In fact, the rates of 0.5 and 1.1 calls/day, for dusky and banded leaf monkeys respectively, underestimate the difference between the species. For *P. obscura*, single calls by one adult male were counted, and for *P. melalophos*, calling rounds involving up to six adult males from as many different groups were counted. The rate of vocal interaction among neighbouring groups is, therefore, several times greater for banded than for dusky leaf monkeys.

The calling pattern of *P. obscura* was non-interactive to such an extent that one might question whether long calls alone can account adequately for the spacing of groups. Another factor might be the conservative nature of pathway use and ranging patterns. Where adjacent groups have stable, well-established territories, these features ensure that inadvertent meetings between groups are avoided. Such mechanisms may be inadequate, however, when a new group is establishing its territory amidst the territories of existing groups (as in early 1970, Chivers, pers. comm.), or in cases of sudden population dislocation. Logging operations near the boundary of the Krau Game Reserve to the south in late 1971 and 1972 may have produced such a dislocation, since MacKinnon and MacKinnon (1978) subsequently reported "fierce clashes and fighting" between groups of dusky leaf monkeys.

RANGING

Home range size

The main study group of *P. obscura*, o-1, covered an area of 33 ha (32 ha used exclusively), and that of *P. melalophos*, m-2, covered an area of 21 ha (14 ha used exclusively), during the period of study.

Home range use

Ranging patterns were analysed for the six-month period from January to July 1971. Day ranges were plotted as completely as possible with reference to tagged and mapped trees for o-1 (fig. 4.6) and for m-2 (fig. 4.7). In the latter half of the study, after several hundred feeding, resting and sleeping trees had been tagged and mapped, it was often possible to represent an entire day range by listing numbered trees in the order in which they were visited.

Full and partial day ranges for each species on 29 days, amounting to 6.8 hr/day for *P. obscura* and 5.0 hr/day for *P. melalophos*, yield rates of travel of 43 m/hr for *P. obscura* and 58 m/hr for *P. melalophos*. Thus, extrapolation to cover the whole active period, yields a mean day range of 559 m for *P. obscura* and of 754 m for *P. melalophos*. While such measures must clearly be treated with circumspection, they indicate the extent of daily travel and the difference between the two species. While days were selected when the group was more tolerant of the presence of the observer, there was still some inhibiting effect which reduced the amount of movement, especially for *P. melalophos*.

Both groups concentrated their activities in the same $2\frac{1}{4}$-ha quadrat - the area bounded by N/S transects 17 and 18 and E/W 8 and 9. Although this area did not differ dramatically from other parts of the forest, it was somewhat unusual in (1) containing more large trees than other quadrats, (2) many of which were members of the family Leguminosae (notably *Koompassia excelsa*, *K. malaccensis* and *Intsia palembanica*). Floristic composition and vegetative structure of this core area for both groups were sampled in a plot of 20 x 200 m (Curtin, 1976b).

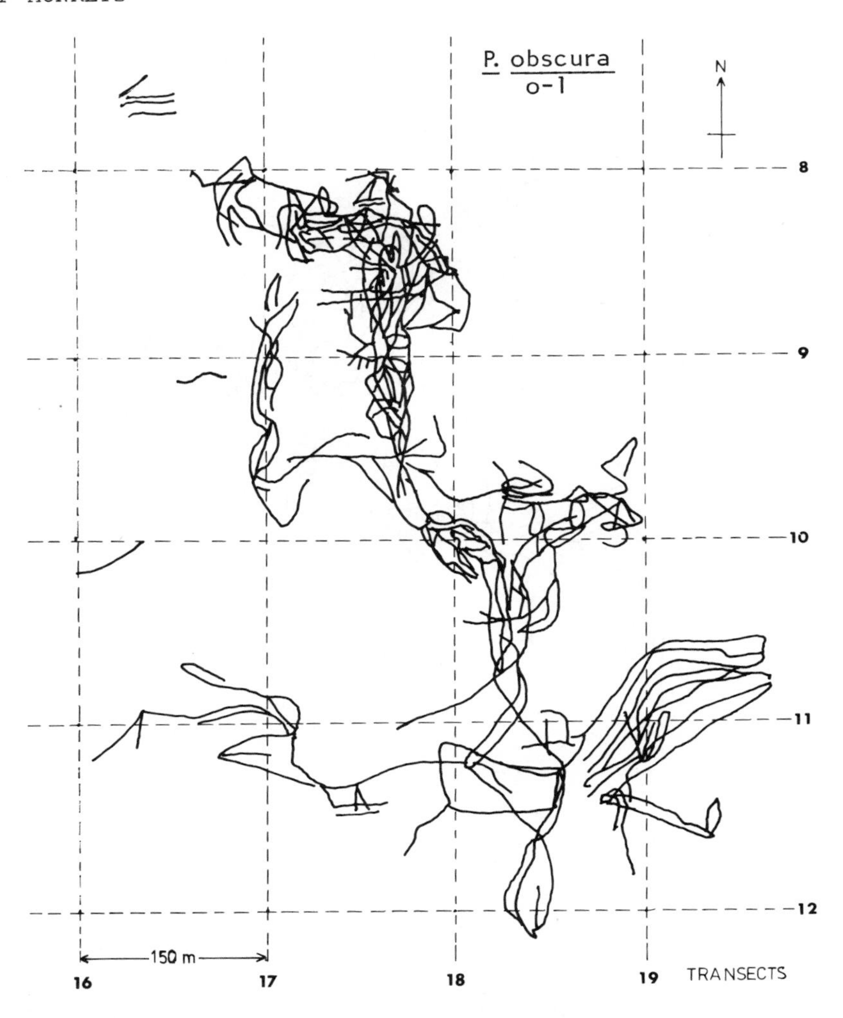

Fig. 4.6. Composite partial day ranges of o-1 (*P. obscura*) on 29 days during the period January 26 - July 27, 1971.

For *P. melalophos* plots of day ranges over several weeks tended to describe a series of irregular loops out of and back to the central core area. The time required to complete each loop was 2-3 days on average; loops were longer in the fruiting season (June-August) and shorter in the dry season (January-March). On several occasions m-2 remained in the core area 1-2 days before moving out on another circuit of its territory.

The importance of the core area as a source of food is illustrated in a map of the distribution of food trees used by m-2 (fig. 4.8). Most of the trees in the $2\frac{1}{4}$-ha quadrat were used not

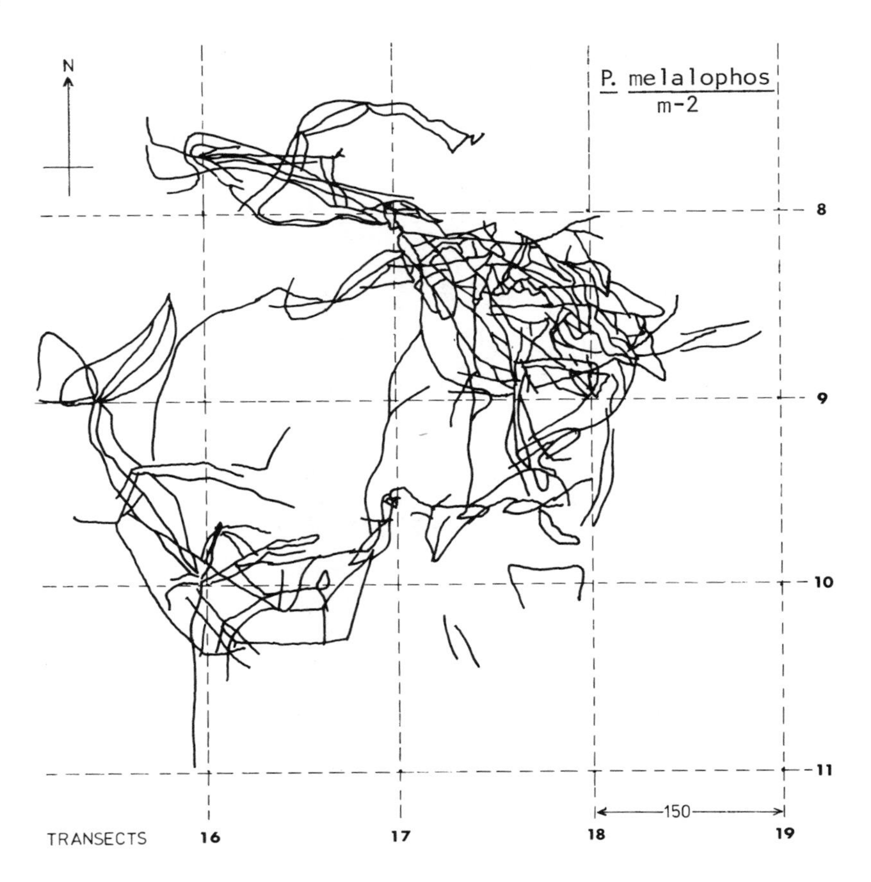

Fig. 4.7. Composite partial day ranges of m-2 (*P. melalophos*) on 37 days during the period January 26 - July 27, 1971.

just once but repeatedly, and nearly every large or medium-sized tree in this quadrat was exploited for food at some time during the course of the year. The pattern of use in the more peripheral parts of the territory was more uneven, particularly in the north-west, where m-2 came into frequent contact with m-4 (fig. 4.1b). It appears that the direction and length of each loop and, ultimately, the overall pattern of exploitation were influenced by: (1) the abundance of preferred food trees, (2) the season of the year, and (3) movements of neighbouring conspecific groups.

The pattern of ranging by *P. obscura*, group o-1, was similar in that two kinds of ranging predominated: (1) loops out from and back to a core area and (2) intensive foraging within the core area. The group circled its territory every 4-5 days on average, however, compared with 2-3 days for m-2. The distribution of feeding trees for o-1 is shown in fig. 4.9.

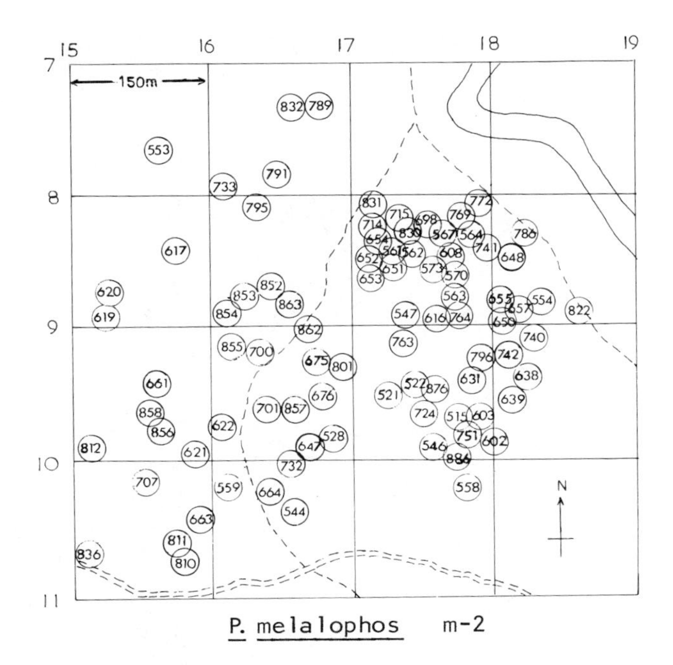

Fig. 4.8. Locations of food trees used by group m-2 (*P. melalophos*).

It is apparent (fig. 4.6) that the movements of *P. obscura* within the core area and the quadrat adjacent to the south were more predictable than for *P. melalophos*. This was a function of the conservative nature of o-1's choice of arboreal pathways. The main pathway in these two quadrats consisted of a chain of large trees, and the nodal points of highest use coincided with three key trees: #515, NAT-1 and #570. The movements of *P. melalophos*, in contrast, were more irregular and less restricted to particular arboreal pathways (fig. 4.7). Possibly this difference between the two species was due largely to inter-specific differences in the use of different levels of the canopy (see below and chapter 6).

The two species differed also in their use of riverine forest. *P. melalophos* was often observed on the margins of the two main rivers in the study area, Sungai Krau and Sg. Lompat, and two groups (Rh-m and m-5) slept frequently in trees bordering the rivers. *P. obscura*, however, appeared to avoid riverine habitats, preferring the mature forest away from the rivers.

Plate XIV. Group of dusky leaf monkeys eating new leaf shoots in a Merbau tree, *Intsia palembanica*, at Kuala Lompat (DJC)

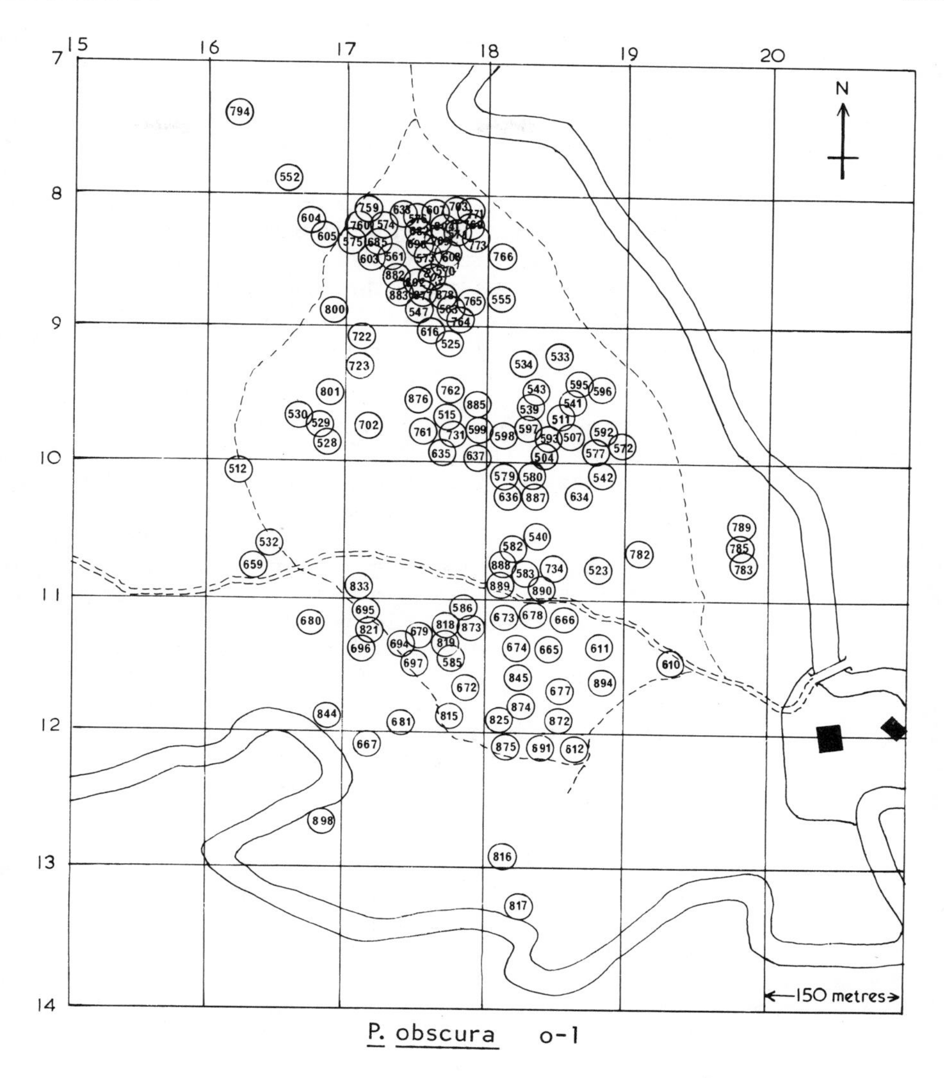

Fig. 4.9. Locations of food trees used by group o-1 (*P. obscura*).

FEEDING

Dietary composition and diversity

The diversity of the diets of the two leaf monkeys differed considerably: *P. melalophos* used half again as many tree species for food as did *P. obscura* (137 species compared with 87, Curtin, 1976b). Since the study area was the same for both leaf monkeys, the floristic diversity available to each must also be the same, and local variation in species richness in the forest cannot be the explanation for the differences.

The study area contained the territories of three *P. melalophos* groups in entirety and of two in part, while only one entire and one partial territory were represented for *P. obscura* (fig. 4.1). Thus the bias would lie in the direction of under-emphasising the differences between the species, since one *P. obscura* group had about three times the potential diversity available to any one *P. melalophos* group (see chapter 2). On the other hand, a factor such as group food traditions (Kawai, 1965) might lead to over-emphasis of the inter-specific difference, since only two groups of *P. obscura* were sampled, compared with five groups of *P. melalophos*.

To what extent was the apparently greater diversity of the *P. melalophos* diet a function of canopy use? Principal food species, defined as trees and lianas fed upon three times or more, are shown for each leaf monkey in table 4.6, together with rankings by height class. Of the 26 species of major importance to *P. obscura*, 25 were classed as emergents - a convincing illustration of the high-canopy habits of this colobine. *P. melalophos*, in contrast, divided their attentions equally between the middle-storey and emergent trees, with only three major food species occurring in the under-storey. The greatest mass and variety of vegetation in this rain-forest is found in the middle-storey, and emergent trees, although individually huge, are fewer in number and poorer in species diversity (Richards, 1952). Inter-specific differences in vertical ranging alone, therefore, may account for the differences observed in dietary diversity.

The dependence of *P. obscura* on emergent tree species was all the more impressive since 1970-71 was an "off" year, when most of the huge dipterocarps neither flowered nor fruited. Although dipterocarps are less abundant in this forest than in many other parts of Malaya and Borneo, their great size makes them potentially important food species where they do occur. And yet, their fruit are clearly designed for wind dispersal and their leaves do not usually appear nutritious (see chapter 2); not one of the food species important to dusky leaf monkeys was a member of this family. The family best represented in their diet was the Leguminosae. All of these species are emergents and some of them, like the mighty 'tualang' (*Koompassia excelsa*), are among the noblest trees in the rain-forest, towering 70 m (230 ft) or more and spreading their canopies over a like distance. The real importance of these legumes lay not only in their vast size, but also in their almost year-round production of food. *P. obscura* fed on *Intsia palembanica* seeds in November, new leaves in January, February and March (Plate XIV), and mature leaves in April. *Koompassia excelsa* provided new leaves in March, mature leaves in August and February; new leaves of *Parkia javanica* were eaten in May, mature leaves in January, March and July

Table 4.6. Principal food species of dusky and banded leaf monkeys.

Family	Species	DUSKY LEAF MONKEY				BANDED LEAF MONKEY			
		No.of trees	No.of occas.	height class	parts consumed	No.of trees	No.of occas.	height class	parts consumed
Anacardiaceae	*Pentaspadon velutinum*	12	14	H	nL mL F	10	10	H	nL mL F
Annonaceae	*Polyalthia* sp.					3	3	L	F
	Xylopia magna	1	3	H	fl				
	Xylopia malayana	1	4	H	nL F				
Apocynaceae	*Alstonia angustiloba*	3	3	H	nL fl				
Chrysobalanaceae	*Maranthes corymbosa*	9	18	H	mL F	3	3	H	F
Combretaceae	**Combretum* sp.	2	3	H	nL mL F	5	6	H	nL mL F
Connaraceae	**Connarus* sp.	4	4	H	mL F				
	**Rourea* sp.	3	3	H	mL F				
Dilleniaceae	*Dillenia reticulata*	3	3	H	nL F fl				
	Dillenia sp.	1	3	H	mL F				
Euphorbiaceae	*Elateriospermum tapos*					4	5	M	nL F
	Pimelodendron griffithanum					3	3	M	F
Flacourtiaceae	*Paropsia veraciformis*					5	8	M	mL F
Gnetaceae	**Gnetum globosum*					2	3	H	F
	**Gnetum* sp.					4	4	M	nL F
Lauraceae	*Beilschmiedia* sp.					2	5	M	nL F
	Dehaasia sp.					2	3	M	mL F
	Litsea sp.					2	3	H	nL mL F
Leguminosae	**Acacia pennata*					3	3	M	nL F
	Cynometra inaequifolia					3	4	H	F s
	**Derris* sp.					7	7	H	nL mL F

Table 4.6 (continued)

Family	Species	DUSKY LEAF MONKEY				BANDED LEAF MONKEY			
		No.of trees	No.of occas.	height class	parts consumed	No.of trees	No.of occas.	height class	parts consumed
Leguminosae (continued)	*Dialium patens*	1	3	H	nL mL F				
	Dialium platysepalum					3	3	H	nL F
	**Entada scandens*					3	5	M	nL mL F fl
	Intsia palembanica	11	12	H	nL mL F	15	18	H	nL mL F fl
	Koompassia excelsa	4	4	H	nL mL				
	Koompassia malaccensis	4	5	H	nL				
	Millettia atropurpurea					3	3	M	nL
	Parkia javanica	6	8	H	nL mL F	3	10	H	F fl
	**Spatholobus* sp.	7	7	H	nL mL F fl				
Linaceae	**Indorouchera griffithii*	2	3	H	nL				
Malvaceae	*Hibiscus floccosus*	3	3	H	nL mL	3	3	H	mL fl
	Hibiscus macrophyllus					3	3	M	mL
Meliaceae	*Dysoxylum costulatum*					3	3	M	F
	Dysoxylum sp.					2	3	M	nL F
Moraceae	*Ficus* spp.	4	5	H	nL mL F	4	5	H	
	**Ficus* spp.	11	12	H	nL mL				
Myrtaceae	*Eugenia* sp.	2	3	H	nL mL	4	4	M	nL F
Olacaceae	*Strombosia javanica*					3	3	H	nL fl
Rhamnaceae	**Ventilago* sp.	4	4	H	nL	4	4	M	nL
Sapindaceae	*Nephelium eriopetalum*					2	6	M	mL F
	Xerospermum muricatum					2	3	M	F
Sapotaceae	*Palaquium hispidum*	3	3	H	mL F				
	Payena lucida					5	6	L	F

Table 4.6 (continued)

Family	Species	DUSKY LEAF MONKEY				BANDED LEAF MONKEY			
		No.of trees	No.of occas.	height class	parts consumed	No.of trees	No.of occas.	height class	parts consumed
Strychnaceae	**Strychnos* sp.					4	4	M	nL F
Theaceae	*Adinandra* sp.					1	3	L	nL mL
Thymelaeceae	*Aquilaria malaccensis*					3	3	M	F
Urticaceae	**Poikilospermum* sp.					2	3	H	nL
Vitaceae	**Vitis* sp.	4	4	H	mL				
Xanthophyllaceae	*Xanthophyllum excelsum*	3	3	M	F fl	8	11	M	nL mL F
	Xanthophyllum scortechinii	2	3	H	mL F	1	7	H	F fl
TOTAL		110	140			139	183		

Total principal food species of dusky leaf monkeys = 26
Total principal food species of banded leaf monkeys = 37

Height class determined from average of estimated tree heights.
Height class of liana (*) = height class of host trees.
H = Emergent (100+ ft.). M = Middle (50-100 ft.). L = Understorey (<50 ft.).

nL = new leaves
mL = mature leaves
F = fruit
s = seeds
fl = flowers

and beans in February and March (table 4.6).

The picture for *P. melalophos* was similar: three leguminous tree species, all emergents, represented less than 10% of important food species, but accounted for almost 20% of feeding occasions (table 4.6 and Curtin, 1976b). *Dialium platysepalum* provided new leaves in August and fruit in August and October. *Intsia palembanica* was eaten at every stage of its vegetative cycle: seeds in October, November and December, flowers in March, and mature leaves in March and October. *Parkia javanica* was mainly important for the beans, which were produced abundantly in the first three months of 1971.

In addition to these legumes, the fruit of *Maranthes corymbosa* (taken mostly unripe), and the ripe fruit of *Ficus* spp. were major food sources for *P. obscura*. The common tree *Pentaspadon velutinum* was exploited by both species, which fed on new leaves, new leaf petioles, mature leaves and fruit.

Dietary proportions

Comparison of the diets of the two leaf monkeys reveals a greater dependence on fruit by *P. melalophos*, and on new and mature leaves by *P. obscura* (Curtin, 1976b). *P. melalophos* consumed 48% fruit and 35% leaves, and *P. obscura* 32% fruit and 58% leaves (fig. 4.10); the former is a "frugivore", and the latter a "folivore", according to terminology commonly used to describe primate diets. These terms should be employed in a relative sense only, however, as one species being more or less frugivorous or folivorous than the other, taking care that the same measures of diet are used - in this case feeding occasions and time spent feeding (Curtin, 1976a). A "feeding occasion" was every instance of feeding observed, whether it involved a juvenile plucking and nibbling a leaf between play bouts or the whole group feeding intensively on a preferred food.

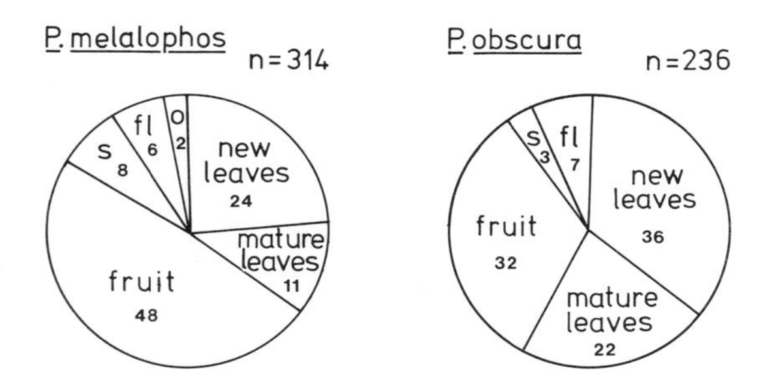

Fig. 4.10. Proportions of food types in the diets of *P. melalophos* and *P. obscura*.

Daily activity

Analysis of feeding occasions derived from observations on five groups of *P. melalophos* shows a bimodal distribution of feeding during the hours from dawn to dusk (fig. 4.11). There was a morning feeding period from 0700-1200 hours, with the most intensive feeding of the day from 0800-1000 hours; a decrease at mid-day, particularly between 1200 and 1300 hours; and an afternoon feeding period from 1500-1900, dusk.

In both morning and afternoon, brief periods of feeding for about 30-60 minutes were interspersed with periods of rest and travel. Major periods of travel occurred: (1) in the early morning between 0600 and 0730 hours; (2) in the early afternoon; and (3) in the hour before dusk. Exceptions occurred when heavy rain fell during the night or in the late afternoon. If there had been heavy rain during the night, the group often remained near the sleeping tree until 0900 or 1000 hours, when the sun had partially dried the foliage and branches, and the group was able to move through the trees with greater ease and safety; under these conditions intensive feeding might not begin until late morning. When heavy rain fell in the afternoon, both feeding and travel were inhibited. If the

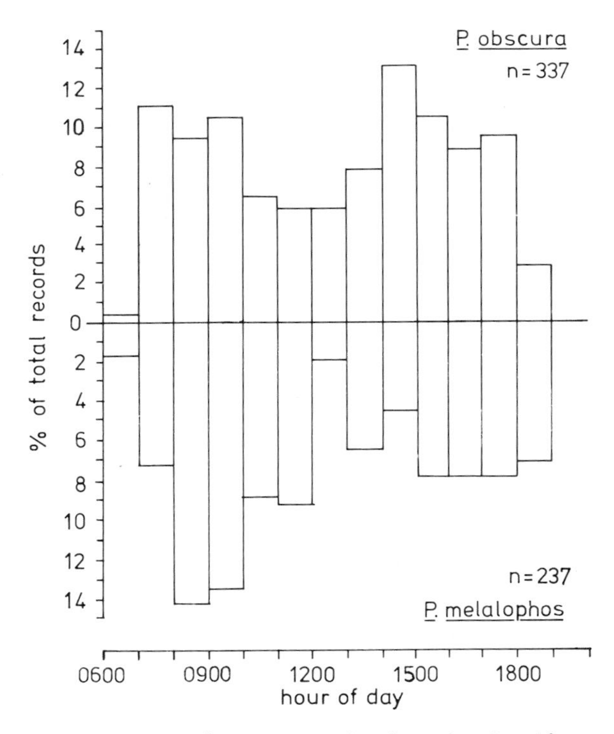

Fig. 4.11. Diurnal variation in feeding.

downpour was both heavy and sustained, and especially if accompanied by strong winds, the group moved only as far as the first suitable sleeping trees.

The data for *P. obscura* also show a bimodal distribution of feeding (fig. 4.11), but the morning feeding extended for only three hours, from 0700-1000 hours, rather than for five hours as recorded for *P. melalophos*. There was a mid-day lull from 1000 to 1400 hours, a feeding peak from 1400-1500 hours, and then a relatively high level until 1800 hours, when feeding decreased markedly.

The differences between the two leaf monkey species may be summarised as follows. *P. obscura* showed (1) a longer mid-day intermission in feeding and (2) an earlier cessation of feeding for the day than *P. melalophos*, and (3) the peak period of feeding in the afternoon, rather than in the morning. On the whole *P. obscura* confined feeding to fewer hours of the day, which illustrates a fundamental difference in feeding patterns between the two species. *P. melalophos* usually moved further each day, sampled more food species, and spent less time in any one food tree. In contrast, *P. obscura* were generally less active: individuals had shorter day ranges, visited fewer feeding trees in an average day, and spent more time in each (Curtin, 1976a). Since feeding bouts were longer, and less likely to be broken between brief periods of travel, feeding thus appears to be restricted to fewer hours of the day than in *P. melalophos*.

Selection of three types of food - fruit, new leaves and mature leaves - defined in terms of the percentage of all feeding occasions in each time interval, are shown for both species in each hour of the day from 0700-1900 (fig. 4.12). There are no marked patterns

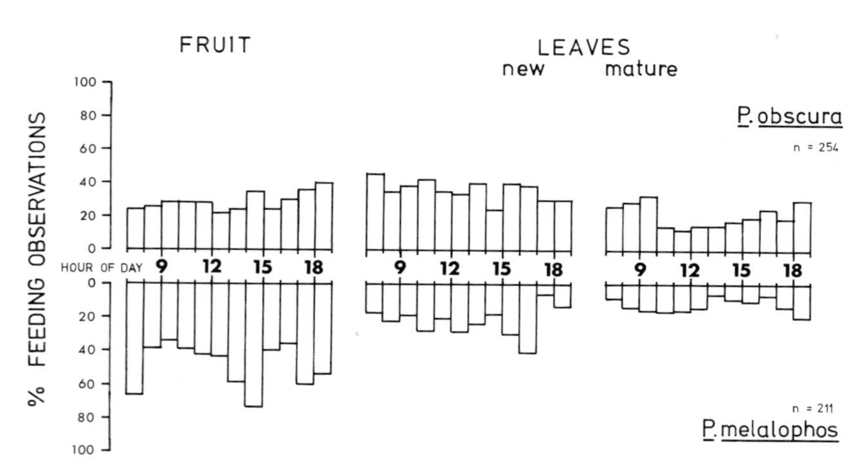

Fig. 4.12. Selection of food types through the day (n = 254 feeding occasions for *P. obscura*, 211 for *P. melalophos*.

of selection for either leaf monkey, although there are slight tendencies for *P. melalophos* to prefer fruit in the early morning, early afternoon and just before dusk, and new leaves in the late afternoon and for *P. obscura* to eat more fruit as the day passes and mature leaves early and late in the day. These results contrast with those of Raemaekers (1978b), in which siamang and lar gibbons select strongly for figs as the first and last foods of the day. The leaf monkeys presumably show less fluctuation in the intakes of different foods, because of their different and steadier metabolic needs based on folivory.

Canopy use

P. melalophos characteristically fed and travelled most in the middle- and under-storeys (7-30 m, 23-98 ft), although behaviours such as resting occurred more frequently in emergent trees. In 299 cases the height of the tree being fed upon was estimated to the

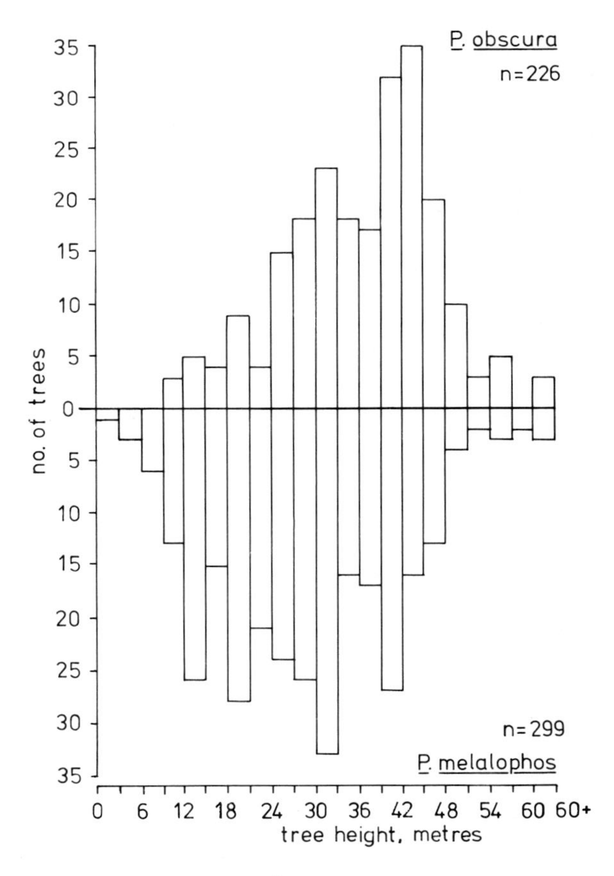

Fig. 4.13. Heights of food trees.

Table 4.7. Heights of feeding trees in relation to food items: *P. melalophos*.

Food item	n	mean height metres	(feet)
New leaves	47	24	(80)
Mature leaves	24	25	(82)
Fruit (tree)	131	25	(82)
Flowers	14	35	(114)
Fruit and beans (vine)	33	33	(109)
Vine leaves	37	29	(95)
Vine flowers	5	24	(81)
Seeds and beans (tree)	17	37	(119)
Other	2	26	(87)
	310	28	(94)

nearest $1\frac{1}{2}$ m; the distribution of heights of these food trees approximate a normal distribution (fig. 4.13), with a mean height of 28 m (92 ft) (Curtin, 1976b). Further analysis according to type of food eaten reveals mean tree heights below the overall average for new leaves, mature leaves and fruit (table 4.7), and above by 8 m for beans and seeds. The latter conforms with what is known of the structure of this forest, in which an unusually large proportion of emergent trees are legumes.

In contrast, the great majority of feeding by *P. obscura* was in the emergent trees and middle-storey; only rarely did they descend to the under-storey, and never were they seen feeding in the sapling level (below 7 m, 23 ft). This is clearly illustrated by the distribution of food tree height from 229 observations (fig. 4.13); most such trees were 24 m (79 ft) or more tall, and the mean height was 35 m (115 ft).

Given these inter-specific differences in food tree height, it is clear why *P. obscura* visited fewer feeding trees each day, and spent more time in each tree than did *P. melalophos*. Generally, the taller the tree, the greater the potential food supply and the less frequent the need to change food trees.

Diet, locomotion and "opportunistic" species

As beak size is a clue to niche differentiation in some birds (Lack, 1971), so relative limb lengths, limb musculature and features of the skeleton, such as vertebral spine length, point to different patterns of locomotion in these two primates (Washburn, 1944; Fleagle, 1976b, c, 1977). These locomotor behaviours are in turn correlated with different strategies of habitat exploitation and vertical ranging in particular. The deliberate quadrupedal striding and leaping of *P. obscura* appear to be associated with the broad, firm supports of the middle and upper levels of the forest. The more saltatory and varied locomotion of *P. melalophos* is in turn associated with the kind of small, flexible supports characteristic of the lower levels of the forest and of the terminal branches of medium and large trees. Of the two species, *P. melalophos* is the more flexible both in its locomotor and vertical ranging patterns (Fleagle, 1977).

The relationship between forest level preference and diet is not obvious, if indeed it exists at all: a high canopy species which is primarily folivorous (*P. obscura*) and a low-canopy species which is primarily frugivorous (*P. melalophos*). One possibility concerns the functional implications of locomotor patterns for access to different types of food sources (Ellefson, 1974; Grand, 1972). In the large discontinuous trees of the high canopy, fruit are borne mainly on the terminal branches, while new and mature leaves are available closer to larger branches. The locomotor demands placed on a primate travelling habitually in this kind of arboreal context may therefore conflict with the efficient exploitation of fruit. A frugivorous high-canopy primate, such as the lar gibbon, provides an exception to this general view (Ellefson, 1974, considers brachiation primarily an adaptation to terminal-branch feeding, rather than an efficient mode of travel through large trees).

On many points *P. melalophos* appears to be a more "opportunistic" species than *P. obscura*. It can be found at all levels of the forest, sometimes foraging on the ground; its diet is more variable, and a captive animal accepts new foods more readily than do its congeners (Curtin, 1976a). Perhaps at least partly as a consequence of this locomotor and dietetic flexibility, *P. melalophos* was encountered in a wider range of habitats outside the study area. It was abundant in secondary forests, where *P. obscura* was more rarely found. Individuals were seen frequently in roadside vegetation or running across roads, while dusky leaf monkeys were never seen moving on the ground outside the Reserve. *P. obscura* was mainly found in mature (though not necessarily primary) forest, although it readily invaded rubber plantations. In all habitats the banded leaf monkey is the more abundant species (Southwick and Cadigan, 1972).

DISCUSSION

Unlike siamang and lar gibbons (chapter 3), the two leaf monkeys at Kuala Lompat do not differ significantly in body size (Curtin and Chivers, 1978). Ecological separation appears to be maintained by a combination of at least four factors: (1) forest level preference, (2) degree of folivory or frugivory, (3) plant species composition of the diet, and (4) rate of activity.

P. obscura is more folivorous, more sedentary, less varied in its diet, and more of a high-canopy species than its congener *P. melalophos*. As with the small and large gibbons, the folivorous species is the less active.

Feeding overlap between the two species is reduced by a number of factors. All tree species fed upon by both leaf monkeys in the same month were large emergents, with the single exception of *Xanthophyllum excelsum* (Curtin, 1976a). Deciduous leguminous emergents, such as *Intsia palembanica, Koompassia excelsa* and *K. malaccensis*, were critical food sources during the dry season months, when their new leaves formed a staple part of the diet of both species. These trees, among the most imposing in the forest, made up a high percentage of all emergents. Species whose fruit were consumed by both leaf monkeys during the main fruiting season, such as *Maranthes corymbosa* and *Ficus* spp., bore conspicuously large fruit crops over unusually long periods compared to other tree species. One *M. corymbosa*, for example, fruited over three months (June-August 1971), and a number of fig trees produced several large crops during the 12-month study.

Inter-specific feeding competition, although relatively frequent (over half *P. obscura*'s food species shared by *P. melalophos*, and over one-third of *P. melalophos*' food species shared by *P. obscura*, Curtin, 1976a), is thus mitigated by at least three factors: (1) tree species shared were nearly all emergents, (2) they were common, and (3) they produced abundant fruit crops.

If adaptive success is measured by biomass, *P. melalophos* must be counted the most successful species of primate, not only at Kuala Lompat (MacKinnon and MacKinnon, 1978), but in Peninsular Malaysia as a whole (Southwick and Cadigan, 1972). One of the factors promoting dense populations of this species is the high degree of overlap among the home ranges of neighbouring groups. Although overlap was considerable in this study (19% for five social groups), an area of exclusive use was maintained by each group (c.f. Chapter 6). The key to the avoidance of conflict is the long calls, which apparently inform neighbouring groups of each other's movements, and allow them to range so as to avoid confrontation. Other factors promoting high density in this species are the very varied diet and the use of all levels of the forest.

The only other colobine in South-east Asia, which has been studied in detail in the field, is the silver leaf monkey, *P. cristata* (Furuya, 1961; Bernstein, 1968a; Wolf and Fleagle, 1977; Wolf, in prep.). The population living at Kuala Selangor is composed of one-male territorial groups, apparently characterised by adult male replacement and subsequent infanticide (Wolf and Fleagle, 1977), behaviours not observed in the leaf monkeys at Kuala Lompat. For the Bornean leaf monkeys, some information is available for *P.*

rubicunda (Stott and Selsor, 1961) and for *P. frontata* and *P. aygula* (Rodman, 1973a). Stott and Selsor report groups of 5-8 *P. rubicunda* and describe a morning display by the "dominant male", which sounds identical to the territorial display of *P. melalophos*. *P. rubicunda* males also displayed to human observers in a manner similar to *P. melalophos* males, and unlike *P. obscura* males, which generally retreated and hid with the rest of the group when disturbed. According to J. MacKinnon (pers. comm.), *P. rubicunda* lives in larger groups than those reported by Stott and Selsor (1961); these groups forage routinely in the lower levels of the forest and frequently come to the ground. Their territorial call is highly reminiscent of the long call of *P. melalophos*. The similarity of calls, and low-level foraging, suggest that in northern Borneo, *P. rubicunda* may be the ecological counterpart of *P. melalophos* in Peninsular Malaysia.

For the Sumatran species, there is almost a complete dearth of information, except for survey data (Wilson and Wilson, 1973, 1975). These species include *P. cristata, P. femoralis, P. thomasi* and *P. melalophos*. In the course of their 15-month survey, the Wilsons distinguished the latter three species on the basis of species-specific long calls. The taxonomic confusion surrounding the classification of South-east Asian colobines is evidenced by the fact that *P. femoralis* in the Wilsons' view is identical to what is called *P. melalophos* in Peninsular Malaysia, while the Sumatran *P. melalophos* is distinct from each. *P. femoralis* is found typically in primary and secondary lowland forests and in rubber plantations, *P. thomasi* in inland primary forest and *P. melalophos* (the most widely distributed of the three) in inland primary forests and in hill and sub-montane forests. Apparently, these three species do not overlap in their distribution. The Wilsons report group sizes of 5-8 animals in primary forest and 10-13 in secondary forest they state that one-male groups appear to be the rule for all Sumatran species of leaf monkey. In contrast, during the study at Kuala Lompat only one out of seven groups had a single adult male, and this was the smallest bisexual group with only 11 individuals.

A satisfactory classification of the Bornean and Sumatran leaf monkeys, particularly those of the 'aygula-melalophos' group, has yet to be achieved. It would seem that the South-east Asian colobines, a group with a great diversity of forms and wide distribution, would repay well further attention from students of primate socio-ecology.

LONG-TAILED MACAQUES

F.P.G. Aldrich-Blake

Department of Psychology
University of Bristol

INTRODUCTION

Long-tailed macaques, *Macaca fascicularis* [RAFFLES, 1821], are small brown monkeys, with paler underparts and often prominent whitish hairs on the face. These 'whiskers' vary greatly in their development between individuals, and are a valuable aid to their recognition. They are often particularly pronounced in mature males (Plate XV). New-born infants are black-furred, with bright pink face and ears; within a week the facial skin fades to pinkish grey, and after about six weeks the black natal coat is replaced by a brown one. Long-tailed macaques are the smallest of the primate species featured in this volume: adult males weight 5-7 kg, and adult females 3-4 kg, so there is marked sexual dimorphism.

Previous studies in the Peninsula have been confined to urban populations, in Singapore (Ellefson, 1967; Chiang, 1968), in Kuala Lumpur (Bernstein, 1967b) and in Penang (Spencer, 1975). Elsewhere in South-east Asia there have been field studies in Java (Angst, 1973, 1974, 1975), in East Kalimantan in the Kutai Reserve (Kurland, 1973; Fittinghoff, 1975; Wheatley, 1976, 1978) and in North Sumatra at Ketambe in the Gunung Leuser Reserve (Schürmann, in prep.). Poirier and Smith (1974) investigated the population that had been founded on Angaur, Micronesia, some years previously.

The present study began in July 1974 and continued until January 1976. The macaques were not the sole subject of study, however; the observer's time was divided between following macaques, collecting data on tree production cycles (with Jeremy Raemaekers), and conducting census walks. The planned monthly routine was to spend 5 days following macaques from dawn to dusk,

Plate XV. Macaques of Peninsular Malaysia

(a) Long-tailed macaque, *Macaca fascicularis* - male and female in Penang Waterfall Gardens (DJC)
(b) Pig-tailed macaque, *Macaca nemestrina* - adult male in Ulu Gombak (DJC)

5 days on census walks (60 days in all), 5 days on phenology and a further 5 days on macaques, although this was not always practicable. Most of the data presented herein were collected between February and May 1975, and September 1975 and January 1976, when the animals were well habituated. Forty virtually complete days of observation were obtained during this period, out of a total of 72 days of attempted following (about 600 hours of observation in all). Even when the macaques were tolerant of an observer at 30 m (98 ft), it was often impossible to keep them under continuous observation, especially in the dense, vine-covered vegetation along the rivers.

Early in 1976 Y.L. Mah of the Universiti Malaya took over observation of the macaques, habituated a second group at Kuala Lompat and a group at the Royal Selangor Golf Club, on the outskirts of Kuala Lumpur and, thereby, was able to make comparisons between groups in natural and modified habitats (Mah, in press, in prep.).

SOCIAL ORGANISATION

Group composition

Accurate group counts were not easy to obtain, since groups commonly fragmented into smaller parties, which might become widely separated. Even at night groups might still be spread over a wide area, sleeping in several different trees. Group size and composition had to be deduced, therefore, from observation of recognisable individuals over several weeks; only rarely were complete counts obtained.

The main study group, group A, appeared to fluctuate slightly in size and composition; the maximum count during 1974-75 was 23. A neighbouring group across the river to the east was counted reliably as having 23 individuals also. Counts of 4 and 11 were obtained for two groups to the west, but these may be incomplete. These figures are comparable with those from Kutai, East Kalimantan (mean size of 12 groups = 18; Kurland, 1973), but smaller than some from Thailand (mean size of 8 groups = 39; Fooden, 1971).

Group A contained 2-3 adult males, 2 sub-adult males, 6-8 adult females, 4-5 juveniles, 2-3 large infants and 2-3 babies (small infants carried by the mother, Plate XVI). Much of this variation is accounted for by changes due to maturation during the study. The uneven socionomic sex ratio and multi-male group structure are characteristic of all macaque species.

The female membership remained essentially stable; recognisable females remained in the group throughout the study. In contrast, adult males moved into and out of the group. At least six, perhaps

Plate XVI. Long-tailed macaques

(a) on the ground and (b) in the trees grooming in the Penang Waterfall Gardens (SPG & DJC)
(c) female and infant walking along barbed wire in the Singapore Botanic Gardens (DJC)

seven, males spent some time in the group between July 1974 and December 1975, although never more than three were seen at any one time. For instance, a male known as Short-tail was seen regularly in July and August 1974, but then disappeared. He may have been the same as an animal of similar appearance seen by the MacKinnons in 1972. Kink-tail, likewise, was seen only in July and August. By August 1974 three further males had appeared - Big-head, Little-big-head and Dark-patch. Big-head disappeared in May 1975, but rejoined the group in November; Little-big-head was not seen after May 1975. A sixth male, New, appeared in November 1975, and possibly a seventh also. Dark-patch was still in the group at this time.

The major changes in male membership in October 1974 and November 1975 appeared to be associated with increased aggression between males and a subsequent increase in sexual activity. The males' testes, also, appeared more prominent at this time; seasonal cycles in testis size, spermatogenesis and body weight have been noted in rhesus monkeys, *M. mulatta*, on Cayo Santiago by Sade (1964) and Conaway and Sade (1965), and in bonnet monkeys, *Macaca radiata*, at the California Primate Research Center by Glick (1979).

Behaviour within groups

Macaques are characterised by a high level of social interactions within groups. Grooming alone took up some 12% of daily activity time at Kuala Lompat (Plate XVI), more than in the other primates. Low visibility meant that few interactions were seen as a whole, but the level of vocalisations heard suggested that agonistic interactions were also more frequent than in sympatric leaf monkey groups of equivalent size.

The adult males are much the most visible individuals; they are usually the first to be seen by an observer, and would appear to play a watch-dog role. Kurland (1973) describes a distraction display by an adult male which successfully led him away from the rest of the group, as described by Curtin (chapter 4) for the banded leaf monkey. Often, however, the group's response to an observer takes on the tone of mobbing, with members of different age and sex facing the observer and calling together.

Behaviour between groups

Very few encounters were seen between groups; two of these occurred at a range of 100 m across the Krau river, and involved staring and calling by both males and females. Such encounters seem to be mild as well as infrequent (cf. the same species at Kutai: Fittinghoff, 1975). There is no regular exchange of loud calls, as occurs between groups of gibbons and leaf monkeys. Indeed, macaques lack an equivalent loud call audible over several hundred metres. Groups would sometimes be aware of the presence of neighbours, however, from their calls or other activities.

RANGING

Home range size

The group often fragmented during the day into parties varying in size from 2-3 animals to a major section of the group. Such parties are typically separated by 200-300 metres, and sometimes more. While they may recombine in the evening before the group settles for the night, this is not invariably the case. Indeed, on one occasion parties slept 600 metres apart. This factor, combined with the limited visibility, means that one can seldom keep in contact with the entire group. Plots of day range, therefore, will often represent the movements of only a part of the group, although they may be complete day ranges for individual animals. A pattern of splitting into foraging sub-groups and recombining at sleeping trees was also found at Kutai (Fittinghoff, 1975).

During the whole study period, July 1974 to early January 1976, group A was seen to range over an area of ca. 35 ha (fig. 5.1) (cf. 80 ha for a group of ca. 20 animals at Kutai, Kurland, 1973). This area was determined by drawing a line around the outermost points of super-imposed maps of day range, and included parts of 50 of the hectare quadrats.

Home range use

Considering only the period between February 1975 and January 1976, complete day ranges were available for 38 days. The distance travelled varied from 150 to 1500 metres, with a mean of 760 m (fig. 5.2), which is comparable to the 400-1000 m recorded by Kurland (1973) at Kutai. The day range can also be expressed in terms of the number of hectare quadrats entered; this varied from 3 to 18 for the parts of the group observed, with a mean of 9.5 (fig. 5.3).

Variation in the patterns of day range can be illustrated by reference to specific observation periods. For example, in late February/early March 1975 two short day ranges of 200 and 600 m centred on the river were followed by long day ranges of 1250 and 1500 m, during which parts of the group travelled widely around the home range; on the fifth day they ranged for 600 m nearer the river (fig. 5.4a). By contrast, in late November/early December distance travelled varied less from day to day, between 700 and 950 m, with most of the movement close to the river (fig. 5.4b). Six days in May 1975 are of special interest, since they illustrate the way in which different parts of the group might range

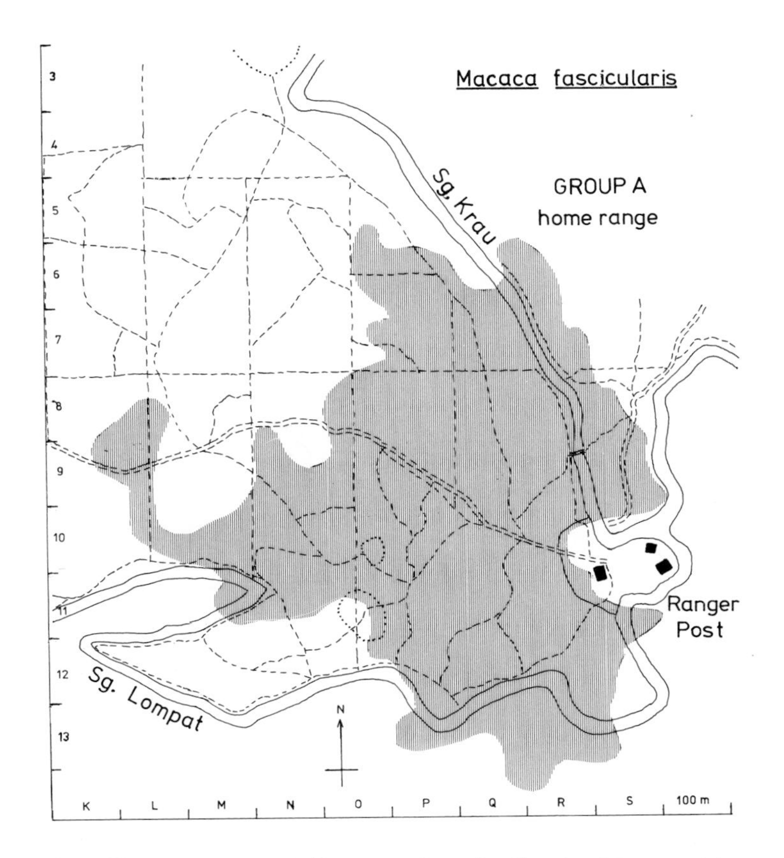

Fig. 5.1. Home range of *Macaca fascicularis* (Group A) at the Kuala Lompat Post of the Krau Game Reserve.

independently. For the first three days, a party of at least 10 were ranging widely (700+, 1000 and 900 m) as a discrete unit (fig. 5.4c). During this time other observers saw other members

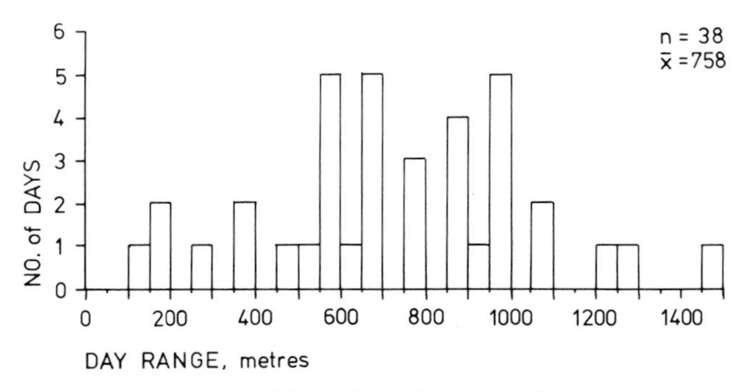

Fig. 5.2. Frequency distribution of day range lengths.

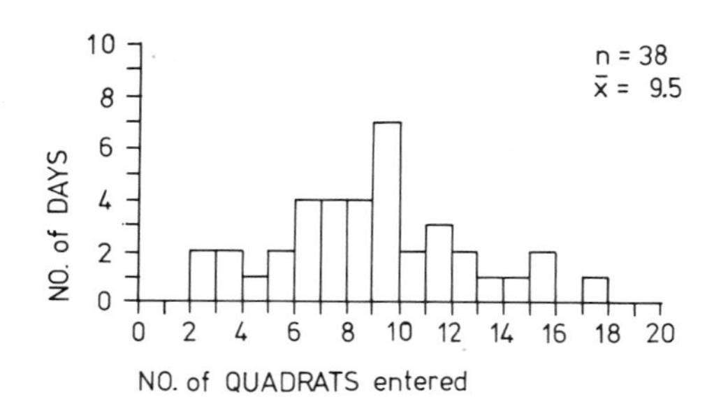

Fig. 5.3. Frequency distribution of number of quadrats entered daily.

of the group more than 500 m from this party. The whole group re-united on the evening of the third day, and on the fourth day they travelled for 1100 m as a compact group; on the fifth day the group

FEBRUARY 26 – MARCH 2 NOVEMBER 30 – DECEMBER 4 MAY 2–7 1975

[a]

	K	L	M	N	O	P	Q	R	S
5									
6									
7									
8				1.2	2.9		4.1	1.0	
9				4.3	0.7	1.9	5.1	2.2	20.3
10				3.6	4.6	1.4	1.2	11.4	22.2
11				1.7	1.2	1.4	1.9	2.2	
12					0.7	0.7	1.9		
13									
14									

[b]

	K	L	M	N	O	P	Q	R	S
5									
6							0.7	0.7	
7								2.0	
8							3.7	6.2	
9						0.2	5.2	10.6	
10						3.9	6.4	19.7	0.5
11						3.2	6.7	17.5	1.5
12						0.7	3.9	1.0	
13							5.7		
14									

[c]

	K	L	M	N	O	P	Q	R	S
5									
6									
7									
8		1.0				1.0	3.9	1.2	
9		0.8				2.7	1.8	0.6	1.0
10		0.6	2.5	1.4	3.7	20.0	4.1	3.1	2.2
11			2.2	1.6	1.6	21.1	9.0	3.9	
12					0.8	5.3	1.2	1.2	
13						2.3			
14									

● food tree
○ night tree

Fig. 5.4. Ranging patterns in three periods in terms of the percentage of observations in each hectare quadrat and day range plots of all or part of the study group.

was more scattered and travelled less far (700 m) and on the sixth day the group was again compact, moving a similar distance (650 m).

The home range was not used evenly. The percentage of all individual records from 10-min scans during the 40-day period (n = 3096) are shown in fig. 5.5. During this period 48 quadrats were entered, but 6 quadrats accounted for 44.6% of all records, with a single quadrat having 15.3%. In contrast, 25 quadrats each had less than 1% of records. One quarter of the home range (12 quadrats) accounted for nearly 70% (fig. 5.6); this may be compared with the "core area" of other species, as defined by Chivers et al. (1975) and the MacKinnons (chapter 6).

Much of the core area lies along the river. A similar preference for riverine vegetation is evident from data collected during census walks covering a wider area. The number of sightings of each species in the 77 hectare quadrats covered by the census route are shown in fig. 5.7. It can be seen that the sightings of macaques are concentrated near the rivers, whereas the leaf monkey species are more evenly distributed and the two apes occur more in the west of the area.

The concentration of activity along the rivers can be explained in part by the distribution of sleeping sites. While they did sometimes sleep elsewhere in their home ranges, the monkeys would often return to the river bank for the night, occasionally reaching it as much as an hour before dark and remaining there for a similar period before moving off in the morning. Group A commonly slept in an area on the banks of the Krau, to the north of the clearing. A few trees, particularly an emergent *Parkia javanica* on the east bank, were especially favoured, although the

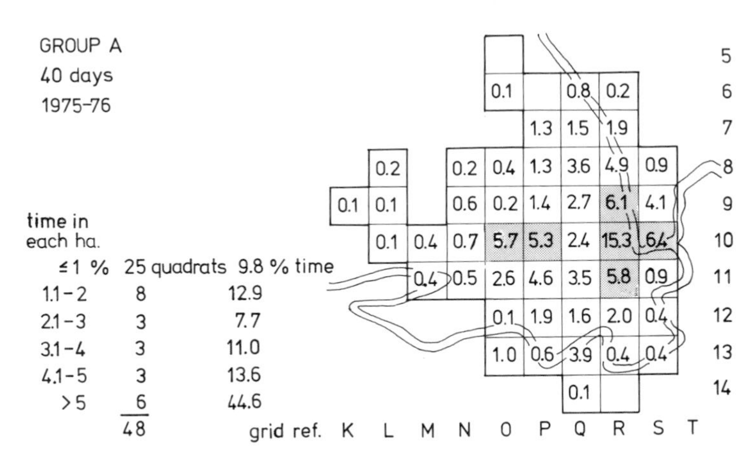

Fig. 5.5. Home range use in terms of the percentage of observations in each hectare quadrat.

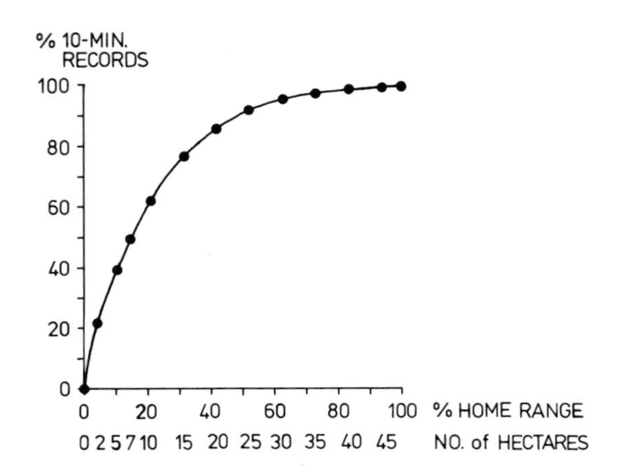

Fig. 5.6. Cumulative ranging time and ranked hectare quadrats.

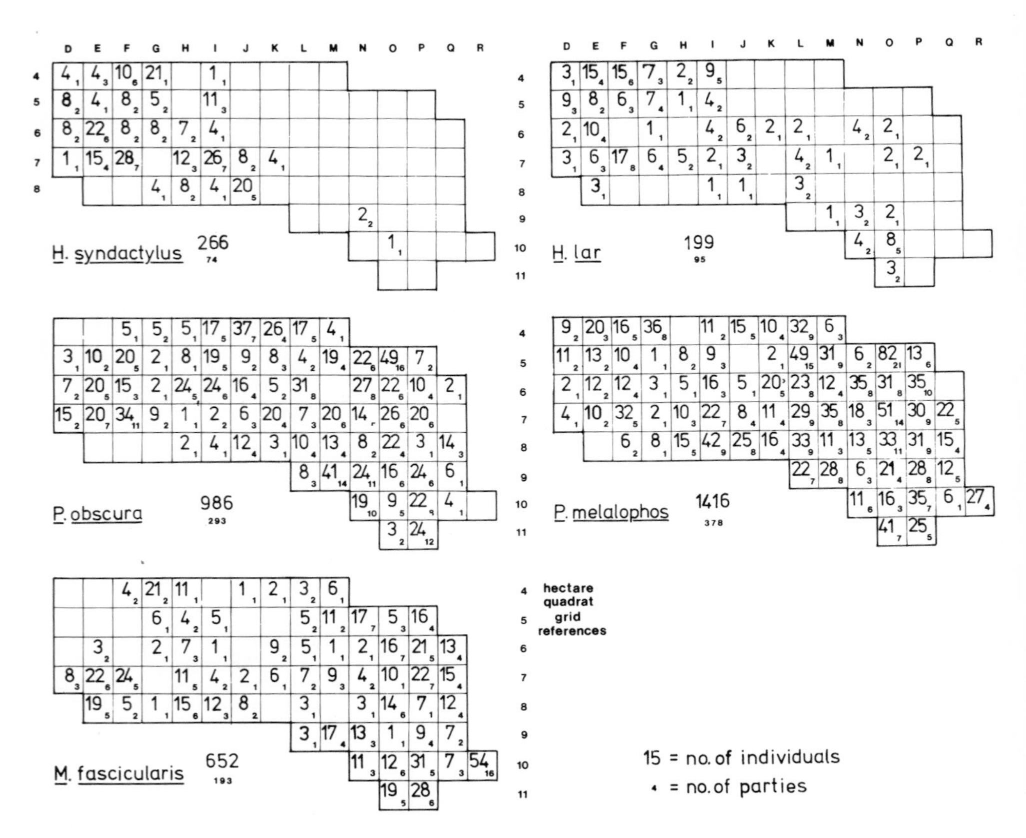

Fig. 5.7. Number of sightings of individuals (and parties) of each primate species in each quadrat during census walks.

group was sometimes scattered over several different trees within that general area. Sleeping trees were by no means always emergents such as the *Parkia*; some were low and covered with dense tangles of vines.

The distribution of sleeping sites does not account, however, for all this concentration of activity. The monkeys would often spend several hours, or even the whole day, within an 100 metres of the river. The riverine vegetation in the core area is quite different to that of the forest as a whole (see chapter 2), and indeed to that of other stretches of river. Its canopy is lower and more broken, and there are fewer large trees and many more vines. The diversity of tree species is lower, and some species are confined to the proximity of the river; this is probably because the soil is waterlogged for much of the year and prone to flooding. There is a further steep bank 50-100 metres from the river, above which the waters seldom rise. Elsewhere the river banks are higher and the vegetation less distinctive.

Concentrations of activity elsewhere in the range can be accounted for by the presence of favoured feeding trees. It is likely that a larger sample of day ranges would show a more even use of the inland parts of the range, since more short-term intensive feeding sites were included in this sample. Continuing observations by Y.L. Mah (pers. comm.) suggest that this is indeed the case.

Fittinghoff (1975) suggests that the species is particularly well adapted to dense vegetation in its locomotion and that, by keeping to riverine forest, it avoids competition from sympatric primates, such as apes, whose locomotor anatomy is better adapted to high forest with less dense and more solid substrate elements. He also suggests that the long-tailed macaques prefer to sleep in leafless trees on river banks because this helps them to mediate avoidance between groups in the absence of loud calls, and thus to maintain a degree of home range exclusiveness. This cannot reasonably be put forward as an argument for keeping to riverine forest for another reason: it would not help in mediating avoidance away from the river. The preference of the long-tailed macaque for riverine forest, and of the pig-tailed macaque for hill forest, might have their origins in Pleistocene climatic changes. For example, the species might have evolved their respective niches during a period of relative dessication, which isolated the pig-tail in higher moist forests, and the long-tail in riverine gallery forests (see Eudey, 1979).

Canopy use

The macaques distribute their time unevenly in the vertical as well as the horizontal dimension. The proportion of time the

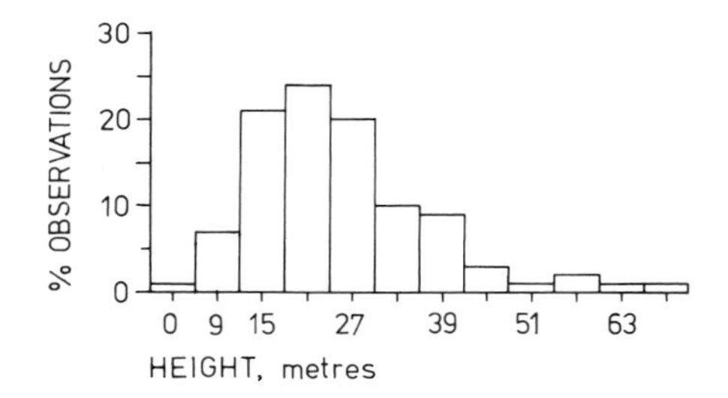

Fig. 5.8. Frequency distribution of sightings of *M. fascicularis* in each height class of the forest canopy.

spent at different levels in the canopy was derived from the number of individual records in each height class in all 10-min scans (fig. 5.8; n = 4427). Only 1.7% of records were of monkeys on the ground; while long-tailed macaques spend much time on the ground in urban or other man-modified habitats, at Kuala Lompat they are predominantly arboreal. More than half the records were below 18.3 m (60 ft), and less than 15% over 30.5 m (100 ft). This could be the result of active selection of the lower levels by the monkeys, or might result merely from there being more branches and foliage at that height. A cumulative percentage plot of the amount of canopy at different heights is shown in fig. 5.9 (see chapter 2) with the data on canopy

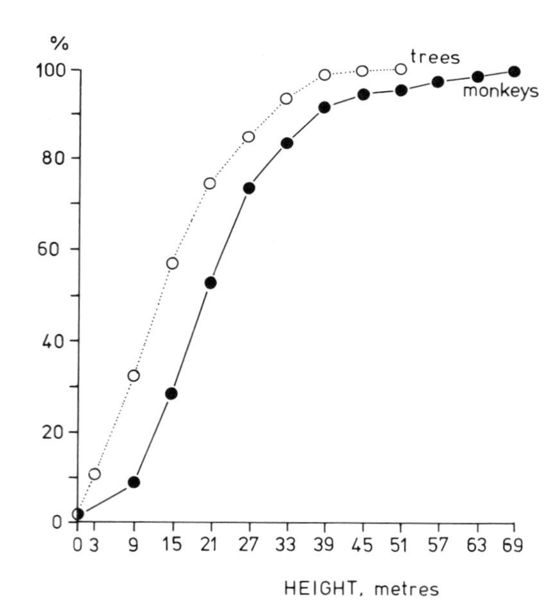

Fig. 5.9. Cumulative percentage plot of the amount of canopy and the sightings of *M. fascicularis* in each height class from ground level.

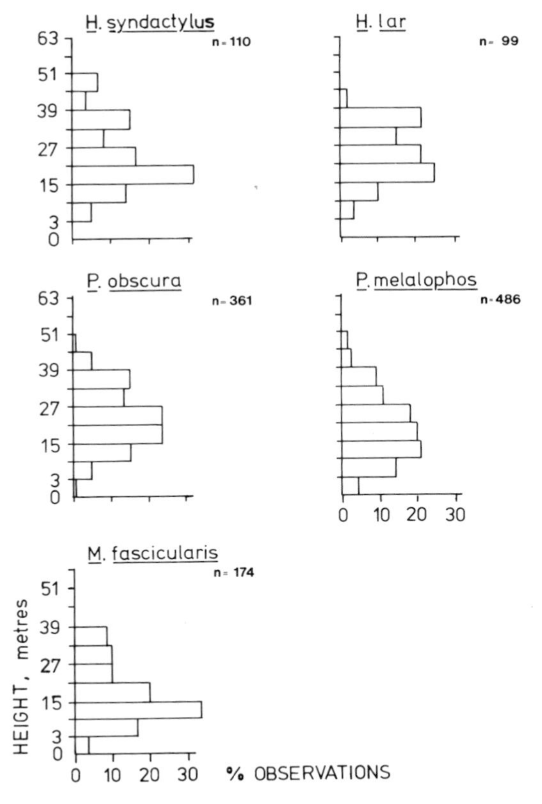

Fig. 5.10. Percentages of sightings of each primate species in each height class during census walks.

use expressed in the same way. From this it can be seen that the monkeys are actively selecting the lower levels. Less than 30% of the foliage, for example, is below the 18.3 m (60 ft) level, and there is more than 25% above 30.5 m (100 ft). The canopy use at Kuala Lompat is very similar to that reported by Rodman (1978) from Kutai in a comparable habitat.

A comparable picture emerges from the data collected during census walks (fig. 5.10). Once again the macaque records are concentrated at the lower levels, but the other species were seen higher (see chapter 6). Cumulative plots of the frequency of each primate species and the amount of canopy at each height (fig. 5.11) confirm that the macaques have an active preference for the lower levels, and suggest that the same may be true for banded leaf monkeys. The reverse seems to be true for lar gibbons, and perhaps also for siamang and dusky leaf monkeys.

Table 5.1. Composition of diets of long-tailed macaques.

FOOD	MONTH				% of IDENTIFIED FOOD FEEDING OBSERVATIONS				
	1975 February	Feb./ Mar.	April	May	Sept./ Oct.	November	December	1976 January	TOTAL
REPRODUCTIVE PLANT PARTS									
Fruit	65.8	89.3	88.0	64.2	32.5	4.9	5.4	16.1	52.4
Flowers	3.2	0	0.5	1.5	7.5	31.7	21.6	1.3	5.4
	69.0	89.3	88.5	65.7	40.0	36.6	26.9	17.4	57.8
VEGETATIVE PLANT PARTS									
Mature leaves	0	0	0	0	2.5	1.2	2.2	5.8	1.5
Young leaves	4.1	2.4	8.2	29.1	2.5	23.2	22.6	9.4	11.2
Leaf stems	0	0	0	0	27.5	0	4.3	10.8	3.4
Roots	0	0	0	0	0	17.1	20.4	0	2.9
	4.1	2.4	8.2	29.1	32.5	41.5	49.5	26.0	19.0
ANIMAL MATTER									
Arthropods	28.8	8.3	3.3	2.5	27.5	22.0	23.7	56.5	23.3
OBSERVATIONS									
identified	222	169	284	134	40	82	93	223	1151
unidentified	22	58	28	76	14	21	23	50	292
TOTAL	244	227	212	210	54	103	116	273	1443

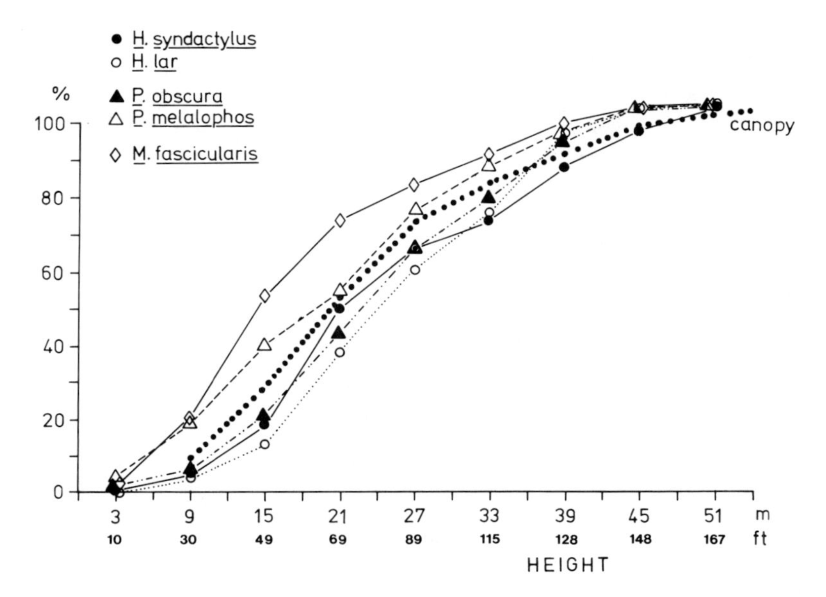

Fig. 5.11. Cumulative percentage plots of sightings of each primate species, and the amount of canopy, in each height class during census walks.

FEEDING

Activity pattern

Activity records collected during 10-min scans suggest that 35% of the time is spent feeding, 20% moving, 34% resting, 12% grooming and less than ½% in other activities. Feeding is generally most intense in the early morning and late afternoon, with periods of inactivity in the middle of the day. Any attempt to express this daily activity pattern graphically, however, might give a distorted picture. Many fewer monkeys are visible during the middle of the day, but those that are visible are more likely to be the active members of the group. Macaques present quite different problems of bias in the collection of data compared to species such as siamang, where all group members can be kept under more or less continuous observation. Care must be exercised, therefore, in any comparisons between species.

Dietary proportions

The composition of the diet was derived from individual records of feeding from 10-min scans during the 40 most complete days of observation in 1975-76 (n = 1443). Eliminating the 20.2% of unidentified items, the diet comprised 57.8% of reproductive plant parts,

19.0% of vegetative plant parts and 23.3% of animal matter (table 5.1). In terms of observation records, 61% of plant matter were taken from trees and 39% from vines. This indicates selection in favour of foods from vines, which comprise much less than 39% of the canopy in the forest as a whole (chapter 2). According to Rodman (1978), the long-tailed macaques at Kutai are about 95% fruit-eating, but his sample size (n = 61) is small. On the whole, the smaller the sample, the more fruit appears to predominate (see Chivers, 1972). Thus the dietary composition of this species at Kuala Lompat judged from census walks (fig. 5.12a, n = 157) includes nearly 20% more fruit and flowers than the composition judged from continuous observation (n = 1443).

Inter-specific comparison

A comparable picture emerges from the smaller sample of feeding records collected during census walks (fig. 5.12a), which can be compared with similar data for the other primate species. The macaques appear to be the most frugivorous of the five species, a feature

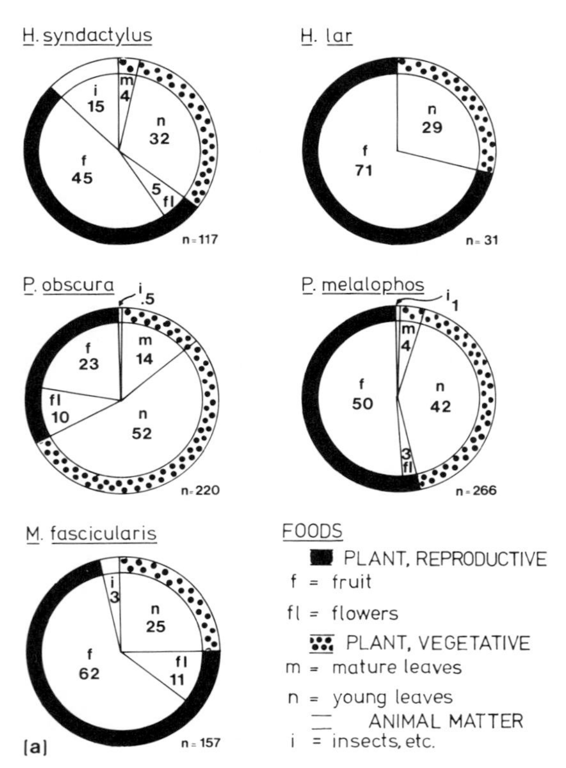

Fig. 5.12. The composition of the diets of each species from observations during census walks during this study (a) compared with results from the 5-day follows in 1973

supported by a comparable set of data collected by the MacKinnons (fig. 5.12b). The lar gibbon comes a close second, but the small sample fails to detect the significant component of animal matter (cf. fig. 3.10). The data on the siamang, from various observers using different methods, are closely comparable, as are those for the leaf monkeys (fig. 5.12, cf. fig. 4.10), with the dusky leaf monkey being the most folivorous.

Perhaps of greater significance than the overall properties of different food items is the contrast in diet between monthly observation periods. For example, the percentage of fruit varies from 4.9 to 89.3, of young leaves from 2.4 to 19.1, and of animal matter (predominantly insects) from 2.5 to 56.5 (table 5.1). There were also contrasts in diet between the different age/sex classes. For instance, in one 5-day period adults ate insects in 18% of observation periods, but juveniles and infants did so in 50%. A young growing animal requires more protein for each kg of body weight than does an adult; if it has similar catch rate, it will obtain a greater proportion of its absolute protein requirement in a given time.

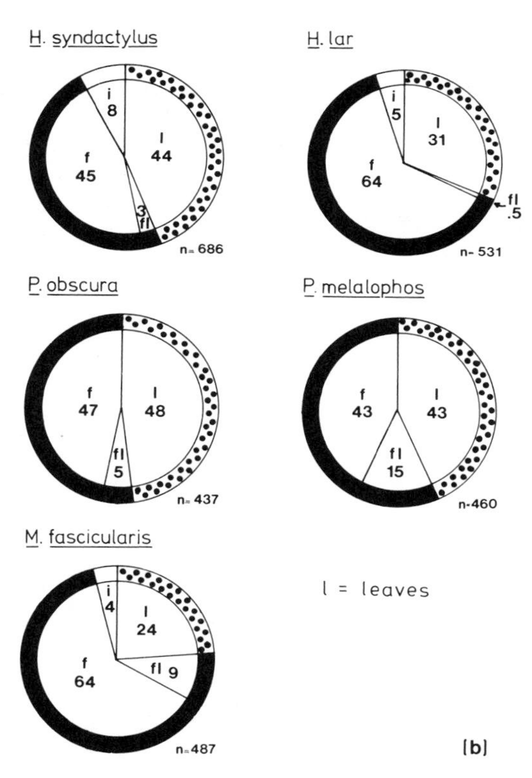

(b) by the MacKinnons (see chapter 6).

Of the other primates in the Krau Game Reserve, at least the gibbons show less variation in diet from month to month, although the leaf monkeys may in some periods show as much variation as the macaques (see table 6.4). The contrast is unlikely to be explicable wholly in terms of sample size, so one may be justified in describing the macaque as an opportunistic feeder, with the flexibility to take advantage of a temporary abundance of particular foods. In this respect the high proportion of roots in the diet in November and December is of interest. At this time a small plot of cassava on the edge of the clearing around the Game Ranger Post was dug up, and the macaques left the trees to eat the exposed roots; this same period also accounted for almost all records of them being on the ground.

Variations in diet from month to month show no clear-cut relationship with the abundance of fruit or young leaves in the forest as a whole, as measured either by the overall number of trees with fruit or young leaves, or by a score weighted according to tree and crop size (fig. 5.13), based on data from plots P10, O8, M8, E7 and E4). This remains true even if one considers only those plots that fall within the range of group A, despite there being contrasts between plots in any one month. This is perhaps not surprising, since a single tree with, say, a heavy crop of palatable fruit might account for a high proportion of feeding records for a particular period, and yet be insignificant in the botanical sample as a whole.

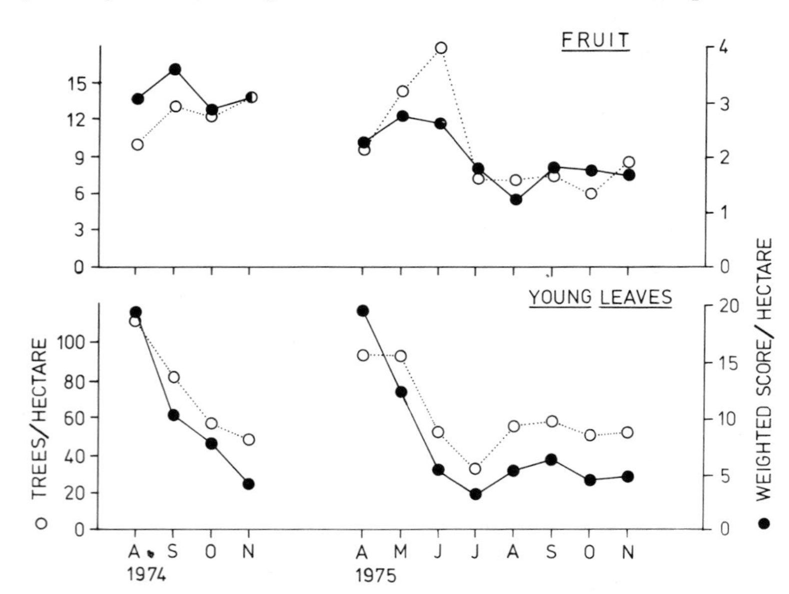

Fig. 5.13. Monthly variation in the production of fruit and young leaves according to the number of trees, with scores weighted for tree and crop size.

Thus we gain some insight into the way of life of this sociable and opportunistic but elusive monkey. The opportunism of this species seems to be typical of the whole macaque-baboon tribe, which has invaded a wider range of habitats than any other non-human primate taxon of equivalent status, from ever-wet forests to dry deserts and mountain tops and to temperate zones with snow in winter. This adaptability is undoubtedly due in large part not just to anatomical and physiological capabilities, but to cognitive and social ones.

J.J. RAEMAEKERS

NICHE DIFFERENTIATION IN A PRIMATE COMMUNITY

John R. and Kathleen S. MacKinnon

formerly Department of Zoology
University of Oxford

INTRODUCTION

The six monkey and ape species in the Krau Game Reserve comprise a primate community, in the ecological rather than sociological sense, together filling the diurnal primate niche in this particular rain-forest. Between them they carve up the primate 'cake' into uneven slices according to the particular characteristics and requirements of each species. To understand better how these species are able to live sympatrically, and what special adaptations permit such coexistence, the authors, helped by two local field assistants, conducted a short but intensive survey of all higher primates in the Kuala Lompat study area between January and July 1973.

Observations were accumulated for a total of 1638 hours, including at least 10 days of dawn-to-dusk observation for each of the five common species - *H. syndactylus, H. lar, P. obscura, P. melalophos* and *M. fascicularis*. Additional information on location, height in canopy, activity and food items was collected during 1433 brief encounters made during the study, including sightings during regular systematic surveys. Since *M. nemestrina* was a very infrequent visitor to the study area, its relationship to the other five species is inferred in part from data on diet and canopy use collected by J.R.M. during 66 encounters with this species in a comparable study area in North Sumatra.

The aim was to gather information for each species on group size and distribution, ranging behaviour (both horizontal and vertical) and dietary preferences. The study was, of course, too short to give a complete picture of the ecology of the primates,

Table 6.1. Comparative ecological data of sympatric primate species.

	H.syndactylus	*H.lar*	*P.obscura*	*P.melalophos*	*M.fascicularis*	*M.nemestrina*
Mean Group Size	3.0	3.5	10.3	9.3	17.0	c35
Ad.♀ body weight (kg)	10.0	5.0	6.0	6.0	3.0	7.0
Biomass group (kg)	26	17	57	51	39	190
Mean daily travel distance (km)	0.64	1.85	0.95	1.15	1.4	2.0
Mean daily travel time (hrs)	1.45	3.10	2.56	2.63	2.59	3.91*
Mean daily feed time (hrs)	5.75	3.40	4.03	3.96	5.17	5.00*
Feed/travel ratio	4.0	1.1	1.6	1.5	2.0	1.3*
Degree of foraging group spread	low	medium	high	low	high	very high
Mean group range size (ha)	28.0	53.0	28.5	21.0	46.2	100+*
Core area size (ha)	16	19	10	7	22	?
Overall density/km^2	4.5	6.1	31.0	74.0	39.0	0.5 (of 20.0*)
Overall biomass km^2/kg	39	30	172	406	89	?
Biomass within core area (kg)	122	67	427	546	133	?
Degree of range overlap between groups	none	almost none	slight	extensive	moderate	?

* data from Sumatra

but this was not the intention (c.f. Chapters 3-5). Our aim was to collect comparative data for the different species in the same area of forest over the same time period, so that any differences found could be attributed to genuine differences in adaptation and preference between the species, rather than anomalies resulting from different weather, locations, food availability or observers.

SOCIAL ORGANISATION

Group size, stability and spread

The first feature to strike an observer was that the six species showed major differences in the groups and social units in which they lived and travelled (table 6.1). The two apes have small, monogamous families, which in siamang are tightly cohesive (Chivers, 1974), but members of a gibbon family may spread out widely when foraging (see Chapter 3). In the leaf monkeys there are small one-male groups, occasional small all-male parties and some large groups (20+), in which there are several large males, with presumably more than one breeding (see also Chapter 4). Groups of *P. obscura* spread out to feed in many trees, but *P. melalophos* groups are more tightly knit, although they may sometimes split completely into separate foraging parties. The two macaques live in large multi-male groups: *M. nemestrina* groups showed enormous spatial spread throughout the day, with animals maintaining contact with one another by means of moaning calls. *M. fascicularis* groups were less widely spread, but, as with *P. melalophos*, the group often splits into two or more separate parties (see Chapter 5). Indeed, the main group of *M. fascicularis* studied spent more time as two separate sub-groups than as a single unit.

Density and biomass

Another conspicuous feature of the Krau primate community was the great disparity in numbers of each species (table 6.1 and Chapter 10), The apes and *M. nemestrina* occurred at low densities; *M. fascicularis* and *P. obscura* were found at much higher densities, and *P. melalophos* was the most abundant of all. Differences in body size are best represented by the weights of adult females, since these are unaffected by selection for sexual dimorphism. The biomass of individual groups and of each species within the study area was estimated from the proportion of adults to young in the population (table 6.1).

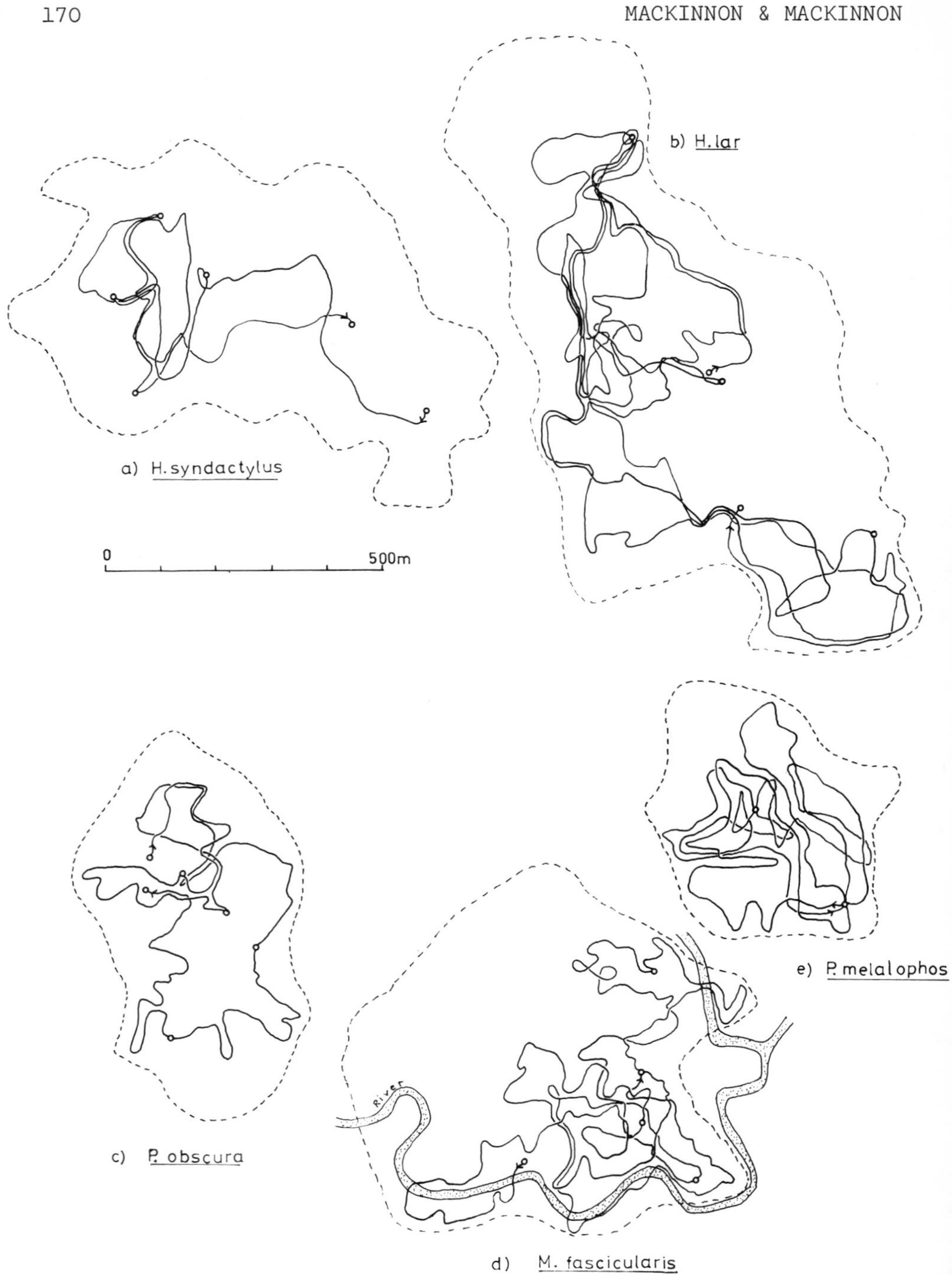

Fig. 6.1. Typical 5-day ranges and home range limits for one group of each species (May-June 1973).

RANGING

Home range size

The size of the home range for each primate species, based on all sightings and travel routes of each group in the study area in 1973, are given in table 6.1. The leaf monkeys consistently had the smallest home ranges; *H. lar* and *M. fascicularis* occupied the largest ranges, and those of *H. syndactylus* were intermediate in size.

Home range use

1. Day range. Actual distances travelled each day also varied markedly between species. The large siamang had the shortest day routes, while the smaller gibbon travelled 2-3 times as far (MacKinnon, 1977). The three arboreal monkeys had day ranges of intermediate lengths, while the semi-terrestrial *M. nemestrina* benefitted from energetically cheap ground travel to clock up the longest routes of all (table 6.1).

If we look at ranging patterns over 5-day periods, other differences emerge (fig. 6.1a-e). *H. syndactylus* made very heavy use of a small part of their range, with travel routes criss-crossing their favourite food trees of the moment; they had many sleeping trees throughout their range, and used whichever was nearest in the evening. *H. lar*, in contrast, covered much more ground and used twice as many, more widely-dispersed, food locations; they returned each evening to sleep within the central part of their range. *P. obscura* progressed steadily around their ranges in wide circuits, with more frequent passages through the core area, again sleeping in many different places. *P. melalophos* had fewer sleeping positions, and the main study group spent more than half the nights in the same location, showing clover-leaf ranging patterns as it covered its range from this tree. *M. fascicularis* had only a few favoured sleeping positions near rivers, and groups generally moved sequentially from one to another in a variable, extensive and dispersed manner.

2. Pattern of use. To quantify more clearly the differences in ranging patterns, day ranges were analysed in terms of the frequency of visits to hectare quadrats during 5-day routes (table 6.2). *H. lar* made far more hectare visits than the other species, with *P. obscura* making the least, but when we examine total visits in relation to the number of different hectares visited, i.e. the tendency to revisit the same hectare, *P. melalophos* emerges as the species visiting each hectare most frequently. *P. obscura* remains the species least likely to revisit an hectare. *P. melalophos* is also the species which covered the greatest proportion of its home range in five days, with the short-routed *H. syndactylus* covering

Table 6.2. Statistics for mean five day ranges of primate species.

	H. syndactylus	*H. lar*	*M. fascicularis*	*P. obscura*	*P. melalophos*
Total visits to hectares	43.6	104.7	61.5	32.7	49.5
No. different hectares visited	20.0	40.2	24.5	18.0	17.0
Mean visits/hectares	2.2	2.6	2.5	1.8	2.9
% range hectares visited	44	68	52	58	85
Mean distance travelled (m)/ hectare visit	80	97	114	126	116

least.

Measurement of the actual distance travelled during each hectare visit gives an indication of the directness of the travel routes in each species. *P. obscura* had the longest routes through hectares, which is consistent with it revisiting hectares least frequently; this thorough use of each hectare visited is probably related to its habit of cropping mature leaves. At the other extreme are the two apes, which took the most direct route through hectares. The need for the heavier *H. syndactylus* to choose efficient travel routes will be discussed below, but it seems that both species had good knowledge and intention of where they were going, and used familiar arboreal pathways to get there; their probably greater intelligence may prove an advantage in this respect.

Cumulative plots of the time spent in each hectare show how one group of each species used their home range at Kuala Lompat in 1973 (fig. 6.2). The severity of curvature indicates the degree of unevenness of use of the range. The species with the smallest home ranges show even more marked core areas than do those with larger ranges, thus exaggerating the differences in local biomass sustained by the forest. The more even use of home range by *M. fascicularis*

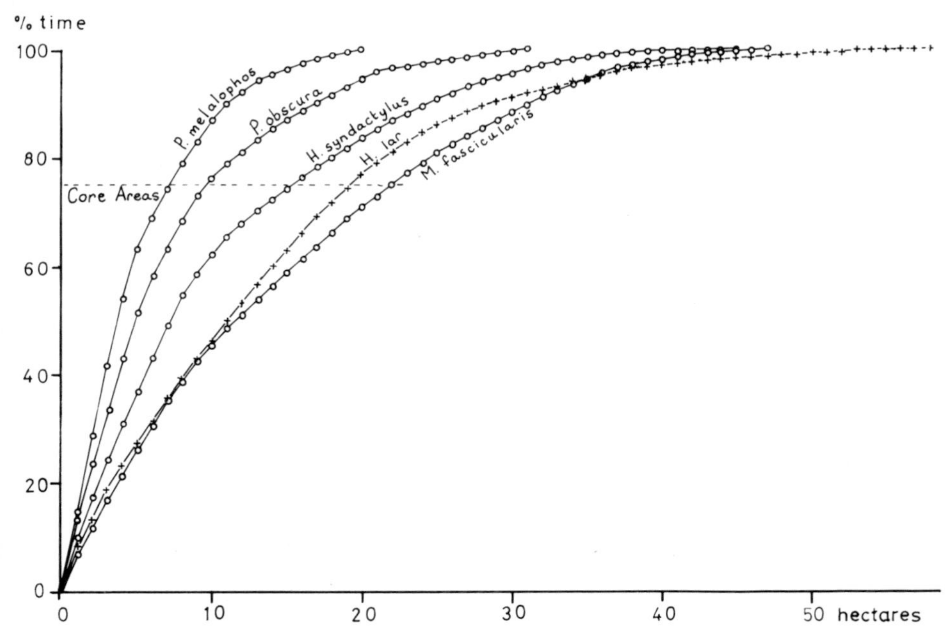

Fig. 6.2. Percentages of ranging time for each species accounted for by cumulative ranked hectares. Core areas are taken to be the fewest hectares accounting for 75% of ranging time.

Table 6.3. Interspecific correlations of spatial use by hectares.

	H.syndactylus	*H.lar*	*M.fascicularis*	*P.obscura*	*P.melalophos*	Breaks in Canopy	Distance from River
H. syndactylus		+.80	-.10	+.57	-.17	-.06	-.52
H. lar	59.1		-.06	+.48	-.12	-.24	-.38
M. fascicularis	26.2	34.2		+.00	+.18	+.03	+.23
P. obscura	36.0	43.8	34.0		+.02	-.18	-.10
P. melalophos	21.6	28.9	53.1	27.9		+.11	+.24

Figures in lower left side of table give actual % of overlap in times spent in each hectare. Figures in upper right side of table give r_s Spearman Rank correlation coefficients. For n = 57 at $p = 0.05$, $r_s = \pm.22$; $p = 0.01$, $r_s = \pm.31$.

results mainly from the strategy of group fragmentation, with different sub-groups using different parts of the group range. For each congeneric pair - gibbons, leaf monkeys and macaques - the species with the greater overall biomass was the species with the smaller home range. Similarly, the species with the greater biomass showed less group spread while foraging.

The horizontal use of space by each species can be compared by analysing their respective use of 57 hectare quadrats in the eastern end of the study area. This area was patrolled thoroughly and systematically for five days each month, from April to July inclusive, to locate groups of monkeys. Frequencies of encounter with each species in each quadrat have been adjusted in relation to the distance travelled by observers, and calculated as percentages of the total encounters for each species over the whole area. Encounters with the apes were too few, but comparable percentages were calculated for each quadrat from the periods of prolonged observation. Similarities and differences in the use of these quadrats was shown by (a) summing the absolute overlap of proportionate time spent in the same hectare by each pair of species, and (b) ranking the 57 quadrats by order of use by each species and calculating correlation coefficients for each pair of species (table 6.3). Quadrat use by each species was also correlated with two physical features that seemed relevant: (a) the amount of canopy break within each hectare, as mapped from ground survey, and (b) the distance from the centre of each quadrat to the nearest river.

The comparison shows a very similar use of hectares by *H. lar* and *H. syndactylus*, and both of these were quite similar to use by *P. obscura*. *M. fascicularis* used the hectares in a manner similar to *P. melalophos*, but both showed marked differences from the other three species. Only *M. fascicularis* and *P. melalophos* showed positive correlations with both the proximity of the river and breaks in the canopy. There seems to be a clear distinction between the preference of the gibbons and *P. obscura* for closed, primary forest, and the slight preference of *M. fascicularis* and *P. melalophos* for more open riverine and secondary forest. These groupings correspond with preferences for canopy levels for travel, for upper and middle/lower storeys respectively (see below).

Canopy use

The structure of the Malayan rain-forest presents a complex three-dimensional matrix of varying locomotor supports, light and humidity levels and plant productivity (Chapter 2). The use of different canopy levels by each species was estimated to see if they are separated ecologically in this way. To avoid observer differences in assessing stratification, data on heights have been taken only from the records of J.R.M. scored (a) at 5-min intervals for each individual during prolonged observations of the apes, and

(b) at the moment of first encounter with monkey groups on a 4-score system (e.g. UUUM means that about 25% of the group were in the middle canopy and the rest in the upper canopy). Comparison of patterns of canopy use - overall, and for feeding and travel - show interesting differences between species in the amount of time spent in the upper (25 m and above), middle (25-8 m) and lower (below 8 m) canopy levels and on the ground (fig. 6.3).

Canopy use by *H. lar* and *H. syndactylus* was extremely similar, and throws no light on their ecological separation, but canopy use by *P. obscura* and *P. melalophos* was markedly different, with the former spending much more time in the upper levels. This corresponds with differences in limb proportions, which render *P. obscura* more suited for quadrupedal locomotion along large boughs among the crowns of the larger trees, while *P. melalophos*, with its higher brachial index, is better adapted for leaping among less sturdy supports (Fleagle, 1976b). The apes also prefer the firmer supports to brachiate beneath or walk bipedally upon (Chapter 8).

The two macaques show even greater differences in canopy use; *M. nemestrina* is essentially a terrestrial quadruped, with a very high inter-membral index, confined mainly to inland hill forest (Chivers, 1971; Rodman, 1973b), whereas *M. fascicularis* is far more arboreal, albeit concentrated in riverine forest. *M. fascicularis* is structurally better adapted for quadrupedalism than *P. obscura*, but, being much smaller, is does not need such sturdy supports and is able to travel quite nimbly along small branches in the middle (and lower) canopy. Thus it can use a similar vertical niche to *P. melalophos*. The two macaques are more similar in their use of canopy levels for feeding than for travel; both species, especially *M. nemestrina*, feed in levels higher than those preferred for travel. This would seem to be less efficient ecologically than the strategies of other forest primates, which have less need to change level when changing activity (except where it may be more economical to travel on the ground).

Inter-group relations

Major differences in range overlap were observed between the different species (table 6.1). No range overlap was observed in *H. syndactylus*, and there were only occasional aggressive incursions into neighbouring ranges by the territorial *H. lar*. The three fights observed among *P. obscura* all appeared to be intra-group aggression; their loud honking calls appeared to have a territorial function, and range overlap was only slight. Range overlap was more marked in *M. fascicularis*, and most extensive in *P. melalophos*, where the range of some smaller groups was overlapped completely by those of larger groups. Loud calls, often given at night, received replies from neighbouring groups, presumably giving good information on locations, but appearing to convey little threat. Groups avoided

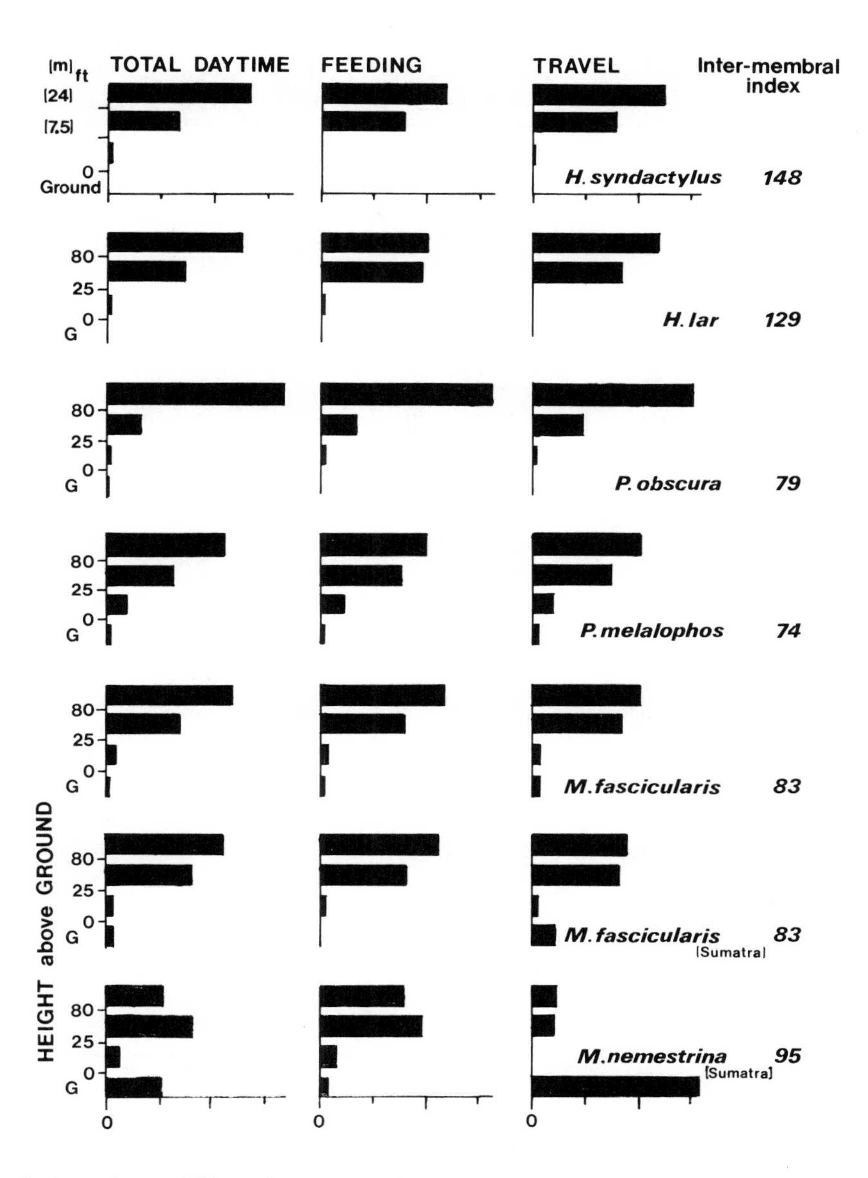

Fig. 6.3. Stratification of primate activities in the forest canopy.

meeting, and one small group was seen hurrying quietly away from a good feeding area on hearing a larger group approach. It is interesting that territoriality seemed so poorly developed, when such

Table 6.4. Percentages of major food types in monthly diet samples of primates.

	H.syndactylus	*H.lar*	*M.fascicularis*	*P.obscura*	*P.melalophos*
Jan/Feb					
fruit	52.4	69.2	80.0	28.6	13.0
flowers	7.5	0	0	0	30.4
leaves	35.4	30.8	20.0	71.4	56.5
insects	4.7	0	0	0	0
Mar					
fruit	37.0	70.4	42.3	50.0	50.0
flowers	5.7	0	30.8	0	15.4
leaves	47.4	22.5	19.2	50.0	34.6
insects	9.9	7.0	7.7	0	0
Apr					
fruit	57.6	64.8	43.5	29.0	39.0
flowers	0.6	1.0	13.5	9.0	11.9
leaves	37.3	31.4	37.6	62.0	49.0
insects	4.5	2.8	5.4	0	0
May					
fruit	30.1	47.8	58.9	36.8	27.7
flowers	0.4	0	2.7	2.5	7.0
leaves	64.5	49.7	35.4	60.7	65.3
insects	5.0	2.5	4.7	0	0
June					
fruit	38.2	57.9	77.0	56.2	47.0
flowers	7.6	1.3	1.6	14.6	17.6
leaves	39.4	34.6	18.1	29.2	35.4
insects	14.8	6.2	3.4	0	0
July					
fruit	52.2	72.5	80.7	83.5	80.1
flowers	0	0	4.3	0.8	5.0
leaves	37.9	19.2	13.9	15.6	14.9
insects	9.9	8.3	5.0	0	0
Total					
fruit	44.6	63.7	63.7	47.3	42.8
flowers	3.6	0.4	8.8	4.5	14.6
leaves	43.7	31.4	24.0	48.2	42.6
insects	8.1	4.5	4.4	0	0
total sample	8630 (686)	3901 (531)	1908 (487)	1450 (437)	1678 (460)

Overall data drawn from 17,567 feeding scores from 2,601 (in brackets) independent feeding occasions.

small home ranges could easily have been defended (but see Chapters 8 and 10).

It appears that the territorial imperative increases with dietary specialisation (see below) and is negatively correlated with density. The extensive range overlap between *P. melalophos* groups may promote a more even use of forest space (see above) than would otherwise be the case.

FEEDING

Diet: food category

Analysis of the diet of each species month by month, classifying food items into four categories - fruit, flowers, leaves and insects - gives little clearer understanding of ecological difference (table 6.4). All species, especially *H. lar* and *M. fascicularis*, had high proportions of fruit in their diet, but the leaf monkeys, in particular, showed marked fluctuations in these proportions from one month to the next. Such fluctuations are reduced, however, most obviously for *H. syndactylus* and *P. melalophos*, by lumping categories of the reproductive parts of plants - fruit and flowers (see Chapter 10). While the primates can be arranged in order of frugivorous tendencies, the differences between species are less than the intra-specific variability between months (fig. 6.4).

Diet: food item

A clearer picture of dietary differences emerges when the feeding data for each species are compared item by item. Considering different parts of one plant species (e.g. fruit, flowers, leaves) as separate items, 376 different food items were recorded during this study. Eighty-three of these items were eaten by at least three of the primate species, and 17 items were consumed by all five species studied. The percentage of total diet of each species accounted for by every food item was calculated. The absolute dietary overlap can be calculated from the total number of different food items taken by each species, and the number of food items shared by species (fig. 6.5); it is the sum of the smaller of the two percentages for each shared food item between a given pair of primate species. For example, if *Intsia* leaves account for 2% of the diet of species A and 3% of the diet of species B, the contribution to absolute overlap for this item is 2%.

The dietary similarities thus revealed show close accord with phylogenetic relationships of the primate species. While analysis by crude categories indicated close similarities in the proportions of leaves and fruit in the diet of siamang and the two leaf monkeys,

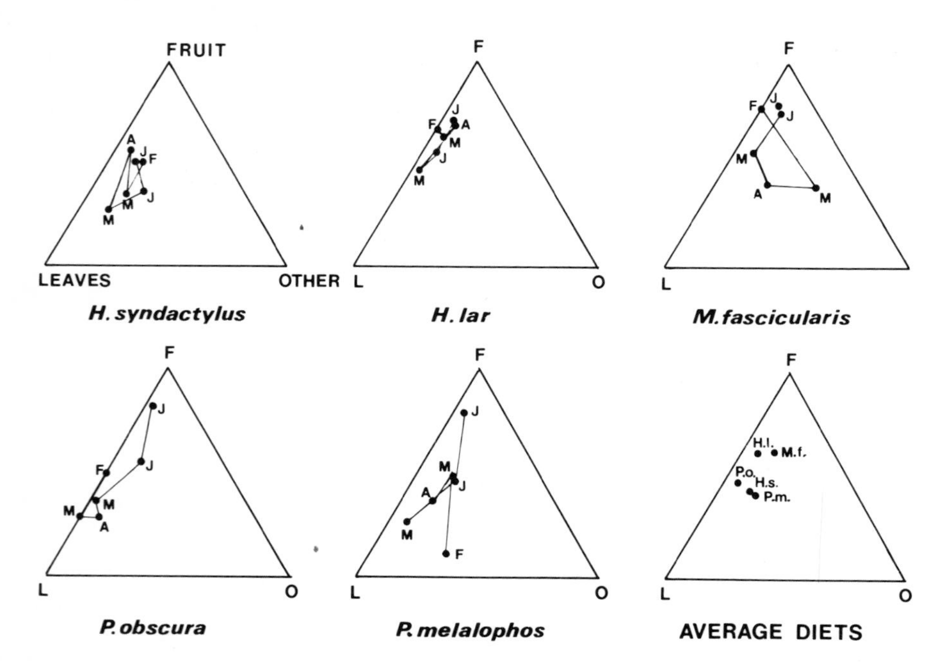

Fig. 6.4. Fruit, leaves and other components as proportions of primate diets, month by month and overall means.

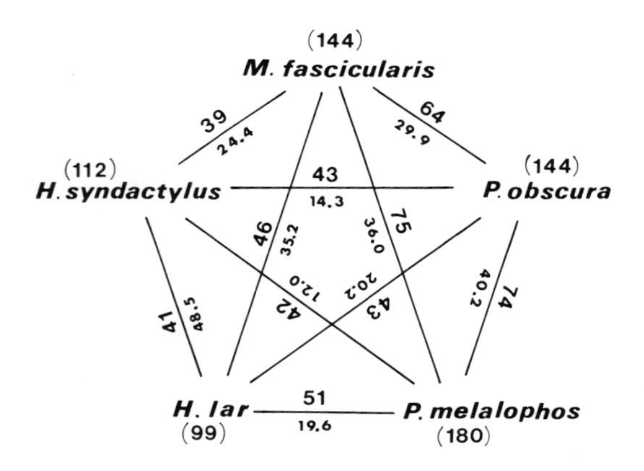

Fig. 6.5. Number of different food items eaten by each primate species, and the number of shared items and the absolute percentage dietary overlap between each pair of species.

analysis item by item shows these similar proportions to be composed of very different plant species. Greatest dietary overlap of food items occurs between the two apes and between the two leaf monkeys, with the diet of the long-tailed macaque intermediate to these two extremes. Insufficient data were collected on the pig-tailed macaque in this study, but observations by J.R.M. in Sumatra indicate closest similarity to its congener. In the Ranun river area of North Sumatra, there was an absolute overlap of 48% between the diets of the two macaques, which is almost as great as that for the two apes in Malaysia.

Diet: plant species

Data for feeding on 20 key food plants were extracted for more detailed analysis (table 6.5); these are the 20 species (or groups of closely-related species) that are highest ranked for their overall use by the primate community during the study period. For each key food plant are calculated (1) the percentage contribution to the total diet of each species, (2) the mean use by all primates, and (3) a measure of each plant's proportionate frequency in the forest. Plant frequencies were determined by a systematic inventory of 2,000 trees (and associated vines) of the middle and upper canopies, conducted in 6m-wide transects along representative grid paths.

The mean of the total proportions of the diets of each primate that are accounted for by the 20 key food plants is 59%. The plant frequencies have been scaled up so that the sum of their frequencies is also 59%, which makes it easier to compare the disproportionate use of these plants by different primates with their respective abundance in the forest (table 6.5).

Several comparisons from these feeding data help us to understand the feeding behaviour of these primates (table 6.5). Comparative measures of the percentage of each primate's diet that is accounted for by its own five most-heavily used food plants gives some indication of the degree of dietary specialisation. The concentration of *H. lar* and *H. syndactylus* on species of fig draws them above *P. obscura* on this index, with *P. melalophos* the least specialised of all (table 6.6). Summing the number of times each primate is the heaviest user of any of the 20 key food plants and the number of times as least-heavy user gives a measure of extremism in dietary preference. The use of the 20 key food plants by each primate can be compared with the overall mean primate use, and with the abundance of the food plants, by summing the overlap percentages.

It will be seen that the rankings for dietary specialisation (table 6.6) are the exact reverse of the rankings for density (table 6.1). *H. syndactylus*, which shows the greatest dietary specialisation, also shows the least correlation with plant

Table 6.5. Relative dependence on key foods.

	H. syndactylus	*H. lar*	*M. fascicularis*	*P. obscura*	*P. melalophos*	$\bar{x}$ primate	f in inventory
Ficus spp. F.L.	31.4	27.1	12.1	2.1	2.2	15.1	2.2
Maranthes corymbosum F.Fl.L	0.4	1.5	4.5	19.5	4.9	6.1	0.2
Sloetia elongata F.Fl.L	12.3	4.9	0.4	1.9	4.2	4.7	7.7
Grewia laurifolia F.L	0.9	6.7	10.6	2.9	1.1	4.4	1.7
Insects	7.5	7.4	4.6	0.0	0.0	3.9	1.0 *
Pentaspadon velutinum F.L	0.02	0.0	0.7	7.1	7.3	3.0	0.9
Derris sp. F.Fl.L	3.0	2.4	1.3	4.6	1.4	2.5	4.1
Intsia palembanica F.Fl.L	0.1	0.7	1.4	3.7	5.2	2.2	3.2
Ventilago sp. L	0.05	1.6	4.7	2.2	1.2	2.0	3.6
Eugenia sp. F.L	0.3	2.7	0.1	1.7	3.7	1.7	4.7
Xerospermum wallichii F.Fl.L	1.5	1.1	3.9	0.1	0.9	1.5	11.9
Gnetum funiculare F.L	1.2	3.7	0.9	0.7	1.1	1.5	0.6
Artocarpus rigidus F	0.0	0.0	5.2	0.2	1.3	1.5	0.4
Sandoricum koatjapi F	0.0	0.0	1.5	3.1	2.4	1.4	0.4
Koompassia malaccensis L	0.03	0.4	0.0	6.0	0.2	1.3	3.6
Xylopia magna F.Fl.L	1.4	0.2	1.9	2.1	0.9	1.3	5.7
Dialium sp. F	0.0	0.0	3.0	1.0	1.9	1.2	0.9
Artocarpus sp. F	0.0	0.0	5.4	0.0	0.0	1.2	0.4
Parkilospermum sp. L	0.5	0.05	0.4	2.1	2.0	1.0	0.8
Oxymitra philipes F.Fl.L	0.8	0.9	0.6	1.2	1.1	0.9	4.7

F = Fruits or Seeds Fl = Flowers L = Leaf parts * estimate

Table 6.6. Comparative feeding indices of sympatric primates.

	H. syndactylus	*H. lar*	*M. fascicularis*	*P. obscura*	*P. melalophos*
% of diet accounted for by 5 preferred food plants	51	46	39	44	23
% of diet accounted for by 20 key foods	61	62	64	62	43
Scores as top user of 20 key foods	3	1	6	7	3
Scores as lowest user of 20 key foods	$5\frac{3}{4}$	$4\frac{3}{4}$	$5\frac{1}{4}$	$3\frac{3}{4}$	$\frac{3}{4}$
Total scores as extreme user of 20 key foods	$8\frac{3}{4}$	$5\frac{3}{4}$	$11\frac{3}{4}$	$10\frac{3}{4}$	$3\frac{3}{4}$
% absolute overlap with primate mean for key foods	34	40	41	33	31
Mean % overlap with all other primate species for total diet	26	32	31	25	22
% overlap with food frequency scores for 20 key foods as proportion of % of diet accounted for by 20 key foods	33	33	35	45	53

frequencies in the forest; it is the primate which most often scores lowest for use of the 20 key food-plant species. *P. melalophos*, which shows the least dietary specialisation, also shows the greatest correlation with the frequency of plant species and scores least often as the lowest user of the 20 key food plants. Moreover, it also comes lowest in overall extremity of dietary preference for the 20 key species, and lowest of all primates in its overlap with the mean use of those plants by primates. The almost perfect agreement between the degree of overlap shown by each primate species with the mean primate use of the 20 key food plants, and the mean dietary overlap with all other primate species for their total itemised diet, confirms that the 20 key plant species do indeed register dietary preferences reflected by the total food lists.

DAILY ACTIVITY

Each species showed a characteristic distribution of its major maintenance activities throughout the day. At one extreme, *H. lar* fed and travelled with little rest until, by about 1600 hours, its day was done, and it had already taken up sleeping positions for the night. By contrast, *P. melalophos* had long bouts of resting at intervals through the day, but often did not retire for the night until about 1900 hours. Different groups of the same species showed minor variations in daily activity patterns, without clouding the basic species pattern (e.g. fig. 6.6a-b).

Although differences in times of activity do not affect the degree of inter-specific competition among primates where their resource requirements overlap, it is possible that different peaks of activity could reduce the actual incidence of conflict at shared food trees. It seems unlikely, however, that different activity routines have arisen for this reason, since one would expect to find the largest differences between species of the same genus. In fact, the reverse is true and, like dietary preferences, similarities in activity patterns reflect phylogeny. The activity pattern of *H. syndactylus* is most like that of *H. lar*, that of *P. obscura* closest to *P. melalophos*, with *M. fascicularis* somewhat intermediate (fig. 6.6).

It seems more probable that activity patterns have evolved to ensure that food is ingested at optimally-spaced intervals throughout the day. The leaf monkeys, which are ingesting large amounts of cellulose requiring bacterial fermentation and, perhaps, considerable quantities of plant toxins from mature leaves and leguminous beans, are probably forced to spread their feeding over a long day so as not to overload their digestive and detoxifying processes at any one time. The apes, by contrast, which ingest few mature leaves but many high-energy, easily digestible, low-toxin fruit, have no such constraints and can complete their daily feeding requirements much earlier.

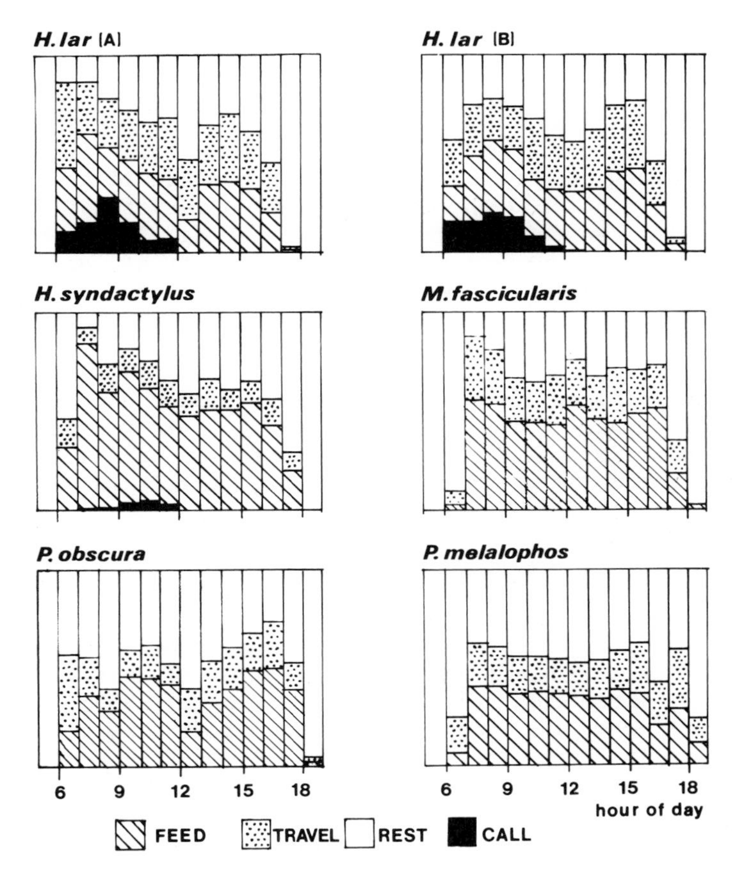

Fig. 6.6. Daily activity patterns for social groups of each species.

DISCUSSION

Crude generalisations about fruit- or leaf-eating specialists may be misleading and of limited use in discussing the Krau primate community; the term 'specialist' itself is open to confusion. All species eat large proportions of both fruit and leaves. It is at the plant species level that dietary preferences separate the primates into three species couplets, in accordance with their phylogenetic relationships. These differences in dietary preference are presumably related to differences in the functional anatomy and chemistry of the gastro-intestinal tract. Dietary overlap between members of different genera is relatively small, and competition for food, therefore, is minimal. Dietary similarities within generic couplets, however, create great potential for competition, and it is probably only because the members of such pairs employ such different foraging strategies that balanced coexistence is possible. In each case different foraging strategies appear to reflect differences in body size and/or limb proportions.

The dietary adaptations of leaf monkeys enable them to exploit the less nutritious and chemically-defended leaves and fruit, which other primates can tackle at most only sparingly. The two species are sufficiently divergent ecologically to coexist, because morphological differences render them suited to travel along different substrates; hence they prefer different parts of the forest, both vertically and horizontally.

The two gibbons appear to rely heavily on the productivity of a single plant genus, the figs *Ficus* spp. Their ranging behaviour, unique mode of locomotion and social organisation all seem specialised towards life as a fig exploiter (or fruit seeker, see Chapter 3). They take the smallest range of food types, show the greatest use of a few plant species, and are by far the least numerous of the diurnal forest primates. Their absence from riverine forest can be explained in part by their being 'outcompeted' there by high densities of long-tailed macaques, which feed heavily on figs but are not so strictly dependent on them. Divergence between the gibbons is clarified only by a consideration of their foraging strategies in relation to body size (MacKinnon, 1977; Raemaekers, 1979). *H. syndactylus* is twice the size of *H. lar*, yet travels only half as far each day and feeds at only half as many feeding sites, but feeds for twice as long; it has to follow direct routes between trees with abundant food. *H. lar*, on the other hand, can afford to exploit smaller, more dispersed, food sources (MacKinnon, 1977).

The macaques show dietary preferences intermediate between leaf monkeys and gibbons. Their contrasting morphology, however, renders them suited to different terrain. *M. nemestrina* is well adapted for long-distance travel along the floor of closed-canopy primary forest, so as to find sufficient good food sources to fulfil the needs of its larger groups; terrestrial travel is particularly economical in hilly terrain, where arboreal travel probably is more costly at times. *M. fascicularis*, by contrast, is closely associated with water; it is a fine swimmer and can cross flooded or swampy areas, and even fast-flowing rivers, with ease, and groups usually sleep near water at night. It shows slight preference for open secondary forest, which is more common in riverine areas, and exploits heavily a few plant species that are more abundant near rivers, e.g. *Dracontomelum mangiferum*, *Xerospermum wallichii*, *Grewia laurifolia* and the vine *Ventilago* sp.

The two commonest primates in the study area, *M. fascicularis* and *P. melalophos*, both operate in similar levels of the canopy and prefer similar parts of the forest. They also show the least specialisation towards particular plant species, but show very different food-finding behaviour.

M. fascicularis is an adaptable opportunist with a high level of short-term selectivity, neglecting some food sources to concentrate on others. One day it may eat almost only one type of flower, another day a single species of fruit, yet at other times it will eat a range of items and travel more extensively. Group cohesion and coordination is as variable, with members concentrated at a large food source or scattered around smaller ones. As conditions within the range change, the group switches its attention to whatever food is the easiest or most available. For example, two days were spent hunting frogs, which were attracted in large numbers to a pool formed by a rain storm.

P. melalophos, by contrast, shows the least extremism in diet, the widest range of food types, and the greatest correlation between the use of foods and their relative frequency in the forest. Members of the group usually forage together as a tight unit, eating whatever food they find, rather than searching more widely for rarer food items of higher value. Since *P. melalophos* is so abundant and uses forest space so evenly, ranges so regularly, and lives in the smallest home ranges with considerable overlap, it should be regarded as the most successful species in this habitat. Its inability to exclude *M. fascicularis* from the riverine areas may be due to the larger group size and greater aggressiveness of the macaque. In several competitive conflicts over feeding and sleeping positions *M. fascicularis*, despite their smaller size, were able to oust *P. melalophos*.

P. obscura, as has been described, has diverged from its congener to show some similarities in habitat preference and use with the apes.

The forest flora are extremely diverse; every hectare of the study area contains about 170 species of tree and 40 species of vine (see Chapter 2). Most of these plants are in leaf all year round, yet little of this vegetation is obviously edible. The mature leaves of forest trees are tough and difficult to ingest; they are difficult to digest because of their high cellulose content and defences of evil-tasting, and often harmful, tannins and other alkaloids. The obvious foods are the sweet fruits of plants that exploit primates as agents of seed dispersal; also the young, tender leaf growth which is not yet high in cellulose nor chemically protected. Such edible products are rare and, for a given plant, seasonal compared with mature leaves (Hladik and Chivers, 1978). The leaves of fast-growing trees and vines of secondary forest are more ephemeral structures than those of the slower-growing primary forest trees; they are usually less well protected and are used more heavily as food by primates.

Specialising within this limited choice of obviously edible foods, which seems to be the strategy of the gibbons, reduces greatly their potential food supply and predetermines a low density of these animals. Broadening the available food supply to include some of the better protected, but more abundant and reliable, food sources, the leaves of primary forest trees, the strategy employed by leaf monkeys permits much higher densities, but involves structural and chemical modifications of the gastro-intestinal tract. Since different plant species use different chemical defences (Janzen, 1975), which require different enzymes for detoxification, the leaf monkeys have had to specialise on particular groups of plants; the ability to exploit these depends on long-term co-evolution between plant and folivore.

Since plants tend to defend their flowers and unripe fruit with the same chemicals used in the leaves, a primate that has broken down the leaf defence of a given plant can usually also exploit its fruit. Hence it should not be surprising that these "leaf specialists" may eat large quantities of fruit. The organic substances most easily assimilated by plants are precisely those most easily assimilated by animals. Fruit, either as a food store laid down to support the young seedling or as a bait for a seed-dispersing agent, constitute a rich, concentrated supply of easily digested food. Thus leguminous beans and some nut kernels are important foods for leaf monkeys.

In those plants which employ seed dispersal by animals, the fruit must be protected chemically from the frugivores until the hard testa is ready. Monkeys that can break the chemical defences, however, can cheat the plant and eat the fruit too early. For example, the fruit of *Maranthes corymbosa* are rarely allowed to ripen for dispersal by gibbons and macaques, because *P. obscura*, which also eats the leaves of this tree, eats the fruit long before they are ripe, and in such quantities that they constitute its main food at this time. It is normal, and sensible, for the primate which makes the most use of the leaves of a given plant, to also make most use of its fruit and flowers (table 6.5); thus it gets maximum food value with least variety of toxins from the smallest area of forest.

The large primate community survives in balanced coexistence, because each primate exploits a different combination of resources. Have these complementary foraging strategies evolved mutually as a result of a long history of sympatry, or is it because the species were preadapted for coexistence that they were able to coexist so successfully when their respective distributions overlapped?

Dietary differences suggest phylogenetic inertia rather than convergence on similar forest niches. Dietary preferences seem to be based on inherited anatomical features, which cannot be changed rapidly to suit new environmental conditions. Morphological differences within genera also seem to guarantee foraging strategies that are sufficiently divergent to permit coexistence.

The two leaf monkeys belong to different groups of an ancient split in the genus *Presbytis*, which some authors (e.g. Brandon-Jones, in prep.) would like to give generic distinction: the *aygula*, or *Presbytis*, group and the *Trachypithecus* group (Pocock, 1934). The split between siamang and the smaller gibbons would also seem to be one of the earliest divisions of the Hylobatidae (Groves, 1972; Chivers, 1977); the two macaques are at opposite ends of the macaque spectrum of limb proportions and tail length. Thus it seems to be ancient differences from separate evolutionary histories that have preadapted the Krau primates to their present coexistence, rather than a development of divergent feeding ecologies in sympatry. Indeed, living in sympatry as they do now, one would expect the different species to converge in their feeding ecology as much as allowed by competitive exclusion. Whatever the origin of the complementary behaviour among these species, it is really the great floristic and structural diversity of habitat that provides so many niches for primates to fill.

Precisely how and why the primate species occur where they do is clearly dependent on innumerable details of food distribution and complex inter-specific competitive relationships. There is something rather predictable, however, about the Krau primate community. It is so typical of the whole Sundaland region, and fits closely to the pattern shown by primate communities in Borneo and Sumatra where we have worked, with apes, leaf monkeys and macaques as the component parts (see Chapter 10).

In all three study areas in Malaya, Sumatra and Borneo, the apes are represented by two or more species of different sizes, but occurring at low density (table 6.7). The same two species of macaque are present at similar densities, with a tendency for the smaller species to be commoner near rivers and coasts, and the larger species to prefer hillier terrain. The leaf monkeys reach the highest densities, no matter whether one or more species are involved (except in riverine habitats where *M. fascicularis* are locally the most numerous). The commonest leaf monkey always belongs to the *aygula* group (*P. melalophos*, *P. thomasi*, *P. aygula* or *P. rubicunda*), and always has smaller home ranges and a wider diet than sympatric members of the *Trachypithecus* group (*P. obscura* or *P. cristata*). The endemic proboscis monkey, *Nasalis larvatus*, in Borneo is a riverine species, whose highly specialised folivorous diet can barely affect the densities of *M. fascicularis* in the same habitat.

Table 6.7. Primate species densities in different Sundaland communities.

	Segama (Borneo)	Ranun (Sumatra)	Krau (Malaya)
Pongo pygmaeus	2.0	2.0	
Hylobates muelleri	11.5		
Hylobates lar		7.0	6.1
Hylobates syndactylus		7.0	4.5
Macaca fascicularis	7.0	50.0	39.0
Macaca nemestrina	9.0	53.0	0.5
Presbytis rubicunda	33.0		
Presbytis thomasi		64.0	
Presbytis melalophos			74.0
Presbytis hosei	5.0		
Presbytis cristata	1.5		
Presbytis obscura			31.0
Nasalis larvatus	9.0		
Totals	78.0	183.0	155.1

Densities expressed as number of animals/km^2.

Although there are considerable floristic differences between the islands of the Sunda Shelf and the Malay Peninsula, it seems that there is something sufficiently constant about lowland rain-forest and its physical structure, in terms of the riverine, secondary, primary and hill sub-habitats, phenological rhythms and types of food available, for the same niches to be available, and filled in a similar fashion. Thus the Krau primate community would seem to be typical of Sundaland, and many of the ecological distinctions discussed here can be applied to the whole region.

LOCOMOTION AND POSTURE

John G. Fleagle

Department of Anatomical Sciences
State University of New York
Stony Brook

INTRODUCTION

Apart from marsupials and edentates, there is no order of mammals, living or fossil, which displays a greater diversity of locomotor and postural adaptations than extant primates. With body weight varying from about 100 grams to 200 kilograms, the Order Primates includes leapers, climbers, brachiators, knuckle-walkers and a variety of arboreal and terrestrial quadrupeds. Despite the interest which primate positional behaviour has engendered in anatomists, anthropologists and zoologists throughout this century, we still understand very little about the relationships between locomotor and postural behaviour on the one hand, and basic aspects of these animals' biology, such as gross diet and use of forest structure on the other.

Are the locomotor and postural adaptations exhibited by extant primates clearly related to gross dietary preferences? Do locomotor differences reflect adaptations to the requirements of moving and feeding in different types of forest or in different parts of the same forest? To what extent does the great diversity of primate locomotion and posture reflect alternate pathways for meeting ecological needs dictated by historical accident?

While hypotheses relating positional behaviour with diet or forest use have been put forth frequently in the past, the information needed to test these hypotheses has always been too scanty to provide a rigorous test (Stern and Oxnard, 1973). Until very recently, the only available information on the naturalistic locomotor behaviour of most species was anecdotal, incidental observations of naturalists, and even when more complete field studies

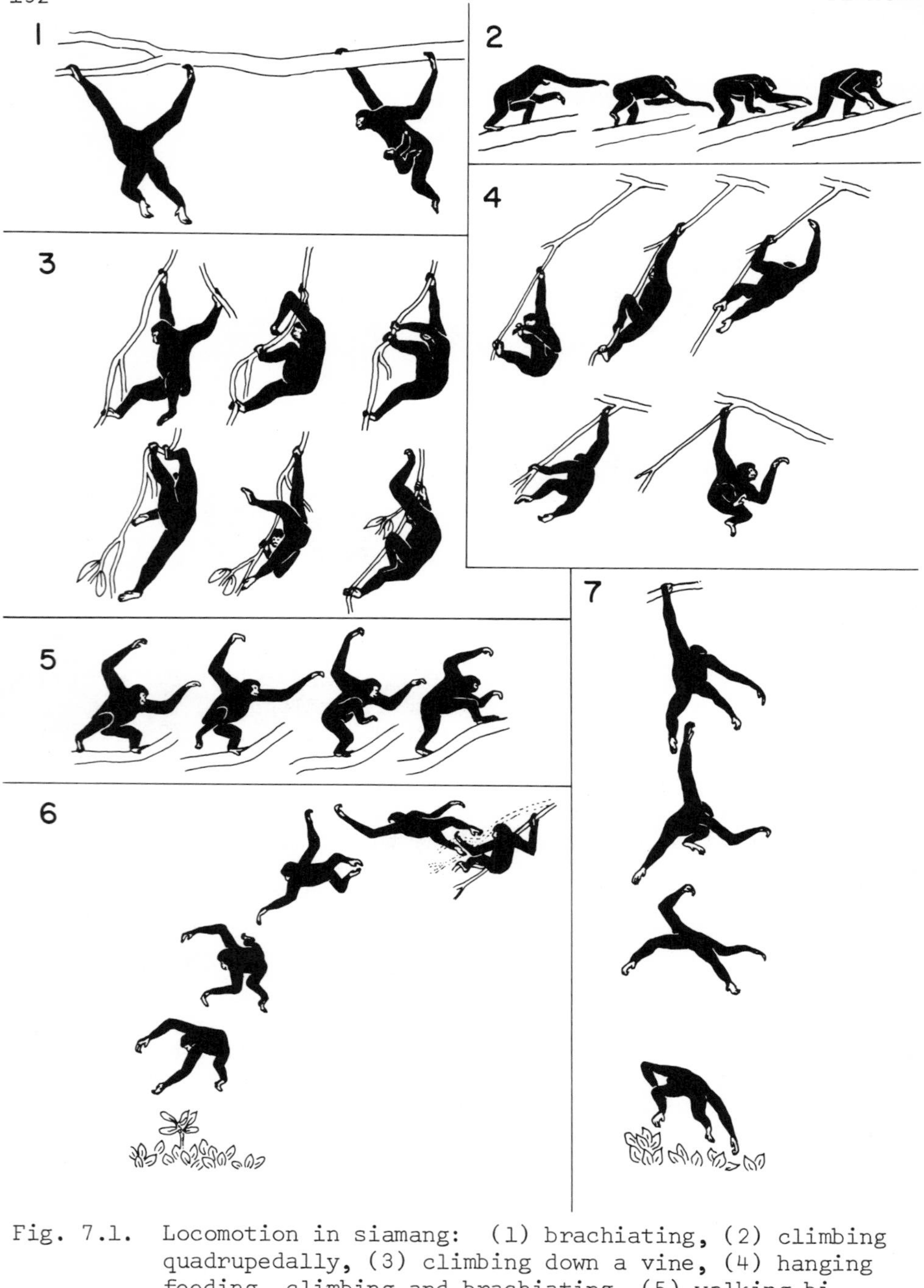

Fig. 7.1. Locomotion in siamang: (1) brachiating, (2) climbing quadrupedally, (3) climbing down a vine, (4) hanging feeding, climbing and brachiating, (5) walking bipedally, (6) leaping and (7) dropping from branch into tree below.

were available on one species, there was never suitable information for comparisons. The purpose in this paper is to examine the relationships between positional behaviour and other aspects of habitat use among the primates of Peninsular Malaysia, for which excellent field data are available, in the hope that any major patterns shown here amongst gibbons, leaf monkeys and macaques will be representative of the order as a whole.

Most of the studies reported in this chapter were conducted according to an uniform procedure described in detail elsewhere (Fleagle, 1976a, 1978; chapter 1). Adults were followed during daily activities, and locomotion was recorded as a series of continuous bouts. For each bout were recorded (1) the type of locomotor behaviour, e.g. leaping, brachiating, (2) the distance travelled and (3) the sizes of arboreal supports used - boughs > 10 cm, branches 10-2 cm, twigs <2 cm diameter. Observations of feeding and resting postures were made by spot samples at 2-min intervals of individuals engaged in those activities.

LOCOMOTOR BEHAVIOUR

Reduced to its simplest, locomotor behaviour is displacement movement from one place to another. Most animals use a variety of different locomotor patterns (such as brachiation, climbing, leaping), which are characterised by particular sequences of limb movement or body orientation. Locomotion takes place within the context of the animals' daily activities. Some locomotion involves movement over fairly large distances, such as between a sleeping site and a food tree, or between two food trees; this can be classified as Locomotion during TRAVEL. Other locomotion involves smaller distances, but more precise endpoints, such as movements between two parts of a food tree, or even between two food items - Locomotion during FEEDING. Finally, locomotion takes place within a structural context imposed by the animal's environment, so that some locomotion takes place on the ground and some on arboreal supports of varying size and orientation and height within the canopy.

Gibbons

The locomotor behaviour of the Malayan siamang and lar gibbon have been described by Fleagle (1976c), and also by Chivers (1974) and Andrews and Groves (1976). Gittins (1979) has discussed the locomotor behaviour of the agile gibbon. The locomotion of hylobatids can be divided into four categories:- brachiation, climbing, bipedalism and leaping (fig. 7.1). Although they occasionally engage in four-limbed climbing sequences, they very rarely show a behaviour comparable to the normal quadrupedal walking or running of most animals, including cercopithecoid monkeys.

Brachiation is the commonest pattern during travel, and climbing while feeding, in all three species (table 7.1). The larger siamang climbs more than the smaller gibbons, which are more adept at richochetal brachiation. Bipedalism and leaping occur occasionally during travel, but almost never while feeding.

All Malayan apes are exclusively arboreal, but in all three species the size of arboreal support used during locomotion varies with both locomotor pattern and behavioural context (table 7.2). Siamang and gibbons tend to brachiate along larger supports and climb among smaller ones. Climbing allows the spread of weight over a greater number of supports in more variable positioning. Bipedalism takes place almost only on relatively large horizontal boughs. Leaping is usually from larger branches into a mass of smaller branches or twigs.

Table 7.1. Locomotion of gibbons.

	H. syndactylus	*H. lar*	*H. agilis*
Locomotion during Travel % of Bouts	n = 643	n = 118	n = 132
Brachiation	.51	.56	
Climbing	.37	.21	
Bipedalism	.06	.08	
Leaping	.06	.15	
Distance per km (in m)			
Brachiating	623	702	74*
Climbing	291	129	14*
Bipedalism	37	58	7*
Leaping	49	110	6*
Locomotion during Feeding % of Bouts	n = 563	n = 93	
Brachiation	.23	.45	
Climbing	.74	.51	
Bipedalism	.03	.02	
Leaping	-	.02	
Distance per km (in m)			
Brachiating	308	480	
Climbing	638	480	
Bipedalism	54	20	
Leaping	-	20	

* Gittins (1979)

Table 7.2. Substrate use by gibbons.

		H. syndactylus	*H. lar*	*H. agilis*
TRAVEL (All)	Bough	.44	.50	.44
	Branch	.41	.41	.43
	Twig	.15	.09	.13
Brachiate	Bough	.57	.62	.32
	Branch	.41	.48	.51
	Twig	.02	-	.17
Climb	Bough	.22	.32	.63
	Branch	.43	.50	.13
	Twig	.35	.18	.25
Bipedal	Bough	.81	1.00	1.00
	Branch	.13	-	-
	Twig	.06	-	-
Leap	Bough	.23	.06	.33
	Branch	.46	.61	.17
	Twig	.31	.33	.50
FEEDING (All)	Bough	.23	.20	
	Branch	.48	.67	
	Twig	.29	.13	
Brachiate	Bough	.42	.29	
	Branch	.50	.69	
	Twig	.08	.02	
Climb	Bough	.13	.11	
	Branch	.49	.64	
	Twig	.38	.25	
Bipedal	Bough	1.00		

Locomotion during TRAVEL occurs on larger supports than does Locomotion during FEEDING. This is not just because feeding requires more climbing and less brachiation; both brachiation and climbing take place on relatively smaller supports during locomotion associated with feeding rather than travel. Furthermore, the average distance travelled in a single locomotor bout is smaller during FEEDING than during TRAVEL.

Leaf monkeys

The locomotion of the dusky and banded leaf monkeys in the Krau Game Reserve is almost totally distinct from that of the lesser apes. Whereas the gibbons are basically brachiators and climbers, the leaf monkeys are predominantly quadrupeds and

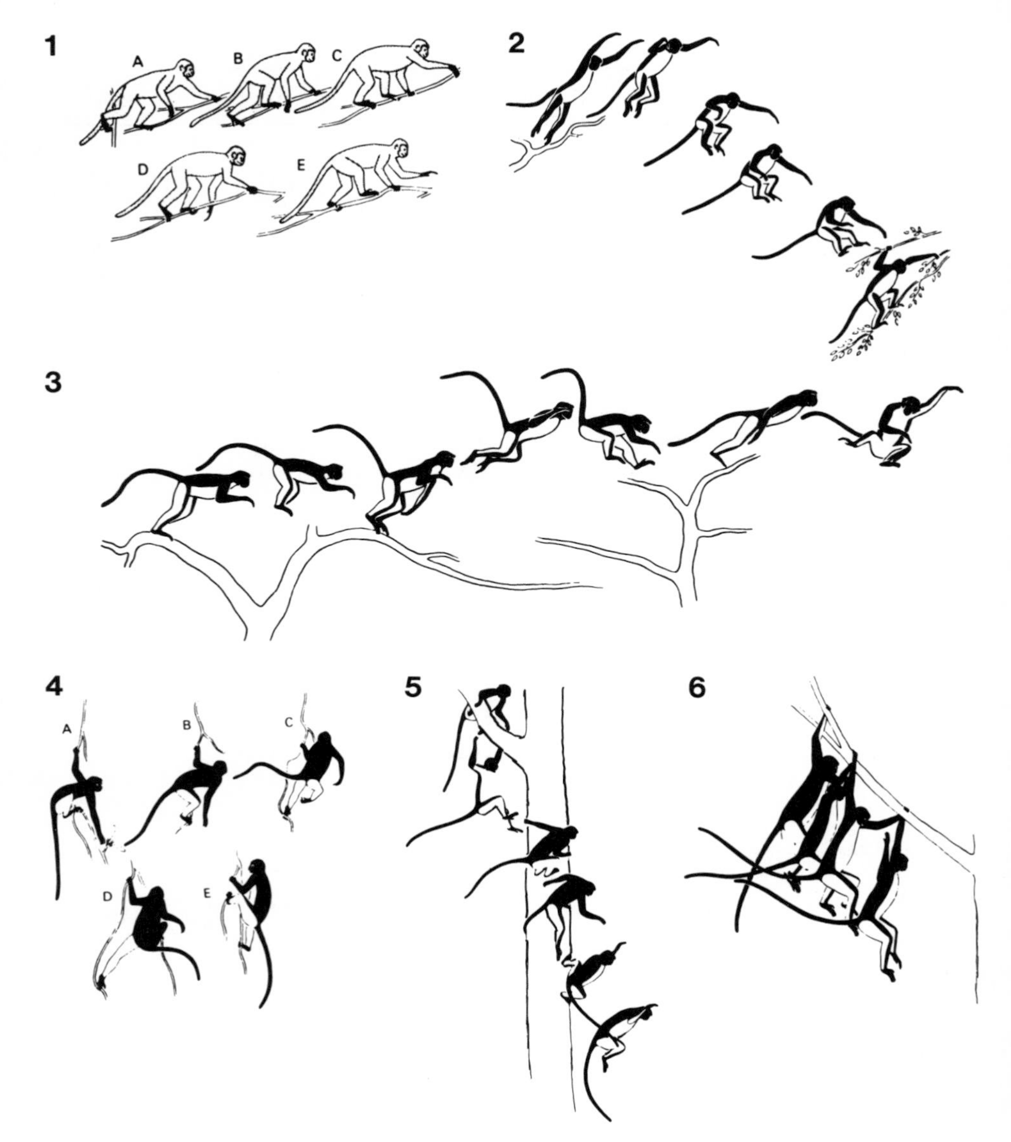

Fig. 7.2. Locomotion in leaf monkeys: (1) dusky leaf monkeys walking quadrupedally, and banded leaf monkeys (2) leaping, (3) hopping, (4) climbing up small vine, (5) climbing down tree trunk, and (6) progressing bimanually.

leapers (fig. 7.2). Even when leaf monkeys climb, it is more of a vertical quadrupedal walking than the quadrumanual climbing exhibited by the gibbons. *P. melalophos*, however, occasionally progress bimanually.

Although the two leaf monkeys at Kuala Lompat use similar locomotor patterns, they do not employ them in the same amounts. *P. melalophos* relies more on leaping, and *P. obscura* more on quadrupedalism (table 7.3; Fleagle, 1978). In both species Locomotion during FEEDING involves more quadrupedal walking and running than does TRAVEL, although the inter-specific differences are still present.

The size of arboreal support used during locomotion varies between species, among locomotor patterns and according to locomotor context (table 7.3). *P. melalophos* tends to move on smaller supports than its congener, even when using the same locomotor mode. During TRAVEL, quadrupedal walking and running takes place on larger supports than does leaping in both species. Leaping is from relatively larger supports to smaller ones. Locomotion during FEEDING takes

Table 7.3. Locomotion of leaf monkeys.

	P. melalophos	*P. obscura*
Locomotion during Travel	n = 731	n = 348
% of Bouts		
Quadrupedal walking & running	20.7	50.6
Leaping	67.5	40.2
Climbing	8.4	9.2
Arm Swinging	3.4	
Distance per km (in m)		
Quadrupedal walking & running	196	660
Leaping	735	287
Climbing	53	53
Arm Swinging	15	
Locomotion during Feeding	n = 235	n = 235
% of Bouts		
Quadrupedal walking & running	35.5	68.6
Leaping	31.5	15.0
Climbing	30.1	16.4
Arm Swinging	.9	
Distance per km (in m)		
Quadrupedal walking & running	346	810
Leaping	410	90
Climbing	244	100
Arm Swinging		

place on smaller supports than does TRAVEL. During feeding, leaping occurs between relatively larger supports, but the other patterns take place on relatively smaller supports in both species. As in the gibbons, the distance travelled during a single locomotor bout is shorter during FEEDING than TRAVEL.

Macaques

There have been no studies dealing specifically with the locomotor behaviour of either long-tailed or pig-tailed macaques. The following description results from a comparison of my notes, photographs and films with brief discussions of the main differences in locomotor behaviour of these cercopithecine monkeys in various parts of South-east Asia by Chivers (1973), Kurland (1973), MacKinnon and MacKinnon (1978) and Rodman (1979).

M. fascicularis is the smallest monkey in Malaysia (see chapter 5); they are predominantly arboreal, but frequently descend to the ground to feed or drink, especially in early morning or late evening The locomotor repertoire is much less versatile than those of the leaf monkeys and gibbons. While *M. fascicularis* occasionally make short leaps and frequently climb up various supports, they are predominantly quadrupedal walkers and runners (fig. 7.3). So extensive is their reliance on this locomotor pattern that they normally walk up and down vertical supports where other species would climb or leap. They appear to be normally palmigrade both in the trees and on the ground. Only Kurland has seen them move "bipedally" along branches while foraging, and this he saw frequently.

M. nemestrina is much larger than its congener; of all Malayan species it is most at ease on the ground and normally travels along the ground, where it moves quadrupedally with the hands digitigrade and the feet palmigrade (fig. 7.3). In the trees it is predominantly quadrupedal, being less adept at climbing among smaller supports and at spectacular leaping than their congeners. They seem to move, however, on larger boughs and even on vertical trunks more than *M. fascicularis*. On larger boughs, as on the ground, *M. nemestrina* is digitigrade.

Comparison and discussion

The six higher primates discussed above display a wide spectrum of locomotor abilities. Despite the differences in locomotor repertoires, the five species that have been studied in detail show a number of similarities in the more general features of their locomotor behaviour. Thus, bouts of locomotion during FEEDING are shorter than bouts of TRAVEL, reflecting the shorter distances between the endpoints in the two types of locomotion - on the one hand different parts of a home range, on the other different parts of a tree. Furthermore, in choosing the arboreal pathways during

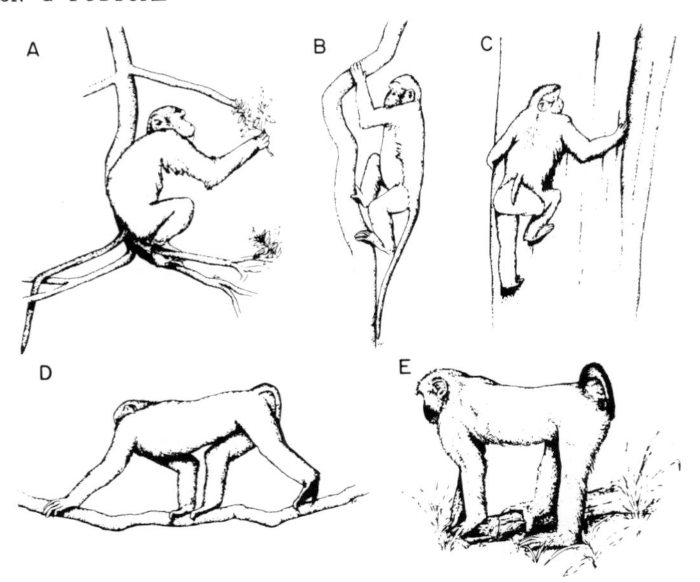

Fig. 7.3. Locomotion and postures of long-tailed (A-B) and pig-tailed (C-E) macaques.

travel about their home range, monkeys and apes have selected routes permitting long stretches of relatively "easy movement", whereas in moving about a food tree routes are determined largely by the location of desired food items. Likewise, the differences in use of locomotor patterns between TRAVEL and FEED, and in the size of supports, are seen in all species and probably reflect the same phenomenon. Locomotion during FEEDING must take place on smaller supports, because this is the location of fruit and leaves, whereas the supports used during TRAVEL are more optional, within restrictions imposed by the forest level in which the animal spends its time. Thus, all species tend to use slower, more cautious methods of locomotion when FEEDING and faster methods of progression, such as brachiation and leaping, during TRAVEL.

Overall, the similarities and differences concur with the taxonomic relationships of the species, a pattern similar to that found for dietary relations (MacKinnon and MacKinnon, 1978; chapter 6). The greatest overlap in locomotor behaviour is between congeners. The next greatest similarity is between leaf monkeys and macaques, who overlap to varying degrees in their use of pronograde quadrupedalism and leaping. The gibbons are not obviously more similar to either leaf monkeys or macaques; although *P. melalophos* occasionally progresses bimanually, this superficial similarity is more than offset by its extraordinary proclivity for leaping, a pattern used rarely by gibbons.

Locomotor differences between congeners can be explained by, or at least correlated with, the feeding strategies of individual species. The rapid, ricochetal brachiation of *H. lar*, permitted by its small size, enables this species to cover a longer day range than the heavier, slower siamang, which, through its greater size, can displace the lar gibbon from preferred food sources. The banded leaf monkey feeds extensively in the more discontinuous understorey of the forest and leaps more than its congener, the dusky leaf monkey, which feeds and moves mainly in the horizontally more continuous canopy above. Finally, the digitigrade hand postures and long strides of the pig-tailed macaque probably permit more rapid terrestrial movement for this wide-ranging species, than the deliberate palmigrade movements of the more arboreal long-tailed macaque with its shorter day ranges.

Although locomotor behaviour can be related rather clearly to the differential strategies of diet and habitat use among sympatric Malayan primates, broad associations between gross aspects of locomotor behaviour and habitat use are much less conspicuous (Fleagle, 1978). It has been argued previously that aspects of locomotor behaviour, such as brachiation, are associated with, and particularly adapted for, use of distinct levels of the forest canopy and specific dietary habits (e.g. Avis, 1962; Andrews and Groves, 1976). In a comprehensive literature review Stern and Oxnard (1973) found no compelling evidence for consistent associations between features of locomotor behaviour and habitat use or diet. The fauna of Peninsular Malaysia provide excellent opportunities to test for such associations.

There appears to be evidence for some general, but not exclusive, associations between locomotor behaviour and habitat use among the arboreal higher primates of Peninsular Malaysia. As MacKinnon and MacKinnon (chapter 6) point out, the suspensory gibbons prefer to travel in the more continuous upper levels of the canopy and to use closed primary forest rather than more open secondary forest. Their preferential use of these zones, however, is also characteristic of the more quadrupedal of the two colobine species, *P. obscura*. *P. melalophos*, the most frequent leaper, seems to prefer the lower levels of the forest and more open secondary forest, which it shares with the quadrupedal macaque, *M. fascicularis*. Finally, the most quadrupedal species of all, *M. nemestrina*, is also the most terrestrial.

Thus, although brachiators seem to be restricted to upper canopy levels and closed primary forest, brachiation is only one way of moving in that general habitat. Likewise, leaping seems to be associated with lower canopy levels and secondary forest, but quadrupedalism also works there. Quadrupedal primates seem able to exploit almost all levels and types of forest to some extent. In summary, the behaviour of Malayan forest primates suggests that for a given forest level or type there is not clearly a <u>best</u> type of locomotor behaviour, but several different types of locomotor behaviour that primates can use to earn a living.

There is also no indication of exclusive associations between locomotor behaviour and gross dietary preferences, such as fruit, leaves, insects. The most frugivorous species include a brachiator (*H. lar*), a leaper (*P. melalophos*), an arboreal quadruped (*M. fascicularis*) and a terrestrial quadruped (*M. nemestrina*). Likewise, the two species that have been reported as the most folivorous (Curtin and Chivers, 1978) are a brachiating ape (*H. syndactylus*) and a quadrupedal colobine (*P. obscura*). Thus among the members of a sympatric primate community, the relationships between locomotor behaviour and dietary preferences are clearly more complicated than the gross associations suggested previously (Fleagle and Mittermeier, 1980).

POSTURAL BEHAVIOUR

Actual movement occupies only a small part of any animal's daily activities; for the rest of the time animals are feeding or resting, and so forth. As with locomotion, however, these activities frequently involve particular patterns of support and limb use that are of significance both for the behavioural ecology of the species and as selective factors in the evolution of their bones and muscles.

Gibbons

The feeding postures of gibbon and siamang can be divided into two categories: (1) hanging, or suspensory, postures, in which the primary support is from above, and (2) seated postures, in which the animal rests on the support while feeding. Both species frequently use suspensory feeding postures, an aspect of their behaviour that clearly distinguishes them from all other Malayan primates. Among both siamang and agile gibbons, seated feeding tends to take place mainly on branches, while suspension occurs from both branches and smaller twigs (table 7.4). There is also a significant association between the use of suspensory postures when feeding on fruit, and seated postures when feeding on leaves (Chivers, 1974; Fleagle, 1976c). There are several possible explanations. Gibbons appear to be much less particular in their selection of individual leaves, than of individual fruit; furthermore, leaves require more chewing and other processing than most fruits.

Thus, the use of several postures during leaf-eating reflects the more sedentary nature of that activity, while the greater use of suspensory postures during fruit-eating may be associated with the more mobile nature of that activity. Certainly this seems to be true of the foraging behaviour of the smaller gibbons, which

Table 7.4. Feeding postures of gibbons.

		H. syndactylus	*H. agilis*	
		Feed	Feed	Forage
Sitting Postures		n = 254	n = 228	n = 91
	Bough	.15	.14	.31
	Branch	.78	.50	.52
	Twig	.07	.36	.18
Hanging Postures (N)		n = 405	n = 94	n = 74
	Bough	.03	.04	.07
	Branch	.49	.35	.46
	Twig	.18	.61	.47

involves much more hanging than does feeding; this association is probably also the optimal energetic strategy for lesser apes. If fruit is the limiting resource in the gibbon's diet, and if suspensory postures are energetically more costly than seated postures (Parsons and Taylor, 1978), then such a distribution of postures represents the most economical strategy for acquiring food.

During daytime resting all gibbons adopt a wide range of postures, most of which can be categorised as sitting, reclining or hanging postures. Seated postures are the most common; both seated and reclining resting postures take place on larger supports than do feeding activities (table 7.5). Reclining postures are almost exclusively on boughs, and seated resting postures are more on boughs than branches.

Table 7.5. Resting postures of Malayan primates.

	H. syndactylus	*H. agilis*	*P. obscura*	*P. melalophos*
Sit (All)	.87	.95	.76	.94
Bough	.58		.74	.74
Branch	.42		.26	.25
Recline (All)	.13	.03	.22	.05
Bough	.96		1.00	1.00
Branch	.04		-	-
Hang (All)	.01	.02	.02	.01

Leaf monkeys

Both *P. melalophos* and *P. obscura* use seated feeding postures virtually all the time (fig. 7.4). Those used most commonly are very similar to feeding postures described by Ripley (1970) for *P. entellus*; the monkeys rest on their ischial callosities with legs spread in front of them and reach through their legs for foliage and fruit. Occasionally a leaf monkey will carry a large fruit to a large bough, where it will sit "hunched" over while it tears the food apart and eats it. *P. melalophos*, in particular, also use a number of suspensory, or more accurately, clinging postures, which involve much less limb mobility than those used by the gibbons. In both leaf monkey species seated feeding postures occur mainly on branches (Fleagle, 1978).

Fig. 7.4. Feeding postures of banded leaf monkeys (A-E) and dusky leaf monkeys (F-H).

Malayan leaf monkeys use both seated and reclining postures during daytime resting (fig. 7.5, table 7.5); both species frequently use tripodial seated postures similar to their feeding postures. *P. melalophos*, more than its congener, rests sitting completely upright straddling a branch, while *P. obscura* tends to sit along the long axis of the support. In both species reclining postures are almost exclusively on boughs (fig. 7.6); seated postures are on boughs for about 75% of the time, and on branches for the remaining 25%.

Macaques

There are no detailed studies of feeding and resting postures of Malayan macaques, but their postural behaviour seems to differ from that of gibbons and leaf monkeys in several ways. Macaques frequently feed in a standing posture, particularly when foraging

Fig. 7.5. Resting postures of banded leaf monkeys.

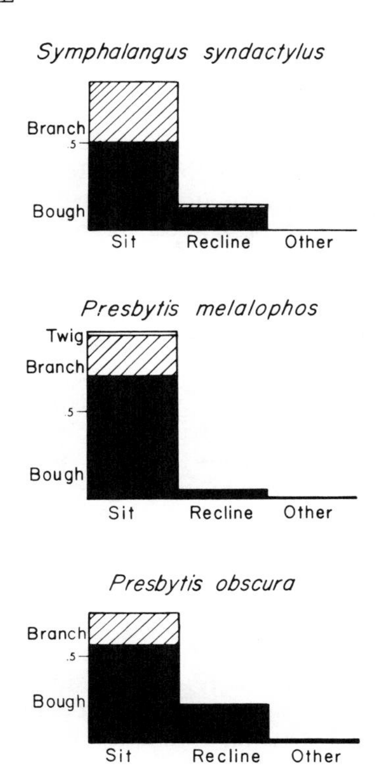

Fig. 7.6. Posture and substrate size in siamang and leaf monkeys.

for insects. They are also more likely to hang by their feet alone to reach food items. Seated feeding postures are more often with the hindlimbs flexed under the body, rather than the more spread postures of gibbons and leaf monkeys. Like the other higher primates, macaques seem both to sit and recline during daytime resting.

Comparison and discussion

As in the analysis of locomotor behaviour, the use of broad categories to describe the postural behaviour of Malayan primates suggests more behavioural similarity between species than actually exists. The variety and extent of suspensory feeding postures of gibbons contrast strikingly with those used by the monkeys. Again, in their more extensive use of seated feeding postures, the two types of monkeys are more similar to each other than either is to the gibbons. Likewise, the gibbons are unparalleled in their extensive use of the smallest supports during suspensory feeding.

The use of larger supports in resting during feeding bouts appears to be a general pattern similar to the support differences found during locomotion. The most appealing interpretation of the resting-feeding differences is that the support context for feeding postures is largely predetermined by the location of most food items at the ends of terminal branches, whereas the animals may choose resting places on the basis of stability or comfort.

Some of the inter-generic differences in feeding postures accord well with other aspects of the animals' feeding strategies. Thus, the use of standing feeding postures by macaques is associated with the need to flush actively insects from the foliage and grab them. This mobile type of foraging for any type of food is further facilitated by the possession of cheek pouches. Likewise, the hindlimb suspension used occasionally by macaques during feeding is certainly more appropriate as a posture while quickly obtaining food for storage in cheek pouches and subsequent mastication than a prolonged feeding posture.

The colobine-hylobatid differences in feeding posture are more difficult to relate clearly to differences in feeding stragegies. Although siamang show a definite tendency to use seated postures while eating leaves, and suspensory postures while eating fruit, banded leaf monkeys manage to eat large amounts of fruit with much less reliance on suspensory postures. As demonstrated by Grand (1972), sit-feeding is certainly a widely-used and successful approach to terminal branch feeding. With such high biomass densities of Asian colobines, and on the available physiological data, there is no support for an argument that hylobatid suspensory feeding is generally more efficient for terminal branch feeding on either leaves or fruit. One might argue, however, that it is a functionally different method of foraging, which gives the lesser apes access to resources that could not be reached so easily by sedentary feeders. The association between fruit-eating and suspensory feeding postures suggests that these resources are most likely to be fruits.

CONCLUDING DISCUSSION

The higher primates of the Malay Peninsula show considerable diversity in locomotor and postural behaviour. The gibbons and two of the leaf monkeys have been well studied in this regard. In contrast, there is no systematic information of a comparable nature for the third leaf monkey or for either of the two macaques. Overall similarities and differences in locomotion and posture appear to lie along phylogenetic lines, with the greatest similarity between congeners, and with the two groups of monkeys showing more similarities among themselves than either shows with the lesser apes.

Although the differences in locomotion and posture between congeners are integrally related to their differences in foraging strategy, there are few, if any, consistent associations between positional behaviour and either use of forest levels or gross diet throughout the six species. Animals with grossly similar positional repertoires frequently occupy different forest levels or have different diets. Likewise, species occupying similar forest levels or having grossly similar diets may show considerable differences in locomotion.

Thus, among Malayan forest primates, and among several other well-studied communities of sympatric primates, it is not possible to associate a particular type of locomotion, such as brachiation or leaping, with a gross dietary preference, such as frugivory or folivory (Fleagle and Mittermeier, 1980). There is no best type of locomotion for either a frugivore or a folivore. The association among positional behaviours, canopy level preferences and diet are much more subtle and complicated by the phylogenetic and competitive histories of the species involved. Thus, while the suspensory abilities of gibbons may give them access to parts of a fig tree that other primates cannot easily reach, or be the best locomotor adaptation for rapid foraging of small, widely-scattered fruit sources, the leaping abilities of banded leaf monkeys gives them access to fruits in the lower levels of the forest, and the terrestrial abilities of pig-tailed macaques enable them to forage over greater distances.

J J RAEMAEKERS

LONG-TERM CHANGES IN BEHAVIOUR

David J. Chivers and J.J. Raemaekers

Sub-department of Veterinary Anatomy and
Department of Physical Anthropology
University of Cambridge

INTRODUCTION

One of the main aims in producing a book of this kind on the ecology and behaviour of a community of primates is to add the dimension of time, which is so often lacking in the published results of field studies, and yet which is so crucial to their interpretation. Having described the behaviour of each species in some detail - in relation to each other and to the environment in which they occur - we should not consider these behaviours, and the weather and trees, over the 10-year period of study and, where possible, try to explain the changes observed.

From the outset we must admit to our dissatisfaction with the extent to which we can achieve these aims or needs. It has not proved possible to sustain observations of key relevance throughout the 10 years. In setting up the study area at Kuala Lompat, it was several years before an objective system for monitoring the leaf flush, flowering and fruiting of trees could be established, and it has not proved possible to obtain observations of sufficient depth on the behaviour of primates other than gibbons. Even the gibbon observations are deficient in their patchiness at crucial times, and in the delay in developing studies of the lar gibbon.

Nevertheless, we feel that we have something tangible to offer; some of the deficiencies that will become apparent are in the process of being remedied during current studies.

The scene is set with a discussion of changes in weather and in the vegetation and its cycles during the 10-year period of study at Kuala Lompat; comparisons are made with similar data from Ulu

Gombak, near Kuala Lumpur, for the preceding decade. Reference is made to the distinctive patterns of calling in each species, and to their significance. The main discussion concerns the social changes that have occurred in the gibbons, with a qualitative description of the social history of the siamang groups TS1 and TS1a and the lar gibbon groups TG2 and GB, with deductions about reproduction, sub-adults and their peripheralisation, pair formation, the role of females in conflicts between groups and group stability. Finally, after brief reference to social changes in the leaf monkeys and macaques, the preliminary results are presented from a quantitative analysis of features of ranging and feeding in the siamang group TS1 on nearly 500 days of dawn-to-dusk observation between 1969 and 1979 inclusive.

We are concerned throughout with two aspects of long-term change, which are closely interwoven: (1) changes within the annual cycle that we believe to be distinct, and (2) changes between annual cycles which, because of the fluctuations that occur in the humid tropics, tend to cloud the annual cycle.

ENVIRONMENT

Weather

Reference has been made to the climate of Peninsular Malaysia and to the weather at Kuala Lompat during the study period in Chapters 1 and 2. In the present context it is worth noting those aspects that introduce seasonality in each year or variation between years.

The influence of the surrounding seas ensures that the Sunda Shelf is one of the wettest regions in the tropics; indeed the success of the tropical forests there - their diversity and tremendous rate of growth - depend on this ever-wet climate (Whitmore, 1975). The lack of prolonged dry seasons, and variations in patterns of rainfall from year to year, lead to doubts that seasonality exists in Peninsular Malaysia, especially among those who have only worked there for 1-3 years. Allowing for some flexibility in timing and intensity, however, the regular annual pattern of north-east followed by south-west monsoons ensures climatic seasonality that is readily apparent over several years. We can consider here the patterns of rainfall reported for the Peninsula as a whole over a period of 25 years or more for some stations by Dale (1959, 1960) and for 20 years (1949-68) by Chia (1977), for Ulu Gombak on the western slopes of the Main Range near Kuala Lumpur from 1961-69 (Medway, 1972), and for Kuala Lompat from 1969-77 (Chapter 2).

Dale (1959) groups Temerloh, the weather station of long-standing nearest to Kuala Lompat, with Kuala Lumpur in the Western

Rainfall Region, with peak rainfall in October, November, December and in March, April, May, although Dale (1960) distinguishes between the Interior and the West Coast. This is a point stressed by Chia (1977) in his harmonic analysis of monthly mean rainfall data by computer; he distinguishes between North-west, West, South-west Coast, East and Interior Regions, with Kuala Lompat belongint to the last. He shows that maximum and minimum rainfalls are lower and earlier in the Interior than on the West. Dale (1960) records October as the wettest month at Temerloh on average with 0.53 in (13.5 mm)/rain day, and July as the driest with 0.21 in (5.3 mm)/rain day; the annual total averages at 70 in (1800 mm). Chia (1977) stresses the primary rainfall peak in October and November, and the secondary peak in April and May.

In producing rainfall probability tables for rubber planters, Wycherley (1967) presents long-term rainfall data for 73 stations in Peninsular Malaysia. Four of them are close to the Krau Game Reserve - Mentekab to the south, Bentong to the south-west, and Raub to the west (all some 40 km from the centre of the reserve), and Tekam about 50 km to the north-east - with records for from 31 to 54 years. The average annual rainfall varies from 2055 mm in the south to 2207 mm in the north-east; the annual patterns at these four stations (fig. 8.1) are highly concordant, with significant fluctuations over the year, thereby providing strong evidence for seasonality of rainfall in the region.

The rainfall in Ulu Gombak averaged 2283 mm (90 in) between 1961 and 1969, with means for successive quarters of the year at 391, 511, 556 and 832 mm (Medway, 1972). Thus the rainfall was spread more evenly over the year than at Kuala Lompat (fig. 8.2), with 172 rain-days on average for 1963-68 inclusive; the average for 5 years at Kuala Lompat was 1948 mm. Medway comments that the typical pattern of major and minor rainy seasons alternating with a short, pronounced dry season and a longer, ill-defined secondary dry season, was best exemplified in 1965; 1963 had an abnormally long first dry season, and 1968 had the next most pronounced dry season, but it was shorter and the subsequent secondary dry season was hardly noticeable (3-month rainfall totals mask these patterns, but they are visible in Medway, 1972, from a system of scoring that precludes the easy computation of monthly totals).

At Kuala Lompat rain usually fell between mid-afternoon and early morning. For 4 complete years (1969-70, 1970-71, 1974-76) annual rainfall averaged 2120 mm (83.5 in), similar to Temerloh (Raemaekers, 1977), but 1977 was unusually dry with only 1434 mm (56.5 in) (Payne, 1979b); rain fell on from 10-85% of days, according to season, in 1976 and 1977, with an average of 53% or 193 rain days/year.

Examination of the rainfall records from Kuala Lompat (fig.

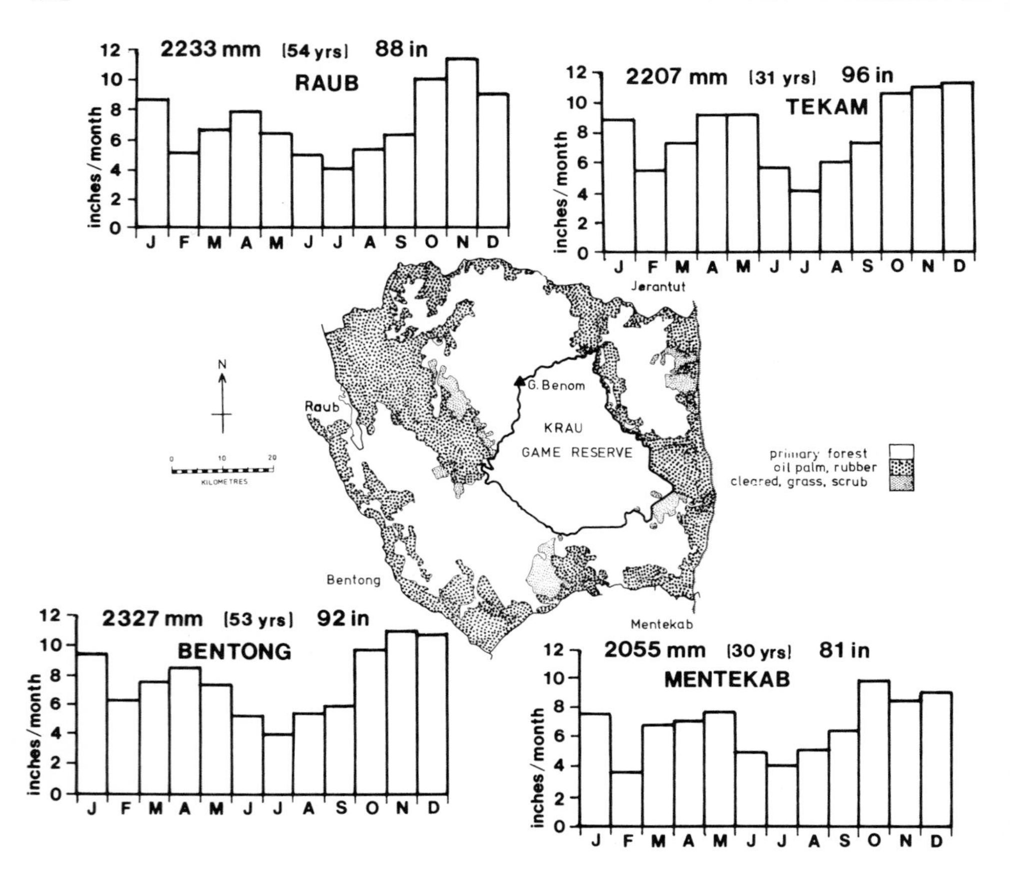

Fig. 8.1. Annual rainfall patterns around the Krau Game Reserve (showing also encroachment on the forest by cultivation, see Chapter 10).

8.3) reveals that in 1969 May and August were the wettest months (although November, December and June were the cloudiest); rain in January 1971 produced the worst flooding in living memory (and extreme discomfort for Sheila Curtin!), otherwise the wettest months in 1971 - May, July, October - had only one-third as much rain. April 1973 was the wettest month of the MacKinnons' study; rainfall was distributed unusually evenly through 1975, with more than 200 mm in March and June, rising to in excess of 300 mm in November. After a distinct peak in April 1976, amounts of rain again increased steadily during the rest of the year to a peak in December. Apart from the low level of rainfall in 1977, the distribution was as expected, with a distinct peak in October and a much lower secondary peak in May and June - a pattern close to the 10-year average (see Chapter 2, fig. 2.3). As the years pass, so the pattern is becoming more, rather than less, distinctive, and the conclusions of Dale

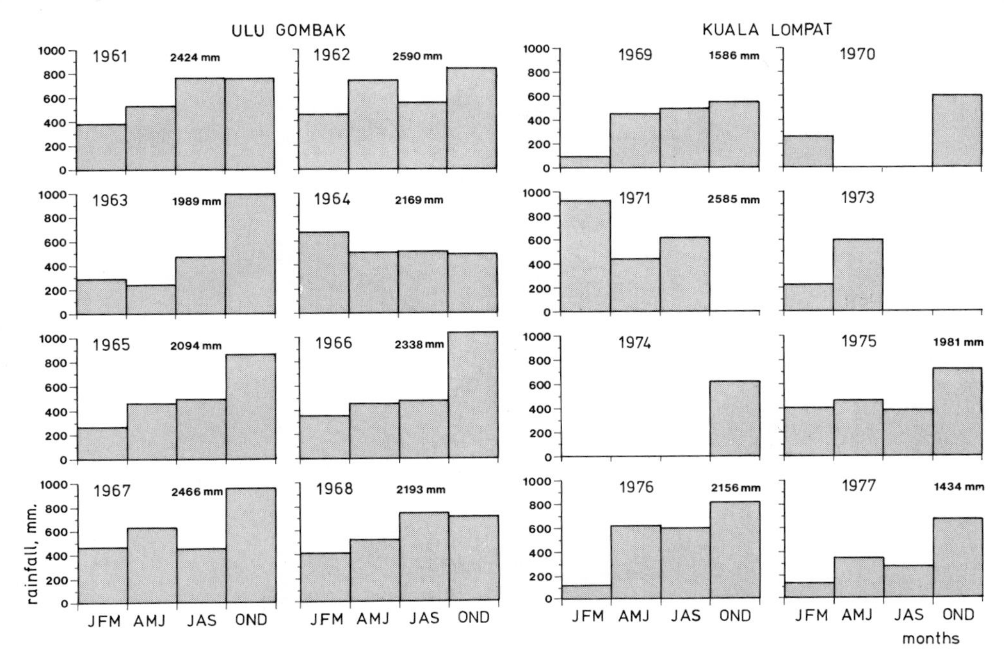

Fig. 8.2. Rainfall in each quarter of the year for Ulu Gombak (left) from 1961 to 1968 (from Medway, 1972) and for Kuala Lompat (right) from 1969 to 1977 (no records for 1972 and quarters with no score).

(1960) and Chia (1977) concerning climatic seasonality are being confirmed.

The seasonal effects of weather on the forest may be masked partly by the different relationship between rainfall and sunshine at different times of year. The wet seasons are characterised by prolonged cloud cover with long periods of light rain in addition to heavy showers; the dry season rain usually falls only in brief heavy showers, with subsequent sunshine exerting a rapid drying effect. In the centre of the Peninsula dry spells rarely exceed 28 days and are usually much shorter (Dale, 1960). Payne (1979b) argues from Schmidt and Ferguson (1951) that evaporation not infrequently exceeds precipitation in the main dry season (first three months of the year), when there is less than 60 mm (2.4 in) of rain/month, but at other times precipitation exceeds evaporation (in months with more than 100 mm or 4 in of rain).

Daily sunshine varied between 5.5 hr/day in January and 6.5 hr/day in July on average in the Interior (Dale, 1964). In 1969-70 there was nearly 9 hours of sunshine daily on average in the main dry season and only 2-4 hours in the main wet season (Chivers, 1974,

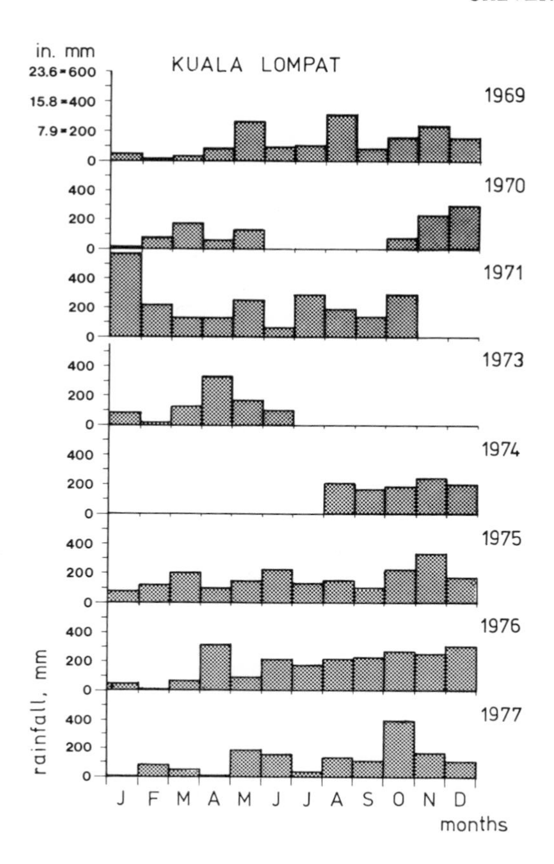

Fig. 8.3. Rainfall in each month of the year at Kuala Lompat from 1969 to 1977 (only 1969, 1975, 1976 and 1977 have complete records - gaps represent no records).

from 15-min samples). Mean daily maximum and minimum temperatures vary little over the year, especially the latter (fig. 8.4). Temperatures are highest during the dry season, sometimes reaching 38°C (100°F), reaching only 32°C (89°F) in the wet seasons, when the lowest temperatures are recorded 19.5°C (67°F). Daily temperature varies between 8 and 13°C (15-21°F), with the lowest range in the wetter months and the highest range in the drier months (Raemaekers, 1977; fig. 8.4). Because of its possible stimulus to flowering (Wycherley, 1973), this measure is of special interest. Temperature changes vary according to height in the canopy; by day the temperature in and above the canopy rises much more than in the lower levels, and by night the air above the canopy cools much more than that inside the canopy. Temperatures presented here refer to shade readings in the clearing at the Ranger Post, but Chivers (1974) shows that these approximate closely shade temperatures at 33.5 m (110 ft) above the ground in the forest canopy.

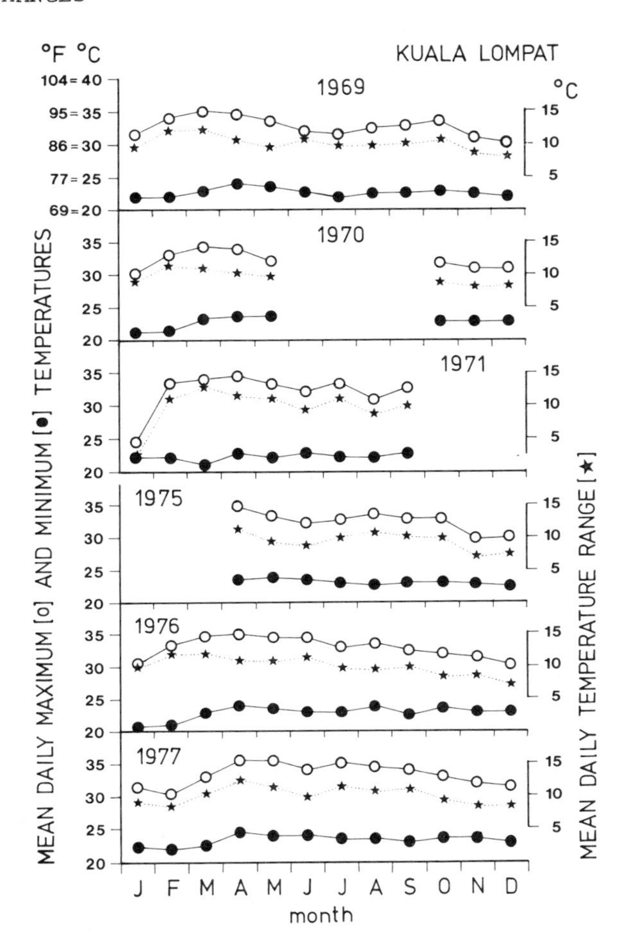

Fig. 8.4. Mean daily maximum and minimum temperatures, and the mean daily temperature range for each month of study at Kuala Lompat from 1969 to 1977.

It will become apparent that it is difficult to relate these patterns of rainfall, sunshine and temperature to changes in primate behaviour, but such relationships become more tangible as one moves from annual through quarterly and monthly to daily comparisons (see final section). Marked dry periods and the subsequent onset of rain, and changes in daily temperature range, seem the most likely trigger to changes in the vegetation and thus to changes in behaviour. Chivers (1974) has drawn attention to the complexity resulting from both immediate and latent effects of such changes in weather. Hopefully we are beginning to develop significant associations between weather, vegetation and behaviour which will eventually lead to clearer explanations. The years 1963, 1968 and 1976 are noteworthy for the pronounced droughts early in the year; these are the years of greatest fruiting (see below). At Kuala Lompat 1969 and 1977 were unusually dry. While the last quarter of the year is usually the wettest, cloudiest and coolest, 1964 and 1971 were wettest in the first quarter because of a shift in the peak of the wet season; the rains were earlier in 1961 and 1968, and the secondary rainy season was dominant in 1962 and 1967. While we are closer to understanding events in the bumper fruiting years, the effects of the latter features are more confusing.

Vegetation

The documentation of cycles of leaf production, flowering and fruiting in the tropical forest is crucial to interpreting the behaviour of primates dwelling therein. Its significance is that it provides a measure of food availability independent of the behaviour of the animals. The first problem is the definition of foods, which of necessity depends on behavioural observation; Chivers (1974) argued that the siamang's knowledge of its home range is so intimate that its ranging behaviour probably provides an accurate indication of food availability; such speculation has at least not been refuted by recent more objective analyses. The supreme problem preventing a speedy resolution of variations in food availability with time is the vast number of species of trees and climber present in the Malayan rain forest; by the end of 1977 395 species of tree had been identified in the study area (about 100 ha) at Kuala Lompat. Of these, 229 species, and a further 93 species of climber, are represented in the 900 or so trees (more than half with climbers) in the 6% of the study area sampled intensively over 29 months.

This large sample investigated over 2½ years can be compared with the impressive pioneering study of McClure (1966) and Medway (1972). From a tree platform in Ulu Gombak they monitored the cycles of trees in its immediate vicinity from 1961 until 1969. From 1963 Medway made fortnightly observations on 61 trees of 45 species (and 17 families); only 10 species were represented by two or more specimens, and only half of these fruited synchronously (some annually, others at longer intervals). Thus we have a small but long-term sample. On average 58% of species flowered, and 47% fruited in each year, but in 1963 and 1968 73 and 88% flowered, and 64 and 79% fruited, respectively; in other years 44-49% of species flowered and 27-40% fruited. In 1963 and 1968 higher proportions of those flowering set fruit - 85 and 89% respectively - whereas the lowest value was 57% in 1966. Most of these species can be assigned to one of several classes of fruiting frequency: less than one year, annual, every two or three years and occasional (in which two groups have been recognised) (table 8.1). Most trees produce new leaves at least once each year; only *Parkia speciosa*, *Koompassia malaccensis*, *Palaquium hispidum* and *Erythroxylum cuneatum*

Table 8.1. Fruiting of trees in Ulu Gombak (1961-1969; from McClure, 1966; Medway, 1972)

BI- or TRI-ANNUAL (2 or 3 times/yr)	ANNUAL	BI- or TRI-ENNIAL (Once every 2 or 3 yrs)
Ficus sumatrana Jan-Feb, May-June, Sep-Oct	*Litsea rostrata* Apr-Aug	*Santiria laevigata*
Ficus ruginerva variable	*Xylopia stenopetala* May-July	*Myristica gigantea*
	Oncosperma horrida May-Dec	*Elateriospermum tapos*
	Parkia speciosa June-Oct	*Artocarpus lanceifolius*
	Sterculia parviflora June-July	*Licania splendens*
	Lithocarpus spicatus Aug-Nov	
	Cyathocalyx pruniferus	
	Melanorrhoea inappendiculata	
	Ficus glabella variable	

OCCASIONAL (every 4-5 years)	every 7-10 years
Shorea curtisii	*Santiria rubiginosa*
Shorea leprosula	*Anisoptera laevis*
Palaquium hispidum	*Hopea* spp.
Dialium patens	*Milletia atropurpura*
Scaphium affini	

had a short annual leafless period; trees of *Santiria rubiginosa*, *Sapium baccatum* and *Cynometra malaccense* flowered regularly without setting fruit. While flowers and fruit were recorded in all months of the year, there was a distinct peak of flowering between March and July and of fruiting between June and October (fig. 8.5).

Although observations at Kuala Lompat span a comparable length of time in succession, they are neither so sustained nor objective until the last 2½ years, when a very large sample of trees was monitored. That only 39% of this sample of 900 trees fruited in the

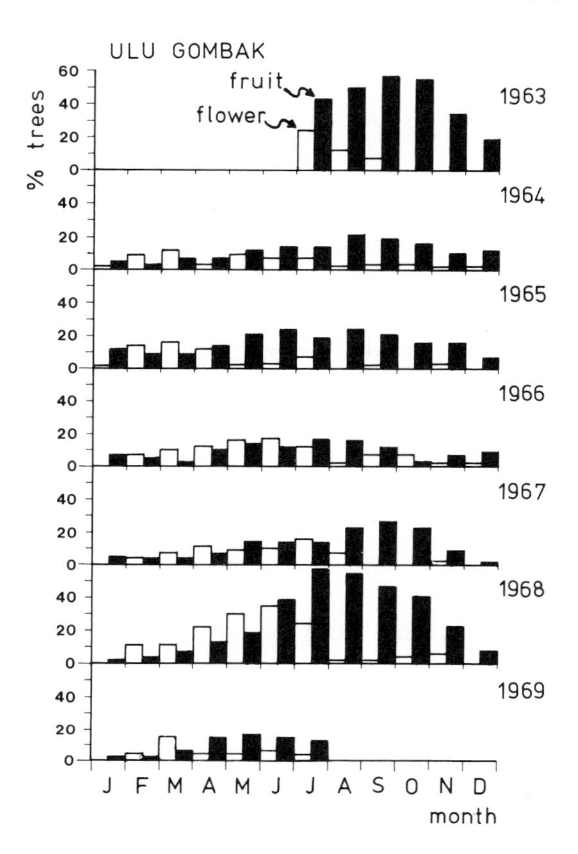

Fig. 8.5. Percentage of trees with flowers and with fruit in each month of the study of 61 trees in Ulu Gombak from 1963 to 1969 (from Medway, 1972).

bumper fruiting year of 1976 may indicate a bias in the small sample of 61 trees in Ulu Gombak - towards the larger trees more exposed to the elements. In most months of the sample at Kuala Lompat less than 5% of trees bore fruit, rising to only a little above that in the fruiting season in 'normal' years, and to 15% in 1976 (fig. 8.6). A discrepancy between observers in the recording of leaf flush was noted in Chapter 2 (fig. 2.9) and the scores subsequent to March 1976 have been corrected upwards (by 17% according to the difference in February-July means - J.J.R. 22.6 cf. J.B.P. 5.3%) yielding a basal level of around 20%. In 1975-76 31% of trees in the sample plots (56% of species) flowered, and 18% of trees (33% of species) fruited. This must give a more accurate picture of fruit (and other foods) available to primates, which use more of the canopy than was sampled in Ulu Gombak.

Data are available on the fruiting of some trees at Kuala

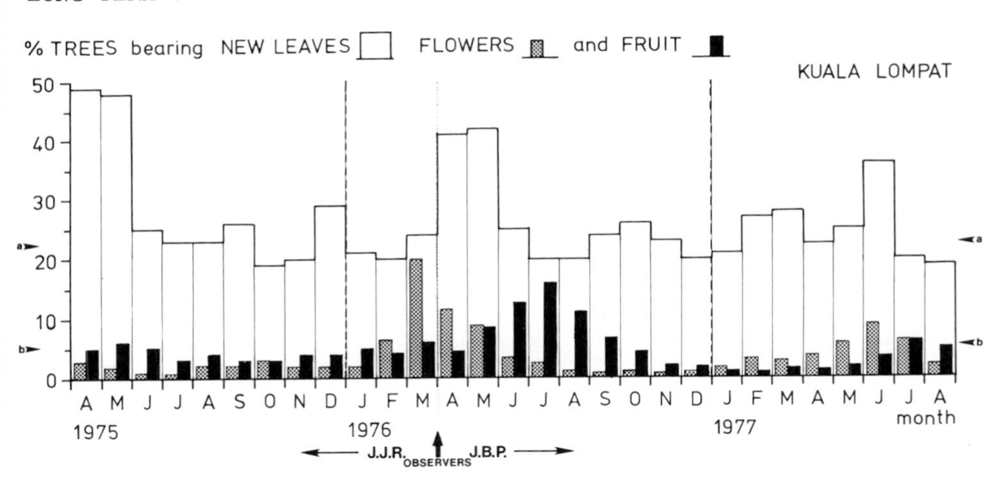

Fig. 8.6. Percentage of trees with leaf flush, flowers and fruit in each month of the study of *c.* 900 trees at Kuala Lompat from 1975 to 1977 (from fig. 2.9 with % leaf scores inflated from April 1976). (a) marks level of mean for leaf flush from July 1975 to February 1976; (b) marks same from July 1976 to February 1977.

Lompat over the eight years (fig. 8.7), but definite conclusions on cyclicity (table 8.2) are impeded by discontinuities in observation. Nevertheless, some interesting patterns of fruiting emerge, especially of fig trees, which are so important to gibbons and other primates as well as to birds; representative specimens of other important food species have also been monitored throughout this period. The Leguminosae are notable for their annual cycles, which include brief leafless periods in *Koompassia excelsa*; *Dillenia reticulata* (Dilleniaceae) also follows this pattern, and *Intsia palembanica* and *Sindora coriacea* have longer leafless periods less often. Species of *Ficus* together exhibit the full range of cycles, but each species tends to be fairly consistent, with some regularly annual and others fruiting two or three times each year.

Thus we are beginning to quantify variations in leaf production, flowering and fruiting in the Malayan forest within and between years, and to identify cycles in many of the tree species important to the primates. There are signs of basal levels of production to which the animals have become adapted, with selection strongest in times of food scarcity and behaviour most diverse in times of abundance (see final section).

Tree species	tree no.	1969	1970	1971	1972	1973	1974	1975	1976	1977
		JFMAMJJASOND	JFMAMJJASOND	JFMAMJJASOND	JFMAMJJASOND	JFMAMJJASOND	JFMAMJJASOND	JFMAMJJASOND	JFMAMJJASOND	JFMAMJJASOND
Ardisia colorata	L366		xx						x	
Artocarpus sp.	S180							xx		
	L880				x			xxx		
Bouea oppositifolia	396		x						x	
	405		x						x	
	M24					x			x	
Calophyllum curtisii	L 54		x							x
	406		x					xxx		
Dracontomelum mangiferum	191	x					x			
Endospermum malaccense	377		x				x			
Eugenia sp.	27	x	x					x		
	1070						x			
Garcinia atrovirides	64						x	xx		
Gnetum sp.	M56					x	x			
	L 99					x	x			
Grewia laurifolia	M148					x		x		
Grewia fibrocarpa	126					x				x
	L199	x					x			
Knema laurina	344		x					x		
Lophopetalum sp.	142	x						x		
Mangifera indica	11	x								x
Maranthes corymbosa	195	xx		x	x	x	x			
Randia scortechinii	212	xxxx						xxx		
	229	xx		x			xx	xxx		xx
	257	x	x				x			
Rubiaceae sp.	230	xx					x	x		
	240	xxx								
Sandoricum koetjapi	M180					x				x
Sapium baccatum	S176	x		x			xxx			
	AB-3						x xx			x
Sarcotheca griffithi	L365		xx			x			x	
Sloetia elongata	17	xx	x							x
	48	xx x	x					xx		
	80	x						x		
	96	x							x	
	104	x						x x		
	186	x						x		
	328		x					x		
Vitex trifoliata	275	x x					x	x		
Xerospermum wallichi	M 40					x		xx		
Xylopia malayana	M 47					x			x	x

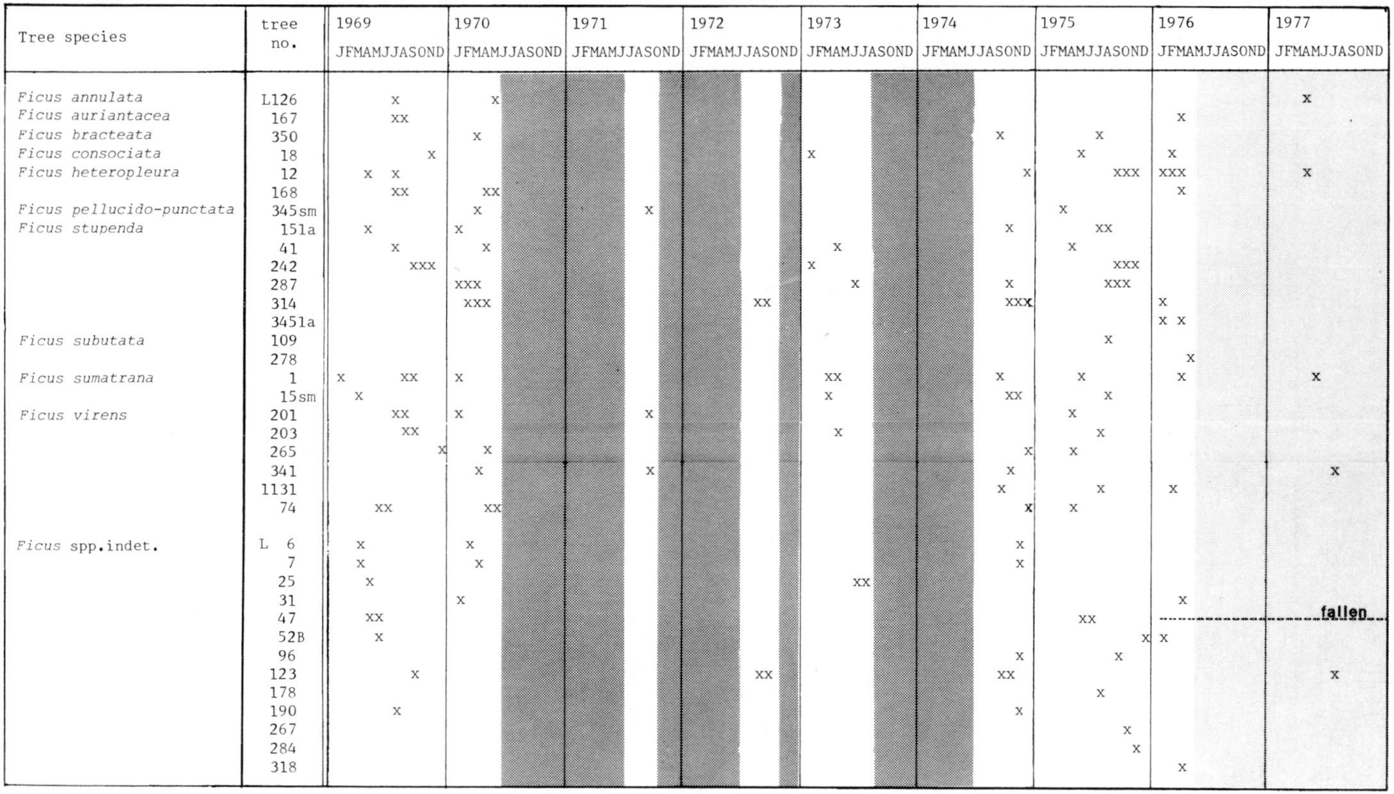

Tree species	tree no.	1969 JFMAMJJASOND	1970 JFMAMJJASOND	1971 JFMAMJJASOND	1972 JFMAMJJASOND	1973 JFMAMJJASOND	1974 JFMAMJJASOND	1975 JFMAMJJASOND	1976 JFMAMJJASOND	1977 JFMAMJJASOND
Ficus annulata	L126	x	x							x
Ficus auriantacea	167	xx							x	
Ficus bracteata	350		x				x	x		
Ficus consociata	18	x				x		x	x	
Ficus heteropleura	12	x x					x	xxx	xxx	x
	168	xx	xx						x	
Ficus pellucido-punctata	345sm		x	x				x		
Ficus stupenda	151a	x	x				x	xx		
	41	x	x			x		x		
	242	xxx				x		xxx		
	287		xxx			x	x	xxx		
	314		xxx		xx		xxx		x	
	3451a								x x	
Ficus subutata	109							x		
	278								x	
Ficus sumatrana	1	x xx	x			xx	x	x	x	x
	15sm	x				x	xx	x		
Ficus virens	201	xx	x	x				x		
	203	xx				x		x		
	265	x	x				x	x		
	341		x	x			x			x
	1131						x	x	x	
	74	xx	xx				x	x		
Ficus spp.indet.	L 6	x	x				x			
	7	x	x				x			
	25	x				xx				
	31		x						x	
	47	xx						xx		fallen
	52B	x						x	x	
	96						x	x		
	123	x			xx		xx			x
	178							x		
	190	x					x			
	267							x		
	284							x		
	318								x	

Fig. 8.7. Patterns of fruiting by certain trees at Kuala Lompat from 1969 to 1977, indicating periods with no observation (heavy stippling) and the final period (light stippling) with incomplete coverage of the sample.

Table 8.2. Fruiting of trees at Kuala Lompat (1969-77).

BI- or TRI-ANNUAL	ANNUAL	BI- or TRI-ENNIAL
Ficus sumatrana	*Ficus annulata*	*Ficus heteropleura*
Ficus virens	*Ficus consociata*	*Artocarpus* spp.
	Ficus stupenda	*Ardisia colorata*
	Maranthes corymbosa	*Bouea oppositifolia*
	Koompassia excelsa	*Dracontomelum mangiferum*
	Parkia speciosa	*Grewia laurifolia*
	Parkia javanica	*Sarcotheca griffithii*
	Saraca thaipingensis	*Baccaurea brevipes*
	Randia scortechinii	*Endospermum diadenum*
	Sloetia elongata	*Sapium baccatum*
	Vitex trifoliata	*Sindora coriacea*
	Xylopia malayana	*Calophyllum curtisii*
	Pentaspadon velutinum	*Aglaia* spp.
	Dillenia reticulata	*Knema laurina*
		Carallia brachiata

OCCASIONAL every 4-5 years	every 7-10 years
Ficus auriantacea	*Ficus subutata*
Mangifera indica	*Shorea leprosula*
Dialium platysepalum	*Dipterocarpus cornutus*
Dialium procerum	
Intsia palembanica	

SOCIAL BEHAVIOUR

Calling

The loud calls characteristic of primates living in the restricted visibility of tropical rain forest, whether given for reasons of group cohesion and/or territorial advertisement and maintenance, yield useful data from the moment the field worker starts his study. These loud calls are important for locating potential study groups, and for speeding the process of habituation. The systematic recording of the frequency and quantity of calling in each species permits analyses that lead to the estimation of population size and to an assessment of the extent to which the timing and amount of calling is indicative of the behaviour and, more important, changes in behaviour in each species. For example, Chivers (1974) based his estimates of the gibbon populations of West Malaysia in 1968 on calling behaviour, and Chivers and Davies (1979) have compared the results of this method with those derived from census walks; furthermore, two years of calling data from Ulu Gombak (Chivers, 1974), indicated an annual increase in calling in both siamang and lar gibbon during the fruiting season in the middle of the year, with an unusually high level of calling during the bumper fruiting year of 1968. There also seems to be a relationship between calling and mating. We shall concern ourselves here mainly with the pattern of calling by each species during the day at Kuala Lompat.

The gibbons lend themselves best to this kind of analysis through their frequent and discrete group calling bout songs (Chapter 3), but it is also applicable to the leaf monkeys whose males emit very loud single calls, often as a series; the long-tailed macaques do not emit such calls. D.J.C. recorded all calls heard on 128 days in 1969 and a further 61 days in 1970; this sample was increased by a further 80 days dispersed in groups from 1972-79 (table 8.3), yielding a total of 2540 call bouts of which 79% were accounted for by siamang and lar gibbons (table 8.4). Thus, lar gibbons call most and dusky leaf monkeys least, even when such call bouts are related to the number of groups in hearing range.

The distribution of calling through the day shows that all species concentrate calling in the morning, and banded leaf monkeys just before dawn, the dusky leaf monkey at dawn, and the two gibbons about two hours after (fig. 8.8). The dusky leaf monkey has a small secondary peak before dusk, and the banded leaf monkey a larger, more protracted one after dusk. In plotting the number of group calls heard in each hour, one sees the differences in quantity of calling, although differences are more marked when one remembers that gibbon calls involve more individuals and last for much longer than those of leaf monkeys. These data are replotted in terms of

Table 8.3. Observations of calling behaviour at Kuala Lompat.

Year	Days	Observer	Months	Subjects
1969	128	D.J. Chivers	Jan-Dec	All groups of gibbon and leaf monkey in and around study area
1970	61		Jan-May	
1972	28		Aug-Sept	
1974	11		June & Aug	
1976	14		Mar-May	
1977	24		July-Nov	
1979	3		April	
	269			
1970-71	243	S.H. Curtin	Oct-Sept	Leaf monkey study groups
1975-76	238	J.J. Raemaekers	Apr-Mar	Siamang (TS1) and lar gibbon (TG2) study groups

Table 8.4. Features of calling at Kuala Lompat during 269 days from 1969 to 1979 inclusive.

Species	No. of call bouts heard	No. of bouts/day heard	Mean no. of groups resident in area	bouts/ group/day
Lar gibbon *Hylobates lar*	1139	4.23	8	0.53
Siamang *H. syndactylus*	868	3.23	10	0.32
Banded leaf monkey *Presbytis melalophos*	331	1.23	7	0.18
Dusky leaf monkey *P. obscura*	202	0.75	8	0.09

the proportion of calls in each 3-hour period for comparison with data from J.J.R.'s study of gibbons (one group of each species, TS1 and TG2) and from Curtin's study of leaf monkeys (see Chapter 4, several groups of each species) (fig. 8.9). Curtin's data show a reversal in size of the peaks in banded leaf monkeys, and a lack of distinct peaks in dusky leaf monkeys.

The observation that most songbirds concentrate singing at dawn and dusk has led to experimental work to determine whether there are

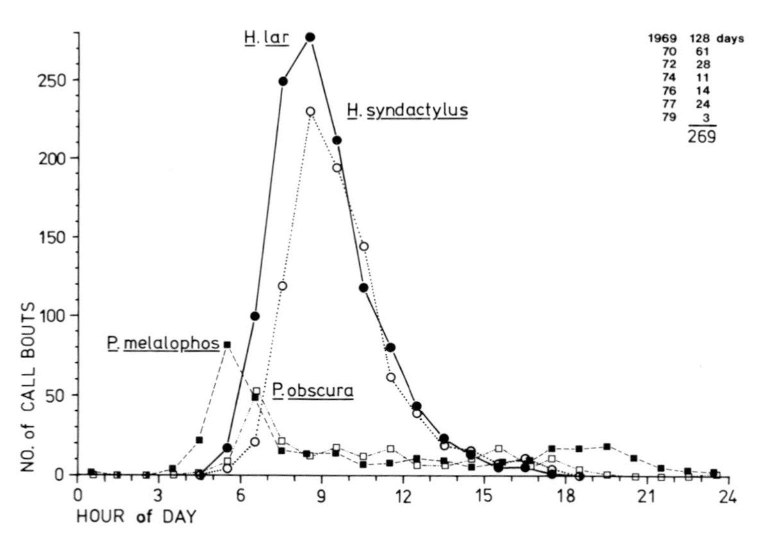

Fig. 8.8. No. of call bouts in each hour of the day by each of the gibbon and leaf monkey species at Kuala Lompat over a sample of 269 days from 1969 to 1979.

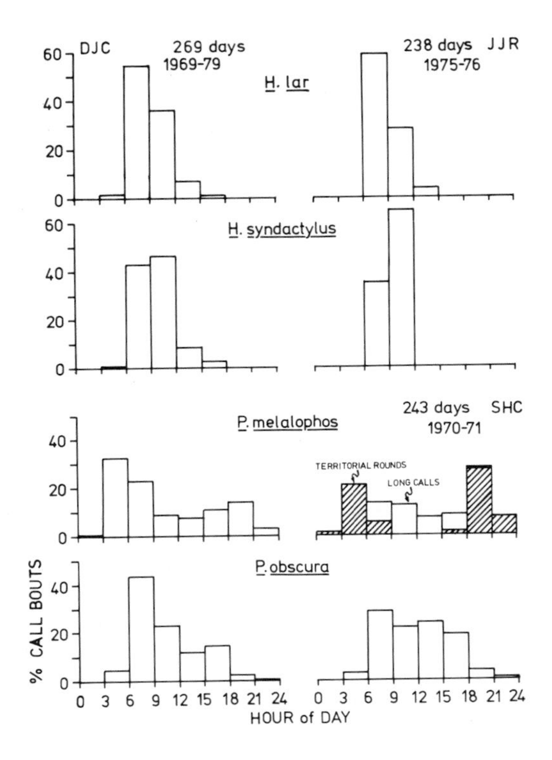

Fig. 8.9. Percentage of call bouts in each 3-hour period of the day for each gibbon and leaf monkey species at Kuala Lompat for 1969 to 1979: sample of 269 days for all groups in the study area for whole period by D.J.C. to left; sample of 238 days from 1975-76 for one group of lar gibbon and one group of siamang by J.J.R. (upper right), and 243 days from 1970-71 for leaf monkey study groups by S.H.C. (lower right).

good acoustical reasons for so doing. It appears that signals do indeed carry furthest through the forest in the early morning, because temperature layering, which tends to reflect sounds, and atmospheric turbulence, which increases attenuation and lowers the signal-to-noise ratio by increasing amplitude fluctuations, are least at this time (Henwood and Fabrick, 1979; Wiley and Richards, 1978; Richards and Wiley, 1980). It may well be then, that one reason why primates call mainly in the earlier part of the morning is because the physical conditions of the environment are then best ... although the cicadas are at their noisiest. Other reasons, of course, might be equally or more important; for instance, if the loud calls of 'group leaders' help to space out groups (as in howling monkeys, Chivers, 1969), it is logical to give them, as do the leaf monkeys, when the animals become active in the morning. The gibbons may delay calling until later, because their much more elaborate and longer calling sessions require more energy, and are thus given mostly after the first feed of the day (although the males in some species call solo around dawn).

While acoustical conditions may favour calling in the early morning, the simultaneous calling of many species of birds and animals, in addition to the loud background noise of insects, would seem disadvantageous. The spread of calling peaks between 0500 and 1000 hours (fig. 8.7) might be the result of selection to avoid inter-specific competition for calling time (in addition to or instead of the differences between leaf monkeys and gibbons mentioned above). Such avoidance has been demonstrated statistically in pairs of sympatric songbird species, both at the level of the local population (Cody and Brown, 1969) and of individual pairs of birds whose territories overlap (Ficken *et al.*, 1974).

MacKinnon (1974, 1977) has shown that sympatric primate species in Sumatra also have asynchronous calling peaks at the level of the local population. He argues that a marked difference in the distribution of calling time between Sumatran and Bornean orang-utans is attributable to greater competition for calling time between species in Sumatra, where there are two, rather than just one, gibbon species. It may be noted that in Sumatra, the calling times of lar gibbon and siamang were better segregated, as they are in other parts of Peninsular Malaysia (Chivers, 1974, 1978). The very elaborate duets of siamang, with considerable information content (Haimoff, in press), may require the later, quieter hour, and the loudness of their calls can cope with any decline in acoustical conditions by mid-morning. At Kuala Lompat, the two species of gibbon are better segregated when individual pairs of overlapping groups, rather than local populations, are compared (fig. 8.9). It must be admitted, however, that there is no hard evidence that species are competing for calling time; the calls of these primates are very different, and one would in fact expect competition to be greatest within species.

Avoiding the masking of one's own song by those of others might also be expected to promote the segregation of calling times within species (Wasserman, 1977). Moreover, if we assume that songs convey genuine information (as opposed to false information, Dawkins and Krebs, 1978), then it should pay to listen to neighbouring conspecifics instead of singing when they do. Wasserman has shown that white-throated sparrows do avoid singing at the same time as their neighbouring conspecifics, and that the closer the neighbour the greater the avoidance. While the peaks of calling in leaf monkeys indicate synchronous calling, the utterance of the short calls in fact alternates between groups. Although lar gibbons tend to have synchronous calling bouts, especially with their closest neighbours, the most elaborate part of the bout - the female great calls - are rarely synchronous. The tendency for groups in a local population to call synchronously seems to apply also to the other small gibbons, e.g. agile (Gittins, 1979), Kloss (Tenaza, 1976; Whitten, 1980; males and females separate). Chivers (1974) also describes the contrasting situation in siamang, where calls pass sequentially around groups in a valley, one stopping when the next starts; this feature, and the higher ratio of songs to inter-group conflicts (Chapter 10), suggests a more careful evaluation of information contained in the song, and a more important role for song in mediating relations between groups.

There is increasing evidence, at least for gibbons, to support previous suggestions (Chivers *et al.*, 1975) that calling is increased when fruit are particularly abundant and when animals are mating, with or without the formation of new groups. Calling is also increased in social upheavals involving more than one group (see below), but is decreased by tragedies within a group and by gross disturbance, such as selective logging within the territory.

Siamang

The main study group, and the only one to persist throughout the study (table 8.7, p.252), is TS1 (fig. 8.10). It was discovered in September 1968 by D.J.C. during his country-wide survey, when it contained adult male and female, sub-adult male, juvenile male and an infant that appeared to be only a few months old (fig. 8.11). A systematic study began in January 1969 (Chivers, 1974), and by April the group was habituated sufficiently for dawn-to-dusk observation; the animals were identified more precisely (Plate XVII) as

adult male, Murgatroyd, at least 17 years of age, with opaque right eye,
adult female, Fenella, also at least 17 years of age (if sub-adult was their first offspring, produced when the parents were 10 years old),
young sub-adult male, Sammy, about 6-7 years of age (if adult by 8-9 years, Napier and Napier, 1967),
juvenile-1, Julian, about 3-4 years old, and
infant-2, Inky, almost 1 year old as just transferred to male for carriage during the day.

During the next 12 months Inky gradually gained his independence of travel, and by early 1970 the peripheralisation of Sammy became marked. His father attacked him more often, and by May 1970 he was preventing Sammy from entering food trees with the rest of the family, by vigorous swinging displays as Sammy approached (Aldrich-Blake and Chivers, 1973). After a quiet year the group's singing increased dramatically in February 1970, and by May Sammy was singing after his parents had stopped in most bouts; his mean distance from Murgatroyd increased. All this coincided with the onset of copulation between Murgatroyd and Fenella in January 1970, rising

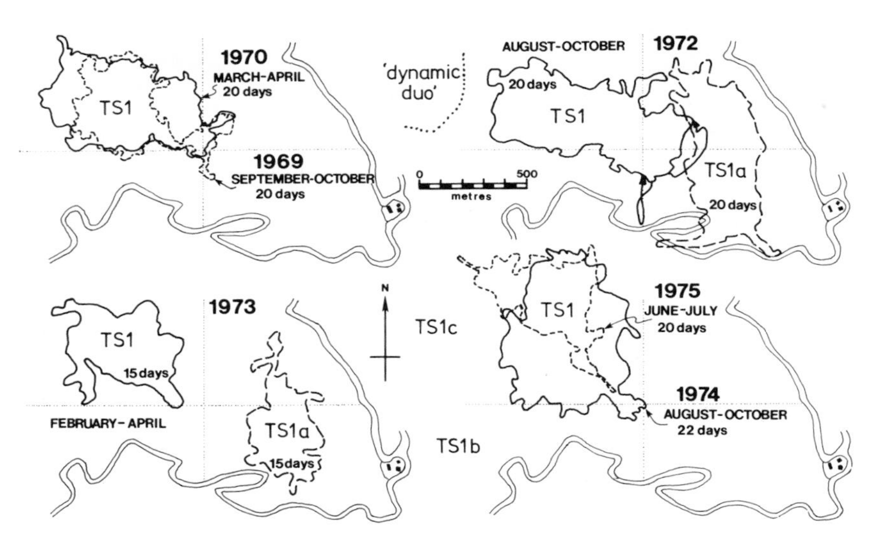

Fig. 8.10. Limits of ranging by the siamang study groups for short periods at intervals through the study period, showing neighbouring groups and N-S transect 15 and E-W transect 10 (dotted lines, cf. fig. 2.2).

to 1.1 copulations/day on average in April.

Observations ceased in May 1970, but when they were resumed incidentally in October of the same year by Curtin, Sammy had left TS1 and was presumed to be the male in the forest adjacent to the east with a young adult female, Sarah, who is believed to be the same animal seen in the south-east of TS1's territory in May 1970, apparently attracted by Sammy's prolonged calling. This new group was labelled TS1a; the forest they occupied contained no resident siamangs. In April 1971, TS1a was joined by an old adult female, Sheila; she left temporarily in July, but returned in September and was still there in November 1971 when Aldrich-Blake ceased his 3 months of observations (Plate XVIII). Sammy played with young Sarah but copulated only with the older Sheila (once ventro-ventrally); in terms of grooming and proximity he was transferring his attention to Sheila.

Meanwhile, Julian of TS1 had become sub-adult and left the group towards the end of 1970, not very long after his elder brother, last seen on October 30th. Early in January 1971 Fenella gave birth, but the infant died later that same year, so that TS1 was reduced to Murgatroyd, Fenella and Inky, now a juvenile. There was a gap in observation until July 1972, by which time TS1a had changed radically; Sammy and old Sheila had gone, but young Sarah remained with a new young male, Imbo. The composition of TS1 had not changed, but a pair of males were located by their calling living to their north-

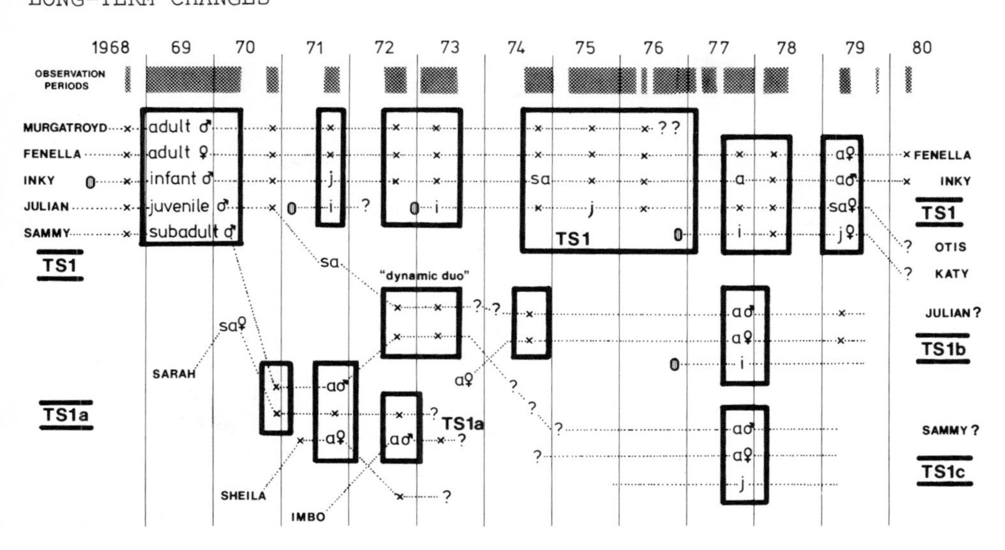

Fig. 8.11. Social history of the siamangs in the study area from 1968 to 1980.

west; their behaviour and appearance suggested them to be Sammy and Julian (Chivers *et al.*, 1975).

From mid-October 1972 until MacKinnon and MacKinnon (1977) started their study towards the end of January 1973 there was another small gap in observation. During this time, probably in December 1972, there was another birth in TS1; the infant, later named Otis, eventually proved to be female (Inky was thought initially to be female!). Sarah had left TS1a, so that Imbo was alone. During February and March 1973 he was often seen travelling and feeding, and often even singing, with a male lar gibbon, Stanley. The partnership ended in April 1973 with the appearance of a female lar gibbon who paired with Stanley, but not before she had associated briefly with Imbo, once duetting with him. Later that month an adult female siamang entered Imbo's home range, three days after his rate of calling had increased. They duetted twice and associated discontinuously, but the female disappeared only five days after first being sighted; by the end of the study in July he had become very lethargic.

A year later, in June 1974, 18-month old Otis was being carried by her father Murgatroyd, as had been the case with Inky in his second year, and she was eating solid food (Chivers, 1975). Imbo had disappeared, and so had the two males located to the north-west of TS1 in 1972, and seen again during 1973, but there was a new group to the south-west of TS1 with a range straddling the Main Trail; the female was of unknown identity, but the male resembled Julian. This group, TS1b, produced an infant towards the end of

Plate XVII. Siamang through the decade (1) 1969-70 (DJC).

(a) TS1 group calling - MURGATROYD and JULIAN in centre, FENELLA to right, SAMMY in upper left (b) TS1 resting - FENELLA above, below from left MURGATROYD, SAMMY, INKY and JULIAN.

(c) Adult male, MURGATROYD and infant, INKY (d) adult female, FENELLA (e) sub-adult male, SAMMY (f) juvenile, JULIAN (h) infant, INKY.

Plate XVIII. Siamang through the decade (2) 1971-72.

(a) TS1, adult male, MURGATROYD, and infant, unnamed in late 1971 (FPGA-B) (b) TS1, adult male, MURGATROYD, and adult female, FENELLA, in 1972 (DJC) (c) TS1, MURGATROYD and FENELLA calling together in mid 1972 (DJC)

(d) RS2 (Ulu Sempam), adult male and female copulating in usual dorso-ventral position in early 1970 (DJC) (e) TSla, adult male, SAMMY, and old adult female, SHEILA, copulating ventro-ventrally in late 1971 (FPGA-B) (f) TSla, adult male, SAMMY, in late 1971 (FPGA-B) (g) TSla, young female, SARAH, and male, SAMMY (right) in late 1971 (FPGA-B) (h) TSla, adult male, IMBO, and young adult female, SARAH, in mid-1972 (DJC).

Plate XIX. Siamang through the decade (3) 1974-79 - TS1 (DJC).
(a) Adult male, MURGATROYD (above) and infant female, OTIS, and sub-adult male, INKY, in mid 1974 (b) infant female, OTIS, in mid-1974 (c) infant female, KATY, with adult male, INKY, in mid-1977

(d) adult female, FENELLA, and (e) juvenile female, OTIS, in mid-1977 (f) adult male, INKY, and (g) sub-adult female, OTIS, with damaged left arm in early 1979.

1976. Sammy may have formed the group seen to their north in recent years (TSlc).

From August 1974 through to the end of J.J.R.'s observation period in March 1976, TS1 remained stable with Murgatroyd, Fenella, Inky and Otis. From the age of 6-8 years Inky was receiving the same kind of peripheralisation treatment from his father that Sammy had early in 1970. Like Sammy he was tending to sleep in a different tree from the others, although of 74 nights in 1974-76, Inky spent only 19 of these in a separate tree from the female and infant, and on 12 of these the male was with him at the start of the night. Unlike Sammy, he stayed with the group, singing in almost every family song bout and not prolonging his song beyond theirs; sometimes he went off alone for a day or two. Murgatroyd and Fenella copulated only rarely (table 8.5).

In July 1976, it was discovered subsequently (Kalang and Payne, pers. comm.), something drastic had happened (reflected also in the change in activity patterns, see below). Murgatroyd disappeared; since he was stiff-limbed and more than 24 years old it is presumed that he died. Sonagraphic analysis of vocalisations from 1974, early 1976, mid-1977 and early and late 1979 by Elliott Haimoff indicate that it is Inky who stayed with Fenella. In late 1979 and early 1980, however, the group was unusually elusive, and we believe it to be just possible that an alien male might have displaced Inky by this time; this problem will have to be resolved by further observation and analysis of calls.

In December 1976 Fenella gave birth again, probably on the 13th (not a Friday!), to a female named Katy (Payne, pers. comm.), who is undoubtedly Murgatroyd's daughter since she must have conceived

Table 8.5. Matings in siamang and lar gibbons at Kuala Lompat and siamang in Ulu Sempam.

			J	F	M	A	M	J	J	A	S	O	N	D
Siamang	RS2	1969	0	0	0	0	0	0	0	2	0	0	0	0
		1970	0	1	6	9								
	TS1	1969												
		1970	1	2	4	3	1							
		1975				1	2	1	0	1	0	1	0	0
		1976	1	2	0									
Lar gibbon	TG2	1975				0	0	1	1	1	0	1	0	0
		1976	0	0	1									

Figures represent number of matings seen; since animals were watched for about 10 days each month, one mating seen in one month represents 0.3 matings/day.

towards the end of J.J.R.'s study. Inky became agitated if an observer approached just after the birth. By February Katy was taking some solid food of her own accord (Kalang, pers. comm.), and from April onwards she played a lot. In May Inky was seen carrying her (Payne, pers. comm.), although she was only six months old.

In October 1977 some siamang were seen briefly in the forest east of TS1, which had lacked resident siamang since Imbo's departure by June 1974; they were not seen again.

In February 1979, TS1 still consisted of Inky, Fenella, Otis (now $7\frac{1}{4}$ years and sub-adult) and Katy (now $3\frac{1}{4}$ years and juvenile) (Plate XIX). In March Otis injured her left arm and was missing for a day or two, apparently resting in the centre of the home range. She rejoined the group and travelled and fed awkwardly with them (Chivers, pers. obs.), but within two weeks she appeared to be back to normal (Mittermeier, pers. comm.).

In August 1979 another dramatic event occurred (Bennett, pers. comm.). When J.J.R. visited Kuala Lompat in October, Otis and Katy had left Inky and Fenella and moved into the old TS1a range. Later that month Otis disappeared, leaving Katy, now nearly 4 years old, alone in the forest adjacent to her natal home range. As with Imbo she was last seen happily munching figs. The details of this development are being followed closely by Bennett *et al.* (in prep.).

Lar gibbon

Although incidental observations were made on the lar gibbon from 1968 onwards, detailed observations and exciting social changes started in 1973 (MacKinnon and MacKinnon, 1977). The main study group TG2 occupied the same tract of forest as the siamang group TS1; several transient lone individuals had been seen over the years up to 1973 in the forest to the east, and there were adjacent groups to the north-west (TG1 - male, female, sub-adult female), to the north (TG1B) and to the south-west (TG4) (fig. 8.12).

TG2 seems to have followed the normal pattern up to 1973, the same two adults remaining paired and sub-adults budding off at intervals (fig. 8.13). In 1969-70 the group consisted of the adult male Gilbert (famous for being two-tone, with a golden 'shirt' and dark-brown 'trousers'), adult female Gertie (dark brown, with much-reduced face ring), a sub-adult male, a juvenile (eventually revealed to be male), and a young infant (subsequently identified as female). By 1972 the composition appeared unchanged, but the two dark-brown immature males spent much time away from their parents, often calling at dawn from the opposite side of the territory; the infant had become a golden-brown juvenile. In 1973 the

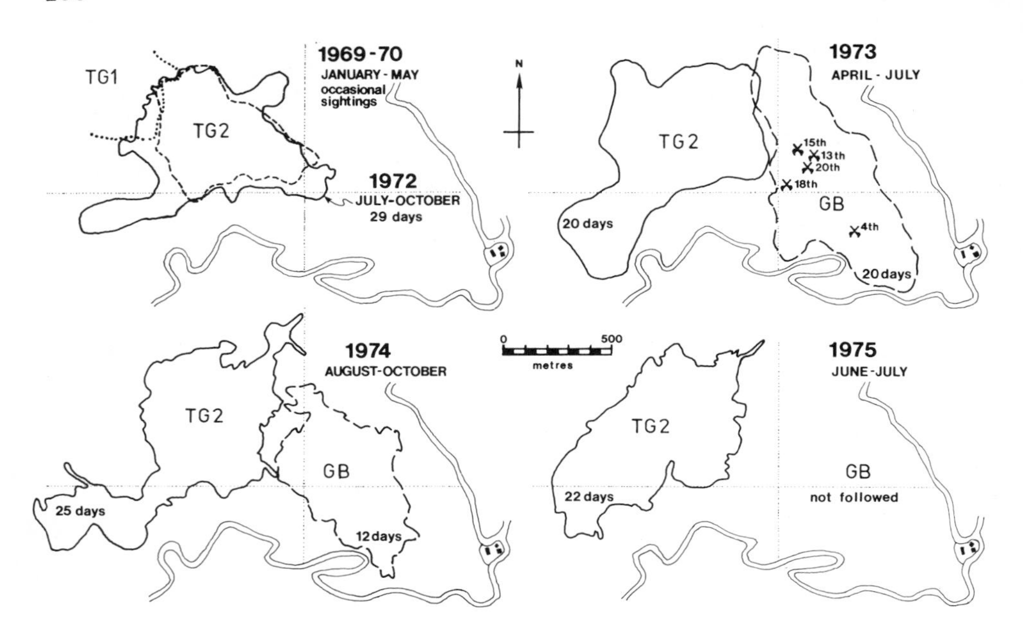

Fig. 8.12. Limits of ranging by the lar gibbon study groups for short periods at intervals through the study period, showing a neighbouring group (present throughout) and N-S transect 15 and E-W transect 10 (dotted lines, cf. fig. 2.2).

two sub-adult males, George and Gordon, were even more peripheral as Gilbert would clearly not tolerate their close proximity; the juvenile female Gail followed her parents closely.

East of TG2 a lone male, Stanley, associated periodically with the male siamang, Imbo (see above), often singing with him; Stanley also sang solo. On April 3rd 1973, while Stanley and Imbo were together, a female lar gibbon appeared, grappled briefly with Stanley and then left, followed by Stanley. On the next two days this pair (GB) duetted, with the result that TG2 rushed over from their territory to attack them. The pair broke up and the female, Sylvia, associated briefly with the siamang Imbo, even duetting with him once; this is the only occasion on which a female lar gibbon has been heard to great call without male lar accompaniment. On April 8th Sylvia and Stanley reunited and duetted, and repeated attacks by TG2 over the next 10 days failed to disrupt them. Of note in these conflicts were the aggressive role of Sylvia, and Gilbert's tolerance of Gordon who assisted in the attacks on GB. Observations ceased in July 1973, when the MacKinnons left.

By August 1974 George had disappeared from TG2, but Gordon remained until November 1975 when he left the group's home range

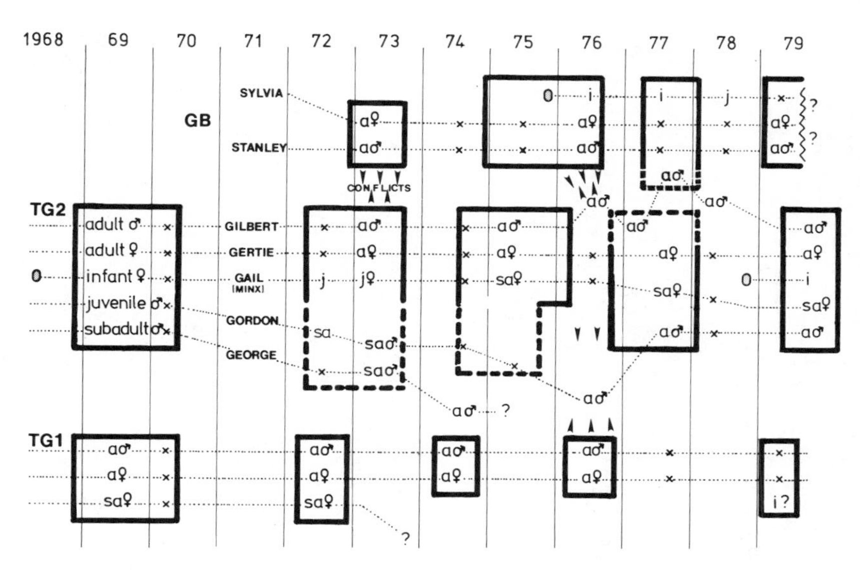

Fig. 8.13. Social history of the lar gibbons in the study area from 1968 to 1979.

(Raemaekers, 1977); over the last 12 months the frequency of observing him in the group had declined gradually from 60 to 3%. Gordon was not usually tolerated close to Gilbert, although he sometimes groomed Gilbert. When with the group (Plate XX) he spent most time with his little sister Gail; he was excluded from small fruit trees, just as were Sammy and Inky in TS1; he never sang with his parents, but gave long solos when they were not singing and he was in some other part of the territory; he sometimes played a leading role in conflicts with Stanley and Sylvia, as he had done in 1973. There was little sexual activity between Gilbert and Gertie prior to Gordon's departure. Gail joined her mother in great calls, at first tenuously, but gaining in strength and timing over the months; learning the female part of the song was obviously a long process.

Throughout this period GB remained stable, and Sylvia bore an infant in December 1975; her great calls were often aborted initially, and only gradually gained in duration over the years.

In July 1976, the month in which TS1 lost its male, TG2 (Plate XX) also entered a period of social flux which still continues. TG2 started fighting with TG1 and GB at an unprecedented rate; 8 conflicts were seen in 5 days (Kalang, pers. comm.). In August a third opponent appeared, a white male to the north of TG2. Conflicts continued at the rate of one a day, with Gertie and Gail participating, just as Sylvia had helped Stanley in April 1973; they were even seen to continue the conflict after Gilbert had retired. In September the social structure collapsed (John Payne and Eamonn

Plate XX. Lar gibbons at Kuala Lompat, group TG2 (DJC).

(a) adult male, GILBERT (b) adult female, GERTIE (c) juvenile female, GAIL (d) sub-adult male, GORDON, in mid-1974

(e) GERTIE (left) and GILBERT being groomed by GAIL (f) GERTIE resting (g) GAIL feeding and (h) GAIL (left) and GILBERT resting in early 1976.

Barrett, pers. comm.); for most of the time Gilbert was ranging alone in his territory. He did, however, take part in one conflict alongside Gertie and Gail against a number of unidentified gibbons inside their territory. He began to drift to the east side of their territory and beyond. The white male and other lar gibbons were seen deep inside TG2's territory, which was now occupied only by Gertie and Gail. In October these two females spent much time in the far noth-west of their territory and beyond, variously associating with the white male and other lar gibbons and fighting with TG1. On the 15th an unprecedented number of gibbons clustered in the centre of TG2's territory, possibly 11 individuals including Gertie and Gail and perhaps GB, all involved in a conflict. Gilbert was seen the next day pursued by a group of four deep inside what had been his territory; he was now at least 22 years old.

A most regrettable gap in the records ensues until January 1977, when Gertie and Gail were observed in the south of their territory, singing in concert with a dark male (Surender Sharma, pers. comm.). In February a group of 4 was seen there, presumably Gertie, Gail, the dark male and possibly Gilbert. This dark male was almost certainly Gordon, and he remained with TG2 (checked by J.J.R. in October 1979). Thus the pre-July 1976 group was reformed, but with Gilbert in a peripheral role, and his son now fully mature and central in the group. In April and May 1977 Gilbert was seen alone several times, although all four were seen together in May.

In September 1977 7 gibbons were seen over the Main Trail in the south of TG2's territory by D.J.C.; Gertie, Gail and Gordon were to the west, and Stanley, Sylvia and their juvenile to the east, calling at each other across Gilbert eating fruit. Gilbert followed GB east, and was seen with them 6 days later in the heart of their territory. In July Gilbert had been seen with a freshly-scarred lip with the other 3 members of TG2; they moved east from the centre of their territory and called. When the other three went away north-west, Gilbert went on east and joined up with GB who were feeding quietly nearby. In November Gilbert was found calling alone on what was now the boundary between the territories of TG2 and GB, the swamp between N/S transects 13 and 14 (fig. 2.2).

Another gap in the records follows until April 1978, when Gilbert was alone, and Gertie and Gail were seen with lar gibbons from the west one day, and fighting with them on the next (probably TG1); Gordon was not seen. There was much singing in May, but the singers were not identified (Kalang, pers. comm.).

In April 1979 Gilbert was not seen, but Gertie had a small, dark brown infant of about 6 months; Gordon and Gail were present. While Gordon seems the most likely father, which would have been incestuous, Gilbert could easily have been the father (especially in view of future events). Stanley, Sylvia and their juvenile were also seen by D.J.C., but there has been no subsequent sighting (Bennett *et al.*, pers. comm.). Gail and Gordon, and Gertie and Gordon sometimes called together, but more often all three called in concert; Gail had only 4-7 notes in her great call, Gertie had about 11.

By April 1979 Gilbert was back with TG2, with Gertie, Gordon, Gail and the unsexed infant; J.J.R. only heard Gail and Gordon singing together; the only known record in gibbons of duetting by brother and sister. In the absence of GB they had extended their range eastwards - back to the 1970 position!

Discussion

Reproduction. Ellefson (1974) suggests that births in a lar gibbon family are spaced every 2-2½ years, with mating taking place for about 6 months at corresponding intervals. Available data on the composition of all gibbon species agree that surviving offspring are spaced at least two years apart. Fenella's unique longitudinal record shows unevenness in the spacing of births: 3 years between Inky in 1968 and the 1971 infant which died, nearly 2 years between it and Otis in 1972, and 4 years between Otis and Katy in 1976. Conception seems to have followed soon after the death of the 1971 infant, and the long interval before Katy's birth might reflect decreasing reproductive capacity with age (in her or Murgatroyd). As far as we know, Gertie of TG2 gave birth only twice during the study period, ten years apart (table 8.6).

The record of births at Kuala Lompat provides convincing evidence of seasonal reproduction. Births are clustered around the turn of the year (table 8.6), during the season of worst weather

Table 8.6. Months of births for siamang and lar gibbons in Peninsula Malaysia.

Siamang	Kuala Lompat	TS1	early in 1968	(Inky)
			early January 1971	(infant which died)
			December 1972	(Otis)
			13th November 1976	(Katy)
	Ulu Gombak[1]		20th October 1968	
Lar gibbon	Tanjong Triang[2]		April 1965	
			August 1965	
	Kuala Lompat	TG2	early in 1968	(Gail)
			early in 1978	

[1] Chivers and Chivers (1975) [2] Ellefson (1974)

and lowest food abundance. Given gestation periods of 7 months for lar gibbons and 7½ months for siamang (Napier and Napier, 1967), conceptions would occur during the flowering-fruiting peaks earlier in the year (fig. 2.9). Chivers (1974) found the correspondence expected from this between mating and the fruiting peak in 1970 in both the Kuala Lompat and Ulu Sempam populations of siamang, as well as a peak of mating in all species in these areas and Ulu Gombak in the bumper fruiting year of 1968. In contrast, the MacKinnons saw no mating in either gibbon at Kuala Lompat at the appropriate time of year in 1973. This absence of reproductive activity in TS1 can be explained by the recent birth of Otis, but TG2 had no offspring for 5 years; they were, however, subjected to considerable stress by the formation of GB. There was no correspondence between mating and fruit abundance in 1974-76 (table 8.5, fig. 2.9; Raemaekers, 1977), but, as in 1973, maybe the level of fruiting was insufficient Katy was born to TS1 after the fruiting peak of 1976, the first bumper year in the Peninsula since 1968, as were infants in the new group to the south-west, and a group on the slopes of Gunung Benong further west.

Weaning may begin as early as 3 months after birth (e.g. Katy in TS1), although it is not usually completed until about one year of age. It can be argued that the reproductive cycle is timed adaptively to ensure that weaning starts during a fruiting peak. The peak nutritional strain on the mother would fall during the flowering peak, some 3-4 months after the infant's birth, when its increasing size demands more milk, and when the mother has reached the peak of stress since conception, towards the end of lactation. The infant will then have an array of soft fruit pulps from which to choose, and the mother will be able to recoup her reserves from these abundant high-energy fruit.

McCann (1933) and Tilson (in press a) report a similar timing of reproductive cycles among hoolock gibbons in the more seasonal environment of Assam. Tilson also argues that cycles may be timed so that weaning coincides with the fruiting peak. The arguments relating to the timing of conception and weaning are not, of course, mutually exclusive, since the former refers to causation and the latter to function.

Sub-adults. The mechanisms whereby sub-adult gibbons split away from their parents, find a mate and establish a territory have been the subject of much theorising since Carpenter's (1940) initial study (e.g. Ellefson, 1974; Chivers, 1974; Brockelman *et al.*, 1973, 1974; Tenaza, 1975, MacKinnon and MacKinnon, 1977; Tilson, in press b). To date Kuala Lompat is the only site to yield long-term observations of known individuals, although Tilson (in press b) has been able to piece together some of the features of group dynamics during his and Tenaza's studies of Kloss gibbons on Siberut. In addition to the advantage of being able to trace the history of individuals over several years, it is important to obtain a reasonably large sample, so as to distinguish between the usual and unusual from the range of possible fates of sub-adults.

Even after 10 years at Kuala Lompat (1) the sample is too small, especially as most observations have been of males, (2) crucial steps have not been observed, (3) there are doubts concerning the ability of the riverine forest in which most observations have been made to support gibbons in the long-term, and, relating to this (4) the forest in which new groups formed have lacked resident groups, and there are doubts as to whether this situation is comparable to gaps that occur between groups of a population.

In the three cases of peripheralisation of a sub-adult described in the chronicle above - Sammy and Inky Siamang and Gordon Lar - the context of aggression against the sub-adult has mostly been food trees, especially rather small fruiting trees. This feature is likely to be universal among gibbons, since it has also been recorded in the agile (Gittins, 1979) and hoolock gibbons (Tilson, in press a), and in other lar gibbons (Carpenter, 1940; Ellefson, 1974); in one case observed by Whitten (1980), the sub-adult male Kloss gibbon drifted slowly and peacefully away from the group. Tilson (in press b) observed three sub-adult males and one female peripheralised during his study of Kloss gibbons, and concludes that maturing females are not excluded to the same extent as males. In almost all cases in siamang and lar gibbons it is the adult male who is aggressive towards the sub-adult male (young males are also evicted by adult males in *Presbytis johnii*, *P. entellus* and *Alouatta villosa*, in contrast to species where they leave of their own volition, Packer, 1979a). Carpenter (1940) suggested that the motivation for this aggression is the male's sexual jealousy of his maturing son, but observations in Ulu Gombak led Chivers (1974) to suggest that the adult male siamang treats female and male sub-adults alike, and that the adult female tends to ignore, and even avoid, both. Moreover, Inky was sometimes attacked by his mother, and once he mounted her without drawing any reaction from his father.

Nevertheless, events in TS1 in early 1970 are indicative of sexual jealousy, because the onset of mating between Murgatroyd and Fenella coincided with increased aggression towards Sammy and to his subsequent departure. The peripheralisation of Gordon from TG2, however, cannot be related to sexual activity in his parents. However much aggressive behaviour there may be between father and son, this is to some considerable extent balanced by friendly behaviour between them, in the form of grooming; this is indicative of the constructive aspect of parental investment to assist their progeny reach breeding status, to which further reference will be made below.

Whatever the reason for the male's aggression, the resultant disadvantage of being excluded from fruiting trees while the rest of the family pick the best fruit would be motivation enough for the sub-adult to leave. It is not the only motive, however, for there is also a clear endogenous socio-sexual urge reflected in solo singing. This is apparently performed only by male sub-adults, never by females. Sammy, Inky, Gordon and Stanley (assuming that he had recently left his parental group) all sang solos to varying degrees, as did unmated males in Ellefson's (1974) lar gibbons and in Tenaza's (1975) Kloss gibbons; the juvenile male of Gittins' DG1 group of agile gibbons did so. Immature siamang males also sing with their parents, as do juvenile and sub-adult females of all species; thereby they might still advertise their presence (Gittins, 1979), but it seems more likely that they are contributing to group cohesion and territorial advertisement. If the immature females were advertising their availability, one would expect them to sing solo as well; instead, sub-adult and lone adult females seem to roam silently around the forest, homing in on the songs of males (siamang: Sarah, Sheila, ?Julian's mate; lar: Sylvia; hoolock: Tilson, in press a; Kloss: Tenaza, 1975).

The only case of group formation reported at Kuala Lompat is that of GB, but the provenance of neither partner is known (although the subsequent relationship between Gilbert and Stanley may be indicative of father and son). For TS1a, the provenance of Sammy is known, but the actual formation was not observed. Both cases may be unusual, as mentioned above, because of the availability of space at the forest edge, perhaps of poor quality for gibbons. It may be more usual to have to force a gap between existing territories. Tenaza's (1975) and Tilson's (in press b) observations on Siberut suggest that this could involve parental help in some cases; Tenaza records the expansion of a Kloss gibbon territory, followed immediately by the increasing independence of the sub-adult male in the newly conquered part. Observations ended before it could be ascertained whether he actually occupied the new part in the end - alone or with a mate - but Tilson found that he had lost it and left by nine months later. To be incorporated into the parental help model, territorial expansion would have to precede aggression resulting in the budding-off of the sub-adult's territory and the contraction of the parental one. Tilson (in press b) reports that a sub-adult male Kloss gibbon was aided by his parents in setting up a territory 150 m from their's, into which he later attracted a female. This was not observed for Sammy at Kuala Lompat; indeed, TS1 clearly tried to limit, and even disrupt, the territory of their offspring Sammy (TS1a) in 1970-71 and in 1972 (Aldrich-Blake and Chivers, 1973; Chivers *et al.*, 1975). Chivers *et al.* argue that such social factors may be even more important than botanical ones for the failure of TS1a, although incompatibility between male and female may have been the decisive factor, especially as it is also suggested that the very loud calls and small territories of siamang create buffer zones into which the group can range in times of food scarcity, and which also afford opportunities for parental help in establishing territories for their progeny. In the case of Gordon, TG2 did contract their territory westwards some 300 metres after the formation of GB, expanding back after its demise, but the contraction was not very amicable and there were other complications.

The third main option open to a sub-adult, apart from occupying empty forest or forcing a gap between empty territories, is to replace an ageing animal, particularly the same-sex parent (Brockelman *et al.*, 1974). For example, a son might replace his father and live (and mate) with his mother. Because of inbreeding depression. incest is not a sound policy and it could only operate for a son maturing towards the end of his father's lifetime. Gittins (1979:281) argues that there would be strong selection against such a move, because the young male would soon have to pair off with another female when his mother died, a lengthy process which would shorten his potential reproductive span. It can be argued more cogently, however, that such a loss in reproduction would be small compared with the saving in cost of establishing a territory, to which a young female could be attracted when the mother died. Population dynamics indicate that high mortality must occur among sub-adults without territories.

Tilson (in press b) observed a sub-adult male Kloss gibbon mate with what was presumably his mother after the father died; the son remained with the mother and she bore his offspring 15 months after the father's death. This incestuous mechanism may have operated twice at Kuala Lompat, in each case when the father was old. In both cases, however, confirmation must await further observation (Bennett *et al.*, in prep). In TS1 Inky has replaced Murgatroyd on his death, and may produce an infant with Fenella; the departure of Otis and Katy is curious, and more indicative of an alien male, unless one thinks in terms of the most selfish of genes! The second case concerns the return of Gordon to TG2 and the peripheral status of Gilbert. Whether or not he did or will mate with his mother will only slightly affect the number of offspring he can sire, but gaining a territory to which he can later attract a younger female by deposing his father in his weakness is a real advantage. Of considerable intrigue is whether that young female could be his sister Gail, who is already present and with whom he already duets.

Availability of space and unmated individuals will be an important determinant of which alternative will be chosen, and such factors might help explain the differences observed at different gibbon study sites.

Slightly altered, this mechanism of incest could be a case of kin selection. If the ageing father were to abdicate voluntarily in favour of his son, then he might thereby increase his inclusive fitness, and success need no longer be oedipal. Evidence for such a kin selection mechanism is weakened by Gilbert's failure to abdicate graciously, but this could have resulted from the aggressive alien contenders for the territory.

Pair formation and group stability. An important part in the establishment of a new group is the formation of the pair bond. The saga of TS1a, the 2½-year gap between pair formation and the first birth in GB and the protracted imbroglios of TG2 from 1976 for at least 4 years, indicate the difficulties which gibbons experience in forming a stable bond. The observation that Sammy mated with the old female Sheila, but not with the young Sarah, conforms with the idea that the absence of peers and strangers during a gibbon's maturation might be a source of difficulty (see Chapter 10). Sheila was presumably an experienced widow guiding the virginal Sammy; Sylvia appeared much older than Stanley in GB. The disadvantages of a much older wife have been discussed above. Difficulties in pair formation may have their advantages in the long run, since a careful choice of mate may be adaptive in a species with relatively stable monogamy.

Although the female usually plays little active part in inter-group conflicts in Malayan gibbons, providing mainly vocal and moral support, at least in the smaller species, exceptions have been revealed at Kuala Lompat. First, Sylvia participated in conflicts against TG2 when she and Stanley were establishing a territory in 1973. It was not a case of Stanley establishing a territory into which she was then invited; the fighting only began when they duetted. At Ketambe in northern Sumatra, a lone male siamang was tolerated by neighbouring groups until joined by a female; conflicts occurred until the female left (MacKinnon, pers. comm.). A pair of gibbons clearly pose a more significant threat than a solitary individual. By contrast, Tilson (in press b) observed that a Kloss gibbon male first establishes a territory and is then joined by a female. Second, Gertie and Gail appear to have been actively involved in conflicts with alien gibbons during the August 1976 debacle. Thus female lar gibbons may take part in conflicts at certain crises in their lives, but refrain from so doing in routine conflicts between established groups. There is doubtless also an element of individual variation; Sylvia may just be an unusually aggressive female. In Kloss gibbons each partner appears usually to defend the pair's territory only against members of the same sex, so that females may be more involved in inter-group conflicts than in other species, although only against other females (Tenaza, 1975). Tilson (in press b) reports that a widowed female Kloss gibbon with offspring defended their territory against neighbouring males, and Srikosamatara (pers. comm.) observed a widowed female pileated gibbon hold her territory for 6 months before disappearing.

The gibbon social system was portrayed in Chapter 3 as being based on permanent monogamous groups. The reader may be a little sceptical of this after reading this social history of gibbons at

Kuala Lompat. We would still argue, however, that gibbon family groups are stable for most of the time. TS1 and TG2 were stable from 1968 to 1976, and probably for at least 8 years before 1968 (based on the oldest offspring), RS2 in Ulu Sempam flourished until 1972, and GS16B in Ulu Gombak until 1977 at least, until disrupted by habitat destruction. The breakdowns that occurred at Kuala Lompat involved adults that were around 25 years of age at least; the longevity record for a captive gibbon is 31½ years (Napier and Napier, 1967). Various observers have commented on their aged appearance, especially Murgatroyd, from as early as 1970; he avoided leaping gaps whenever possible in 1975-76. Thus these two groups may simultaneously have reached a crisis that occurs only once every 10-20 years, when one or both parents become vulnerable to conspecifics or sicken and die.

Leaf monkeys

In comparison with the gibbons there is very little that can be said about long-term changes in the behaviour of monkeys at Kuala Lompat. Some changes with time have been detected, however, among the leaf monkeys, which help to explain apparent contradictions between the observations of Sheila Curtin in 1970-71 and those of the MacKinnons in 1973. These are distinct from any differences attributable to sampling method, intensity of samples and the pattern of sampling over the year (see Chapter 10); they relate to long-term fluctuations in food supply. In Ulu Gombak during the 1960s Medway (1972) demonstrated that in his sample of trees about 40% fruited in most years, but in certain years (referred to above as bumper fruiting years) about 80% may fruit; these peaks result from the coincidence of fruiting in most trees of different cycles - short-term endogenous, annual, biennial, long-term sporadic.

The fluctuations in food supply are widespread over at least the central part of the Peninsula, and are reflected in the abundance of wild or domestic fruit consumed by man, such as durian, rambutan, mangosteen, and preceded by the conspicuous flowering of the dipterocarps during a prolonged dry spell. In recent times years of such distinction have been 1963, 1968 and 1976. D.J.C. started his study in 1968 and towards the end of the survey and during the start of intensive study in Ulu Sempam and at Kuala Lompat, with continuing road-side observations in Ulu Gombak, infant monkeys and apes were very conspicuous; there might be 2-3 orange infants in groups of dusky leaf monkeys, whereas subsequently one rarely saw more than one, as Sheila Curtin discovered. Thus 1969 marked a period of increase in group size, and Curtin was essentially observing the maturation of these groups. By 1973 these groups were top-heavy with mature or nearly mature-animals, and the MacKinnons were observing their fragmentation, quite aggressive at times, into more, smaller groups. This seems the most likely explanation for their observations of groups two-thirds of the size recorded by Curtin, and for their rather smaller home ranges.

Thus, the cycle of group expansion restarted, slowly at first but gaining momentum after the 1976 glut of food with another peak of births early in 1977 (fig. 8.14). The scores for young infant dusky leaf monkeys are taken as indicative of this long-term pattern, but they should be viewed with caution: the 1969-70 sightings were obtained while following siamang from dawn until dusk for up to 20 days each month at Kuala Lompat and in Ulu Sempam; those for 1970-71 relate only to births at Kuala Lompat; orange infants were seen only occasionally in the 1972-76 observation periods: John Payne saw virtually no young infants until 1977. There are indications that mating coincides with the peaks of rainfall that may or may not follow prolonged droughts, with many more births several months after the marked droughts, which are followed by great abundance of fruit. While gibbons seem to respond initially to an abundance of flowers and then to fruit, changes in leaf monkey behaviour seem at times a response to peaks of leaf production, although the major changes described here clearly relate to the occasional years of super-abundance of fruit.

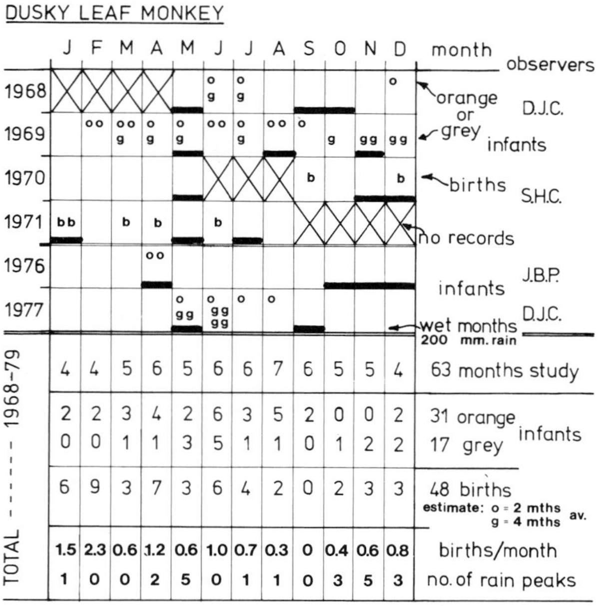

DUSKY LEAF MONKEY

	J	F	M	A	M	J	J	A	S	O	N	D	month / observers
1968						o g	o g					o	orange or grey infants; D.J.C.
1969		oo	oo g	o g	o g	oo	o g	oo	o	g	gg	gg	
1970									b			b	births; S.H.C.
1971	bb		b	b		b							no records
1976				oo									infants; J.B.P.
1977					o gg	o gg gg	o	o					D.J.C.; wet months 200 mm. rain
TOTAL 1968-79	4	4	5	6	5	6	6	7	6	5	5	4	63 months study
	2	2	3	4	2	6	3	5	2	0	0	2	31 orange infants
	0	0	1	1	3	5	1	1	0	1	2	2	17 grey infants
	6	9	3	7	3	6	4	2	0	2	3	3	48 births; estimate: o = 2 mths, g = 4 mths av.
	1.5	2.3	0.6	1.2	0.6	1.0	0.7	0.3	0	0.4	0.6	0.8	births/month
	1	0	0	2	5	0	1	1	0	3	5	3	no. of rain peaks

Fig. 8.14. Sightings of young infant dusky leaf monkeys, and records of births, from 1969 to 1971 and from 1976 to 1977 at Kuala Lompat, indicating months of more than 200 mm rainfall and estimating pattern of births through the year.

Such long-term changes in behaviour, of which we seem to have clear indication and explanation, have a special significance for the usual short-term field studies of primates and the interpretation of their results.

Macaques

Unlike the gibbons and leaf monkeys, there is no indication of marked changes in the size and behaviour of groups of macaques during this 10-year period. While pig-tailed macaques were perhaps seen too infrequently to be sure of this, there has always been a group of long-tailed macaques of between 20 and 30 individuals around the Ranger Post, and a rather smaller group along the river south of TS1's territory (often ranging north into it). Perhaps their opportunistic behaviour (Chapters 6 and 10) renders them less dependent on the occasional bumper fruiting years, so that recruitment and loss of individuals is more steady, or at least closely balanced with the maintenance of group integrity (in terms of size and space). This behaviour in a stable environment should contrast with their behaviour in disturbed habitats, where they are reputed to be unusual among primates in their ease and speed of adaptation, which must involve a marked increase in recruitment rate when such novel situations present themselves.

MAINTENANCE BEHAVIOUR

This section relates only to siamang, from whom daily scores for measures of the activity period, ranging and feeding are available for TS1 for 487 days spread over 65 months between April 1969 and April 1979 inclusive, for consecutive periods of 2-14 days, usually 5 or 10 and once 30 (table 8.7; fig. 8.15). The measures available and the mean values in each of these 65 samples are given in Appendix II, where more dètailed plots, including the range of observations in each sample, are given for TS1 in comparison with TS1a for 1969-73. The weather has been discussed in an earlier section, but the mean daily scores for 15-min samples of sunshine, cloud cover and rainfall in each observation period, and the monthly amounts of rainfall, are given here for comparison with behaviour (fig. 8.15). With odd exceptions, such as April 1973, the weather is fairly consistent during the day-time observation periods, with a tendency for more sunshine early in the year and more cloud toward the end.

Activity period

The onset of daily activity is one of the least variable aspects of siamang behaviour. While they may stir from the sleeping positions at dawn, they usually leave the night sleeping tree just after sunrise, about 30 minutes later; they leave later in the cloudier and wetter months, but usually leave the night tree before

Table 8.7. Observations of siamang group TS1.

Year	Months	Days	Observers
1969	9	99	D.J. and S.T. Chivers
1970	5	49	D.J. and S.T. Chivers
1971	2	18	F.P.G. Aldrich-Blake
1972	4	20	D.J. Chivers, J.J. Raemaekers, J.G. Fleagle
1973	6	34	J.R. and K.S. MacKinnon, Kalang
1974	2	8	D.J. and S.T. Chivers
	4	48	J.J. Raemaekers, Kalang
1975	9	88	J.J. Raemaekers, Kalang
1976	3	30	J.J. Raemaekers, Kalang
	6	22	J.B. Payne, E.B.M. Barrett, Kalang
1977	6	25	Kalang, S.K. Sharma, D.J. Chivers
	4	17	A.G. Davies
1978	4	20	A.G. Davies, Kalang
1979	1	9	D.J. Chivers, Kalang
	65	487	

sunrise in the drier months at the start of the year (fig. 8.16). The cessation of daily activity is much more variable, thereby determining the length of the active day. Entering the night tree became later as TS1a came into existence, and then earlier after its disappearance. Thus we see steady increases in the length of the alert period, from about 9 to 11 hours, in 1970, 1974, early 1976 and to 10 hours in 1977, with abrupt reversions in 1973, 1974-75 and late 1976. The longer active days can be related to peaks of food abundance, and the shorter ones to food scarcity and social upheaval, as in late 1976 to early 1977 (the loss of the male and the birth of an infant).

Thus, there is evidence of longer term cycles in addition to the short-term fluctuations, which may or not be seasonal. Fluctuations in the clock time of leaving and entering night trees are reduced somewhat by plotting in relation to minutes before or after sunrise and minutes before sunset, respectively. The main feature concerns the length of the active period. The low level of activity in 1969, presumably following hyper-activity in the bumper fruiting year of 1968 (as evidenced by the dramatic calling heard in Ulu Gombak and other survey sites), was followed by a marked increase early in 1970 to a level apparently maintained until 1974, since when there has been a steady but variable decrease. Calling by TS1 has

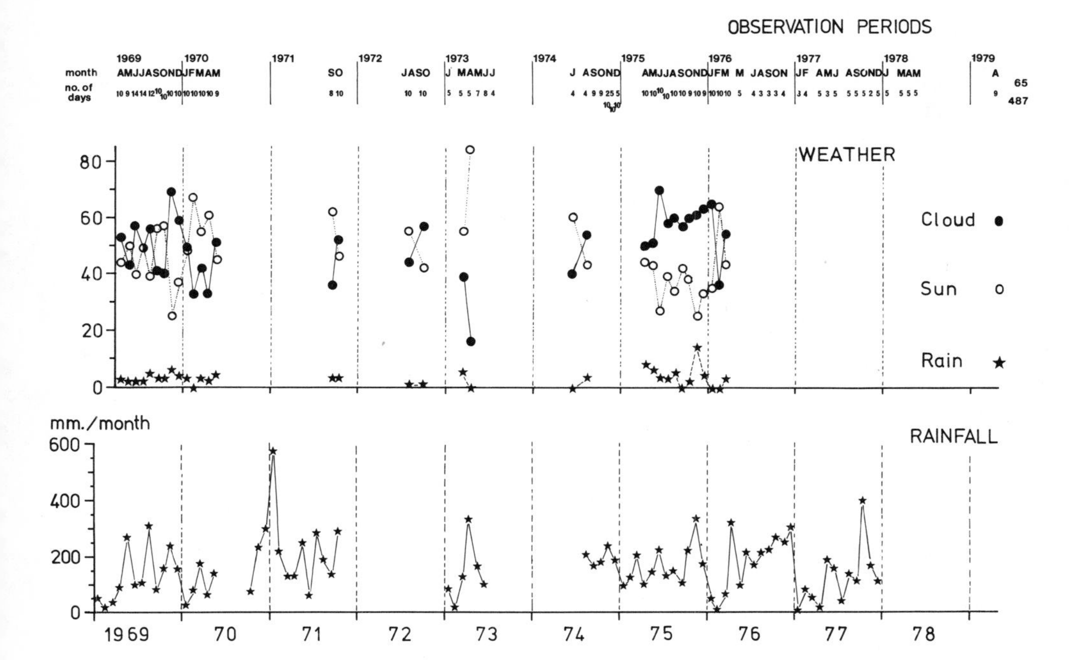

Fig. 8.15. Long-term quantitative observations of siamang behaviour at Kuala Lompat: sample size (days) in each month of observation of TS1 from dawn to dusk from 1969 to 1979, mean monthly score of sunshine, cloud cover and rain (from 15-min samples from 0600-1745 hours), and amount of rain in each month.

been greatest when the activity period is longest (fig. 8.16).

Ranging

Use of the home range has been recorded in terms of the total number, and number of different hectare quadrats entered daily on average in each sample, the mean day range and direct distance between night trees, the proportion of 10-min samples in different levels of the canopy, the proportion of the activity period spent travelling and the mean number of minutes/day for travelling in each observation period (fig. 8.17).

The areal and linear measures of day range are very highly correlated (Chivers, 1974). Day range increases in the middle months of every year, except 1969 (remarkably constant throughout) and 1975. At this time, TS1 annual range to and from the seasonally-swampy area (now dry) in the south-east of their home range on a 2-3 day cycle from their core area on slightly higher, drier ground; fruit seems to be more abundant at such times, especially in this south-

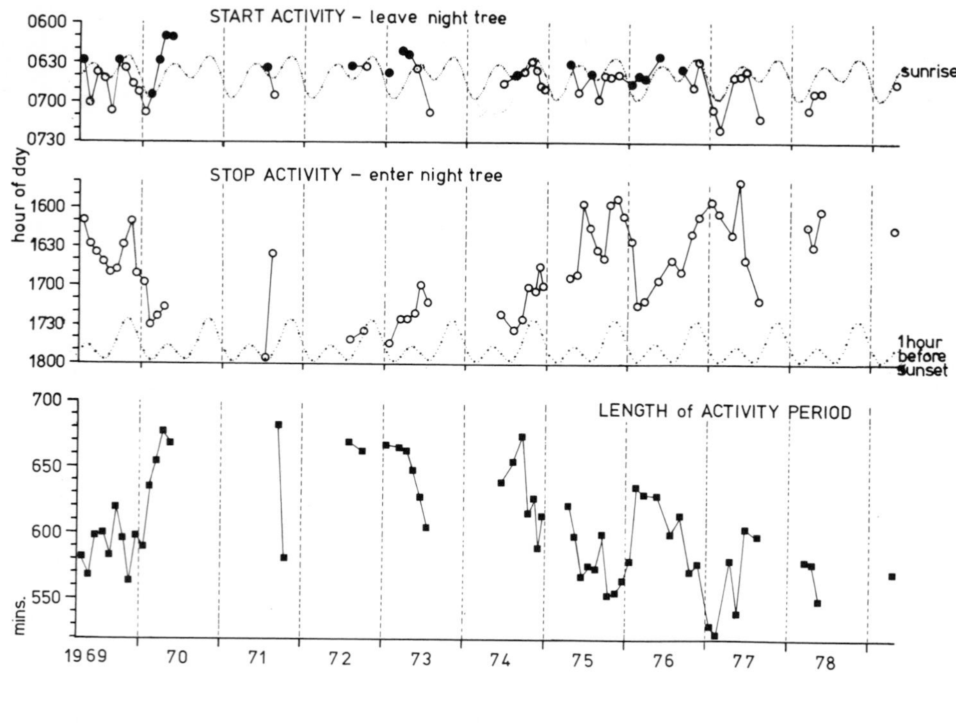
START ACTIVITY - leave night tree
sunrise
hour of day
STOP ACTIVITY - enter night tree
1hour before sunset
LENGTH of ACTIVITY PERIOD
mins.
1969
70
71
72
73
74
75
76
77
78

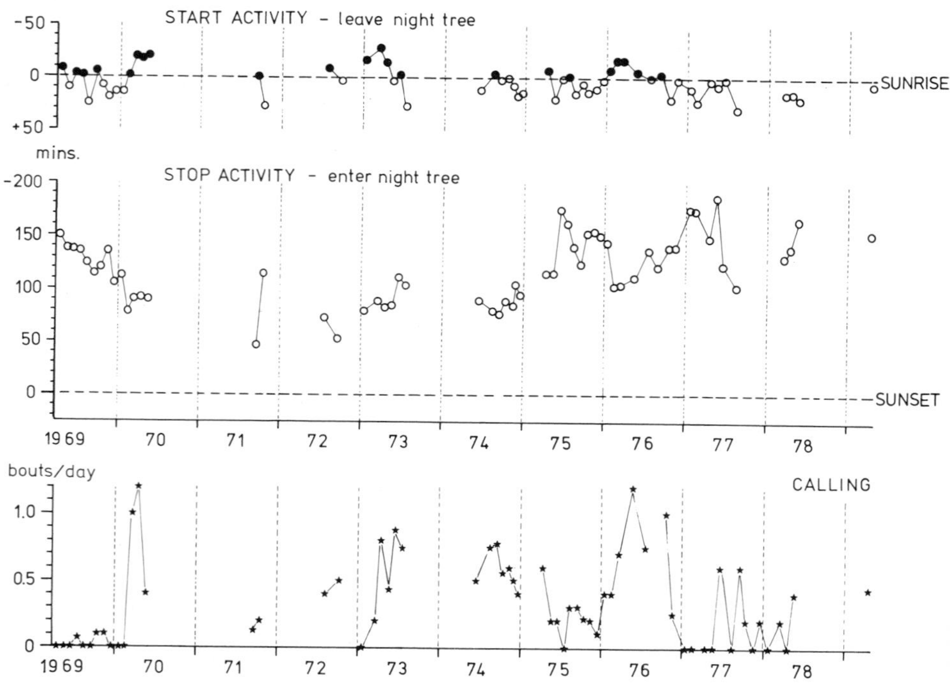
START ACTIVITY - leave night tree
SUNRISE
mins.
STOP ACTIVITY - enter night tree
SUNSET
bouts/day
CALLING
1969
70
71
72
73
74
75
76
77
78

east part.

Night position shift is much less variable, usually between 200 and 300 metres. Except in 1976, they change little with the marked increase in day range, indicating the increased circularity of these longer daily travels.

The siamang usually sleep in emergents, and spend most of the day in the main canopy layer of the forest, descending into understorey in the hottest part of the day when food is available there. Similarly, those months with unusually high activity in the emergents reflects abundant food at this level, most notably the annual leaf flush of *Koompassia excelsa* and fruiting figs on large dipterocarps.

While daily scores for travel and feeding fluctuate considerably between observation periods, they are much less variable when presented as proportions of the activity period: feeding for 40-50%, travel for 15-20% (fig. 8.17). Travel time does not vary as markedly as do sctivity period and day range, suggesting that travel is more rapid in these active times. The low level of travel and high level of feeding throughout 1969 reflect the increased intake of leaves following the scarcity of fruit following the bumper fruiting year of 1968 (see below). The conspicuous decrease in feeding in 1976 coincides with the disappearance of the adult male.

Feeding

To elaborate on the general points made above about feeding, mean daily intakes in each observation period of fruit and flowers, leaves, animal matter, with the fruit and leaves of *Ficus* species also shown separately, are presented as no. of mins/day and % of daily feeding time (fig. 8.18). Fruit consumption exceeds leaf-eating in 33 (57%) of the 58 observation periods for which data are available. The considerable fluctuation between samples reflects the considerable day-to-day variation that has been described by Chivers (1974), Chivers *et al.* (1975) and Raemaekers (1977). Nevertheless, mid-1973, late 1974, early 1975, 1976, early and late 1977 and early 1978 stand out as periods where fruit-eating predominates. While there are good reasons for believing that the unusual behaviour in 1976 relates to Murgatroyd's disappearance, it is possible that the low feeding time, mostly on fruit, relates, at least in part, to a super-abundance of highly nutritious foods. Observations were not sufficiently systematic or protracted during this period, how-

---

Fig. 8.16. Mean monthly scores for time of starting and stopping activity by siamang group TS1 (by clock time and from sunrise or sunset), for length of activity period, and for bouts of calling each day from 1969-79.

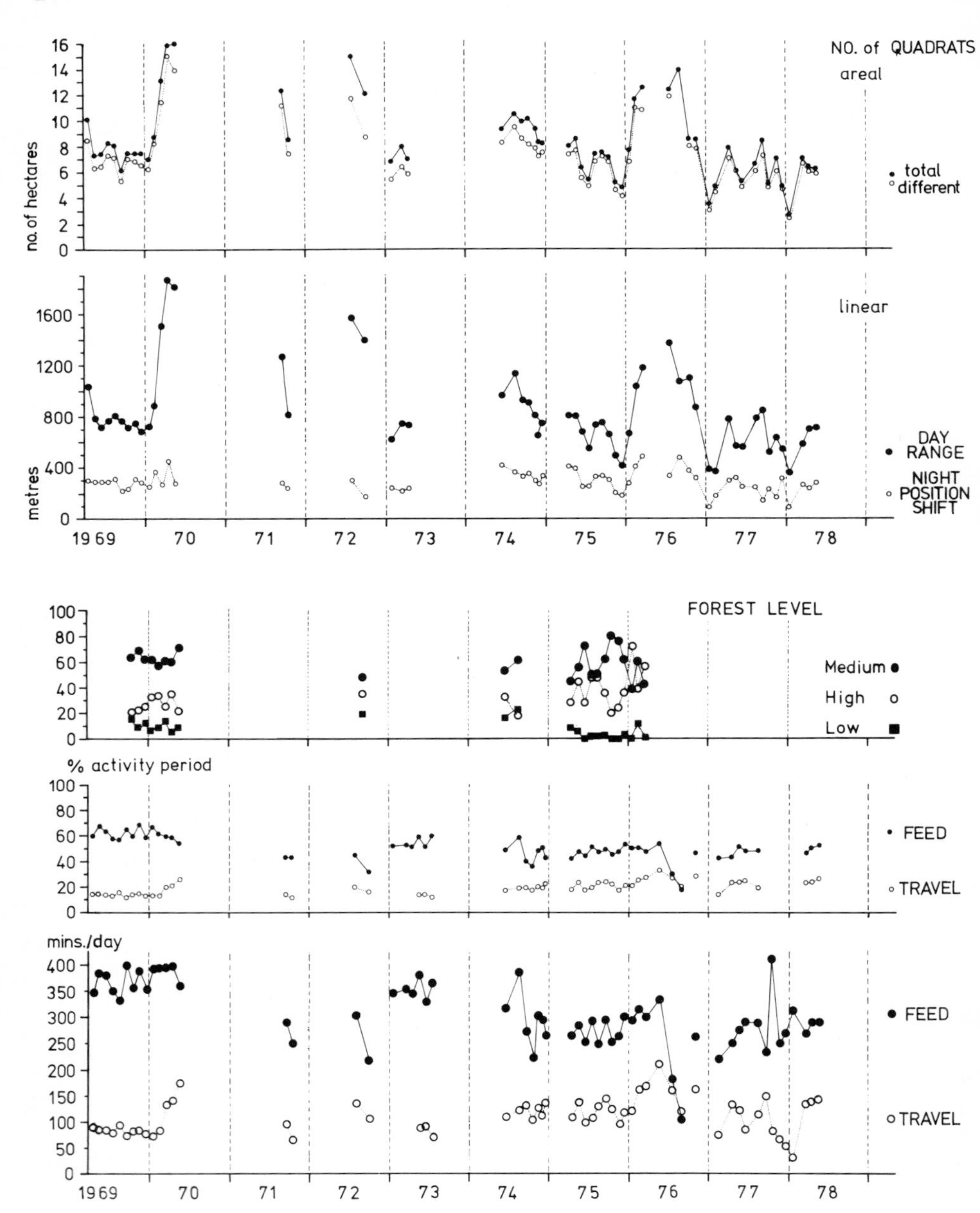

Fig. 8.17. Mean monthly scores for the number of quadrats entered each day by siamang group TS1 from 1969-79, for the mean length of day range and night position shift, for the proportion of time spent in each canopy level, and for the amount of travelling and feeding each day (as no. of mins and as % of activity period).

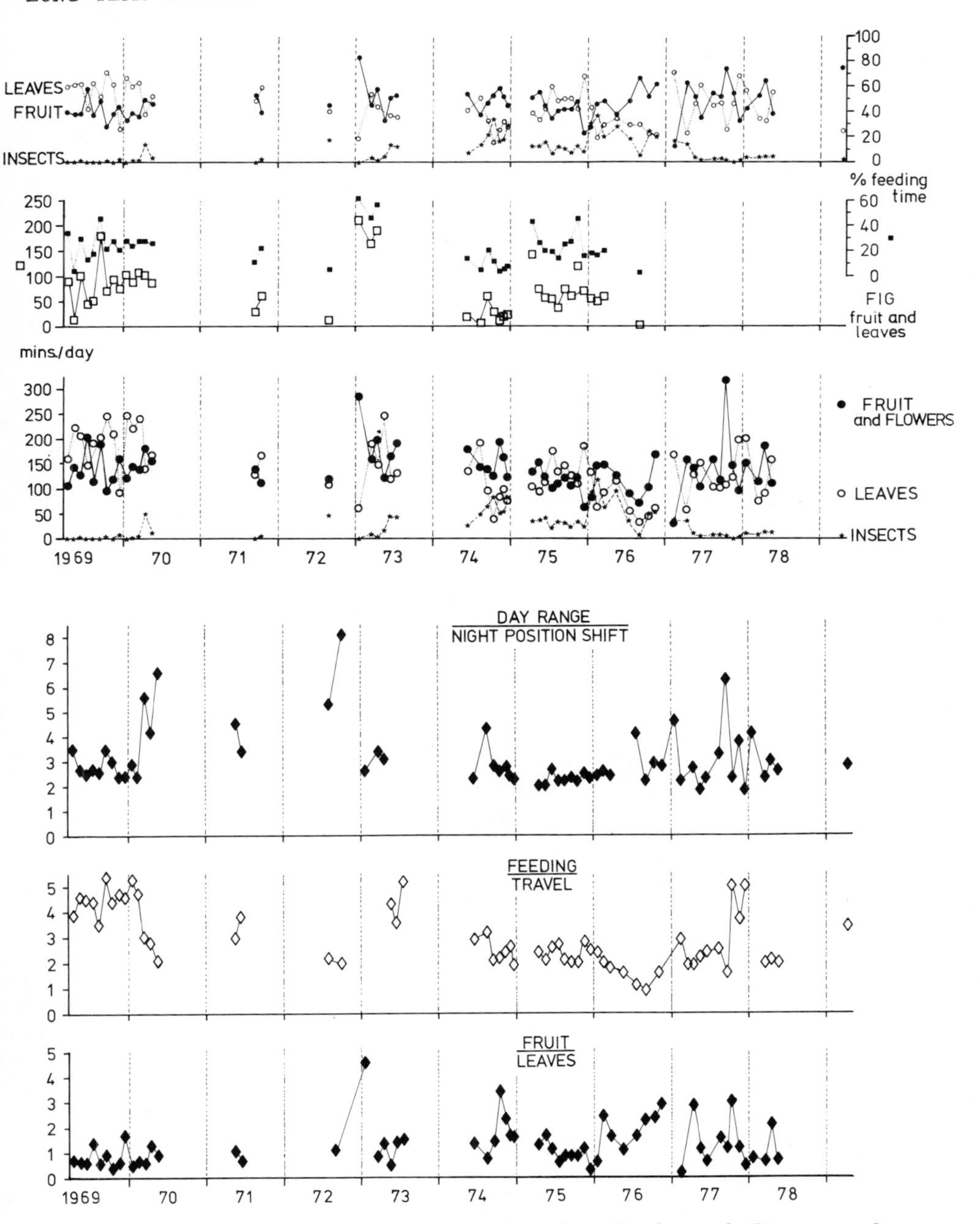

Fig. 8.18. Mean monthly scores for eating fruit and flowers, leaves and insects by siamang group TS1 from 1969-79, as no. of mins/day and as % of daily feeding time, with comparable scores for figs (above); monthly indices of ranging, activity and diet (below).

ever, to be sure of the explanation of the observed changes in behaviour.

Leaf consumption peaks in 1969-70, 1975 and early, middle and late 1977. The consumption of animal matter, mostly termites and caterpillars, persists at a low level throughout with peaks in mid-1970 and mid-1973, late 1975 and early 1976. Fig consumption is maintained at a high level throughout, at about 20% of feeding time, except for periods of occasional scarcity or glut.

Discussion

The calculation of certain ratios highlights certain behavioural patterns mentioned above (fig. 8.18). Day range is usually 2-3 times the length of night position shift; there are five periods when the ratio reaches 4, and three when it reaches or exceeds. In 1970 the disparately long day range follows a relatively quiescent period in TS1's life, in 1972 it coincides with a peak of border conflicts with TS1a, and the brief peak in 1977 has no obvious explanation.

The relation between time spent feeding and travelling seems to operate on two levels. Apart from the depression to equality during the unusual circumstances of 1976, the amount of time spent feeding is usually 2-3 times that spent travelling. There are three periods when it rises to 4-5 times - through 1969, mid-1973 and late 1977. While travel time was slightly depressed, especially in 1977, feeding time was increased - on leaves in 1969 and on animal matter in 1973. The ratio between the consumption of fruits and leaves usually fluctuates around equality, with no obvious pattern other than that described above.

It is too early to present a detailed multi-variate analysis of the daily scores summarised above. This is clearly essential for a full description and explanation of the relationships mentioned, and such an analysis is in progress. For the time being we can only report the results of a Principal Components Analysis conducted by Yarrow Robertson on the sample means of the variables described above. This showed that 51% of the variance is explained by the first two components and 71% by the first four. The first principal component, explaining 29% of the variance, separates out major climatic variables from those concerned with behaviour, confirming our belief that weather is a direct major influence on ranging and feeding (fig. 8.19). The second principal component separates travelling from feeding variables, refuting suggestions that the choice between more fruit or more leaves is of major importance, although the amount of animal matter is separated from amounts of plant matter by this second major component.

Analysis beyond this is tentative, since several of the

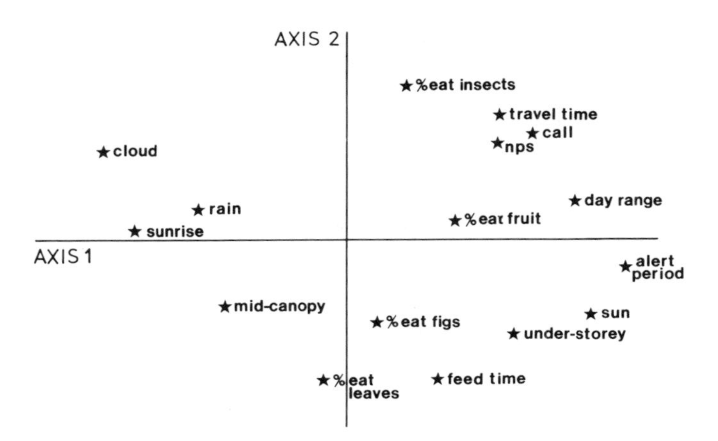

Fig. 8.19. The first two coordinates of a principal components analysis of the monthly means of some of the variables measured in the study of the behaviour of siamang group TS1 during 487 days in 65 months from 1969 to 1979.

variables need to be transformed to make them more normal (less skewed, less kurtosis), but the major component should still hold due to the robustness of Principal Components Analysis to slight deviations from normality. Several other variables are important to understanding the relations between ranging, feeding and weather, such as the indices derived last, but their highly skewed nature may prevent their use. Seasonality in the forest is hard to assess from measures of feeding behaviour (Chapter 2), especially when primates such as siamang seem to be ironing out environmental fluctuations in food production through their food selection (Chivers *et al.*, 1975)....and yet, some changes in behaviour such as breeding seem to relate to changes in the forest as a whole. With leaves almost as important as fruit in their diet, more data from measures independent of forest activity are needed (as used in Chapter 2), especially as the weather variables used are not direct measures of seasonality in the forest, and as there are great cyclical variations between years when measured over a period as short as 10 years.

An independent measure of forest activity is available for a shorter 12-month sample (Raemaekers, in press). In a simple correlation analysis weather again emerges as a major determinant of the monthly median of day range length, which is negatively correlated with rainfall. Day range length is equally strongly, but positively, related to the combined abundance of all gibbon food types (figs, other fruit, flowers, young leaves of species eaten by gibbons) weighted for the gibbon's selection of each type. Partial correlation shows that these two straight correlations are boosted by coincidental correlation between rainfall and food abundance in the same month, but are still high when this is held constant. In line with the results concerning feeding and travelling in the multi-variate analysis, no strong relationship emerges between route length and the abundance of any one food type. Neither is there any relationship between feeding time and food abundance.

COMPETITORS

John B. Payne

Sub-department of Veterinary Anatomy
University of Cambridge

INTRODUCTION

Most of the research carried out at Kuala Lompat has been concerned with the primate community. The importance of the trees in which primates live, and on which they feed, has been emphasized. The main features of the forest were described in Chapter 2, and the continual reference to plant species supplying food in subsequent chapters emphasizes further the importance of studying plants as well as animals. The main foods eaten by primates have been classified as fruits, leaves and animal matter (mainly arthropods). Various birds and other mammals share foods eaten by primates. Since competition is assumed to occur when the availability of a food resource required by one species is reduced by the activity of another, any partially or wholly arboreal bird or mammal taking food consumed by primates may be regarded as potential competitors.

Arthropods, in varying proportions, figure in the diets of most arboreal birds and mammals of the Malayan rain-forest. Not many mammals, and no birds take leaves as a major dietary component. Apart from the monkeys and apes, only the colugo (*Cynocephalus variegatus*) and some of the flying squirrels (sub-family Petauristinae) are significantly folivorous (Medway, 1969; Muul and Lim, 1978), but little is known of their ranging behaviour and food preferences. Fruit is an important food for many arboreal birds and mammals; it is likely that the ranging patterns of some species are determined mainly by the dispersion of fruit sources. This is thought to be so for the smaller gibbons (Chapter 3), the frugivorous pigeons, the giant squirrels (*Ratufa* spp.; Payne, 1979b) and probably the common palm civet (Bartels, 1964).

This chapter is concerned with the animals which compete with primates for food in the Malayan rain-forest. Since little is known either of the food species of nocturnal folivores or of the arthropod component of primate diets, attention is focused on fruit-eating animals, and the kinds of fruit that they eat. Some of the main characteristics of various taxa of fruit-eating animals are discussed, with particular emphasis on the arboreal and diurnal tree squirrels. As in Chapter 2, an important distinction is made between those animals which eat or otherwise destroy seeds, and those which eat only the pulp or rind around the seed, thus acting as seed dispersers. While both kinds of animals are frugivores, those which generally eat seeds or drop them beneath the parent plant are named seed destroyers, and those which generally eat fruit pulp and excrete or spit out the seeds are named seed dispersers.

BIRDS

The ability to fly aids access to widely-spaced food sources. Since plant species occur at low density in tropical rain-forest, many foods, especially fruit, are widely scattered and rare. Perhaps for these reasons, the most specialised frugivores have evolved among flying animals: medium to large birds and bats. Species of bird identified at Kuala Lompat, which are known or believed to be partially or wholly frugivorous are listed in Table 9.1; records of food items for these birds are rare.

Pigeons are believed to be exclusively frugivorous (Medway and Wells, 1976), but they concentrate on "unspecialised" fruit, such as those of species of *Ficus, Eugenia, Calophyllum* and Euphorbiaceae, whose seeds the pigeons disperse. The parrots and parakeets are fruit specialists, probably unique among families of Malayan forest birds in being primarily seed destroyers. All these birds range widely for food; there are no highly specialised exclusive frugivores in South-east Asia, such as occur in Neotropical forests (Snow, 1976), and which have a close co-evolutionary relationship with certain tree taxa.

The hornbills are most often observed feeding in strangling figs, but the kinds of fruit which have evolved more specifically for dispersal by hornbills bear dry nutritious pulp with a large seed. The advantage to the tree of a large seed is that the large food store contained therein increases the changes of successful germination under conditions of low light intensity and, often, poor soil.

There are big differences in body size between the common species of taxonomically-related birds which occur at Kuala Lompat (King *et al.*, 1975). This allows specialisation on fruits of different sizes and on feeding in different parts of tree crowns

Table 9.1. Species of birds known to occur at Kuala Lompat, which are known or believed to be partially or wholly frugivorous. Sources of records: MW, Medway and Wells (1971); M, MacKinnon (unpublished data).

Family and Species Names	English Name	Source
PHASIANIDAE (terrestrial feeders)		
Rollulus roulroul	Crested wood partridge	MW
Lophura erythropthalma	Crestless fireback pheasant	MW
L. tignita	Crested fireback pheasant	M
Polyplectron malaccense	Malayan peacock pheasant	MW
Argusianus argus	Great argus pheasant	MW
COLUMBIDAE		
Treron capellei	Larger thick-billed pigeon	MW
T. curvirostra	Lesser thick-billed pigeon	MW
T. olax	Little green pigeon	MW
T. vernans	Pink-necked pigeon	M
T. bicincta	Orange-breasted pigeon	M
Ptilinopus jambu	Jambu fruit pigeon	MW
Ducula aenea	Green Imperial pigeon	M
D. badia	Mountain Imperial pigeon	M
Chalcophaps indica	Emerald dove	MW
PSITTACIDAE		
Psittinus cyanurus	Blue-naped parrot	MW
Loriculus galgulus	Blue-crowned hanging parakeet	MW
BUCEROTIDAE		
Berenicornis comatus	White-crowned hornbill	M
Anorrhinus galeritus	Bushy-crested hornbill	MW
Anthracoceros malayanus	Black hornbill	MW
A. convexus	Pied hornbill	MW
Rhyticeros undulatus	Wreathed hornbill	M
Buceros rhinoceros	Rhinoceros hornbill	MW
Rhinoplax vigil	Helmeted hornbill	MW
CAPITONIDAE		
Megalaima chrysopogon	Gold-whiskered barbet	MW
M. mystacophanos	Gaudy barbet	M
M. henricii	Yellow-crowned barbet	MW
M. australis	Little barbet	MW
Calorhamphus fuliginosus	Brown barbet	MW

Continued/.....

...../Table 9.1 (continued)

Family and Species Name	English Name	Source
PYCNONOTIDAE		
Pycnonotus melanoleucos	Black-and-white bulbul	MW
P. atriceps	Black-headed bulbul	MW
P. melanicterus	Black-crested bulbul	M
P. cyaniventris	Grey-bellied bulbul	M
P. zeylanicus	Yellow-crowned bulbul	MW
P. finlaysoni	Stripe-throated bulbul	MW
P. goiavier	Yellow-vented bulbul	M
P. plumosus	Olive-winged bulbul	M
P. brunneus	Red-eyed bulbul	MW
P. simplex	White-eyed brown bulbul	MW
P. erythropthalmus	Spectacles bulbul	M
P. eutilotus	Crested brown bulbul	MW
Criniger bres	Grey-cheeked bulbul	MW
C. finschii	Finsch's bulbul	MW
C. phaeocephalus	White-throated bulbul	MW
Hypsipetes criniger	Hairy-backed bulbul	MW
H. viridescens	Crested olive bulbul	MW
AEGITHINIDAE		
Aegithinia viridissima	Green iora	MW
Chloropsis cyanopogon	Lesser green leafbird	MW
C. sonnerati	Greater green leafbird	M
C. cochinchinensis	Blue-winged leafbird	MW
Irena puella	Fairy bluebird	MW
ORIOLIDAE		
Oriolus xanthornus	Black-headed oriole	MW
CORVIDAE		
Platylophus galericulatus	Crested jay	MW
P. leucopterus	Black crested magpie	MW
Corvus enca	Slender-billed crow	MW
C. macrorhynchus	Large-billed crow	MW
DICAEIDAE		
Dicaeum cruentatum	Scarlet-backed flowerpecker	M
D. sanguinolentum	Fire-breasted flowerpecker	M
D. trigonostigma	Orange-bellied flowerpecker	M
Prionochilus percussus	Crimson-breasted flowerpecker	M
P. maculatus	Yellow-throated flowerpecker	MW

Continued/.....

...../Table 9.1 (continued)

Family and Species Name	English Name	Source
STURNIDAE		
Gracula religiosa	Hill myna	MW
TIMALIDAE		
Pomatorhinus montanus	Chestnut-backed scimitar babbler	
CAMPEPHAGIDAE		
Pericrocotus divaricatus	Ashy minivet	MW
P. flammeus	Scarlet minivet	MW

(Medway and Wells, 1976). The green pigeons, *Treron* spp., forage and feed in flocks, covering great distances. Many food species are shared by different bird species, but the larger pigeons can eat larger fruits, while the smaller ones can feed from smaller supports. The jambu fruit pigeon and imperial pigeon were seen rarely at Kuala Lompat; they are known to cover great distances to feed, even in one day. Unlike the *Treron* spp., they do not seem to take many figs, and need to cover wide areas to find other ripe fruits of the upper canopy. The emerald dove feeds mainly on the ground and in small trees; at Kuala Lompat it took fruit from several species of Euphorbiaceae.

Among the hornbills, the three large species (wreathed, rhinoceros and helmeted) forage by flying over the forest, settling to feed only in tall trees. The black and pied hornbills resemble each other in body size and individual spacing. At Kuala Lompat, both species appeared to live in family groups of two adults and offspring (more than one generation), foraging in overlapping home ranges, which were smaller than those of other hornbills. Pied hornbills were absent from forest away from rivers, while the black hornbills were most abundant where riverine forest gives way to "inland" forest. The bushy-crested hornbill foraged at all canopy levels in larger, more closely-knit groups within larger home ranges than the black or pied. The white-crowned hornbill was seen either alone or paired, with a single offspring, foraging almost entirely in the lower canopy levels; it was the only hornbill never seen eating figs at Kuala Lompat. All species were seen searching for, and occasionally eating, animal food, including small vertebrates.

Table 9.2. Species of Malayan mammals known to occur at Kuala Lompat which are partially or entirely frugivorous.

Family and Species	English Name	Feeding Habits Noct.	Diur.	Arb.	Terr.
PTEROPODIDAE					
Cynopterus brachyotis	Malaysian fruit bat	N		A	
C. horsfieldi	Horsfield's fruit bat	N		A	
Peuthetor lucasi	Dusky fruit bat	N		A	
Balionycteris maculata	Spotted-winged fruit bat	N		A	
Macroglossus sp.	Long-tongued fruit bat	N		A	
Eonycteris spelaea	Cave fruit bat	N		A	
TUPAIIDAE					
Tupaia glis	Common tree shrew		D	A	T
T. minor	Lesser tree shrew		D	A	T
LORISIDAE					
Nycticebus coucang	Slow loris	N		A	
CERCOPITHECIDAE					
Presbytis obscura	Dusky leaf monkey		D	A	
P. melalophos	Banded leaf monkey		D	A	
Macaca fascicularis	Long-tailed macaque		D	A	
M. nemestrina	Pig-tailed macaque		D	A	T
HYLOBATIDAE					
Hylobates syndactylus	Siamang		D	A	
H. lar	Lar gibbon		D	A	
SCIURIDAE					
Ratufa bicolor	Black giant squirrel		D	A	
R. affinis	Pale giant squirrel		D	A	
Callosciurus prevostii	Prevost's squirrel		D	A	
C. notatus	Plantain squirrel		D	A	
C. caniceps	Black-banded squirrel		D	A	
Sundasciurus hippurus	Horse-tailed squirrel		D	A	
S. tenuis	Slender little squirrel		D	A	T
S. lowii	Low's little squirrel		D		T
Lariscus insignis	Three-striped ground squirrel		D		T
Petinomys sp.	Flying squirrel	N		A	
Iomys horsfieldis	Horsfield's flying squirrel	N		A	
Petaurista petaurista	Red giant flying squirrel	N		A	

Continued/.....

...../Table 9.2 (continued)

Family and Species	English Name	Feeding Habits Noct.	Diur.	Arb.	Terr.
MURIDAE					
Rattus tiomanicus	Malaysian wood rat	N			T
R. muelleri	Mueller's rat	N			T
R. bowersii	Bower's rat	N			T
R. cremoriventer	Dark-tailed tree rat	N		A	T
R. surifer	Red spiny rat	N			T
R. rajah	Brown spiny rat	N			T
R. whiteheadi	Whitehead's rat	N			T
R. sabanus	Long-tailed giant rat	N		A	T
HYSTRICIDAE					
Hystrix brachyura	Malayan porcupine	N			T
URISIDAE					
Helaractos malayanus	Malayan sun bear	N	D	A	T
VIVERRIDAE					
Viverra malaccensis	Little civet	N			T
Viverra sp.	Malay civet	N			T
Paradoxurus hermaphroditus	Common palm civet	N		A	
Paguma larvata	Masked palm civet	N		A	
Arctictis binturong	Binturong	N		A	T
Arctogalidia trivirgata	Three-striped palm civet	N		A	
Cynogale bennettii	Otter civet	N			T
ELEPHANTIDAE					
Elephas maximus	Asiatic elephant	N	D		T
TAPIRIDAE					
Tapirus indicus	Malayan tapir	N	D		T
SUIDAE					
Sus scrofa	Wild pig	N	D		T
TRAGULIDAE					
Tragulus javanicus	Lesser mouse deer	N	D		T
T. napu	Large mouse deer	N	D		T
CERVIDAE					
Muntiacus muntjak	Barking deer	N	D		T
BOVIDAE					
Bos gaurus	Gaur	N	D		T

MAMMALS

Mammalian frugivores at Kuala Lompat are nocturnal and diurnal, arboreal and terrestrial (table 9.2). Of the arboreal forms, only the primates and tree squirrels are diurnal, although the binturong sometimes feeds during the day.

Six species of fruit bat (Pteropodidae) have been identified at Kuala Lompat (Medway and Wells, 1971); all are believed to be seed dispersers. The large *Cynopterus* species have not been reported in the area during recent primate studies.

The tree shrews forage largely at tree fall sites, among low vegetation and on the ground, searching for arthropods and small fruits, but they have been seen entering strangling figs for fruit. They are most active just after dawn and before dusk.

The flying squirrels are all noctural feeders, although they often become active sometime before dusk, and may not return to their tree holes until after dawn. Muul and Lim (1978) found that the smaller species are mainly frugivorous, and the larger ones more folivorous; they found flying squirrels to be more abundant than indicated by collecting records, which come largely from shooting diurnal animals and trapping nocturnal ones. All are probably seed destroyers. The giant flying squirrels have been seen eating *Durio* fruit and acorns of *Lithocarpus*; the smaller species raid orchards for cultivated fruit.

The sun bear and *Viverra* civets seem to overlap little with primates in diet and habitat use. The nocturnal palm civets and binturong, however, take large amounts of fruit, and are agile, arboreal feeders. The palm civets seem to favour those species of fruit with rather small, hard seeds, surrounded by a thin layer of pulp; the pulp is digested and the seeds excreted, usually on exposed places such as tree falls, trails and landslips.

The large, terrestrial mammals feed on all manner of fallen fruit. The mouse deer are said, by Orang Asli (the indigenous people), to be largely frugivorous. Several individuals congregate at night to feed under large trees on fallen fruit, displaced by heavy exploitation during daylight.

The lowland forests of Malaya support eleven species of diurnal squirrel, two of which are ground-dwellers. All occur within the Kuala Lompat study area, but most observations have been made on the six most common and easily seen arboreal species: *Ratufa bicolor, R. affinis, Callsciurus prevosti, C. notatus, Sundasciurus tenuis, S. hippurus* (MacKinnon, 1978; Payne, 1979b) (fig. 9.1).

The two terrestrial species, *Lariscus insignis* and *Rhinosciurus*

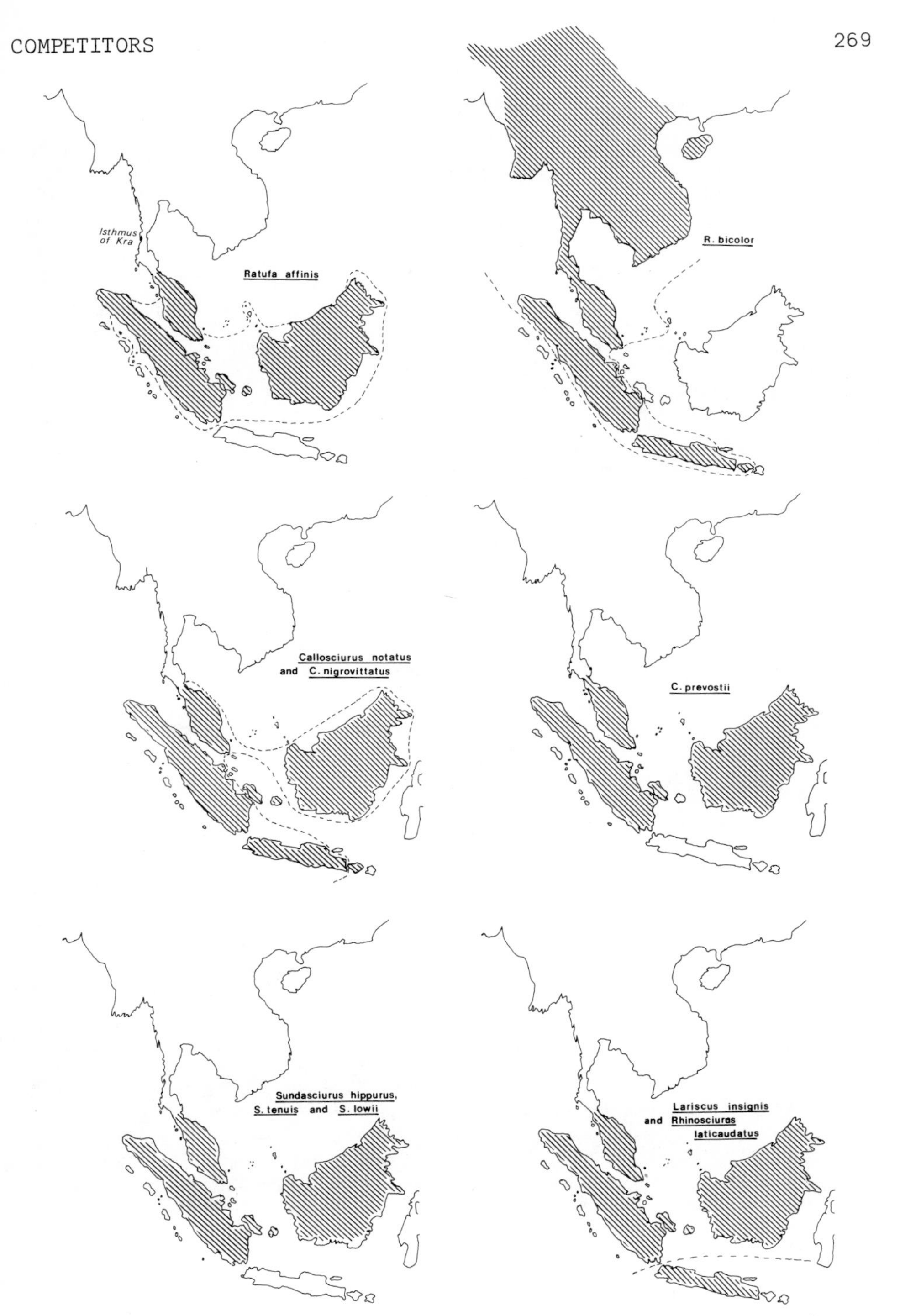

Fig. 9.1. South-east Asian distribution of Malayan tree squirrels

Plate XXI. Squirrels at Kuala Lompat (JBP).

Adult *Ratufa bicolor* resting in *Maranthes corymbosa* tree (above). A 5-month old *R. bicolor* in the usual feeding posture of *Ratufa* (inset). Adult *R. affinis* feeding on sap in an *Intsia palembanica* tree (below).

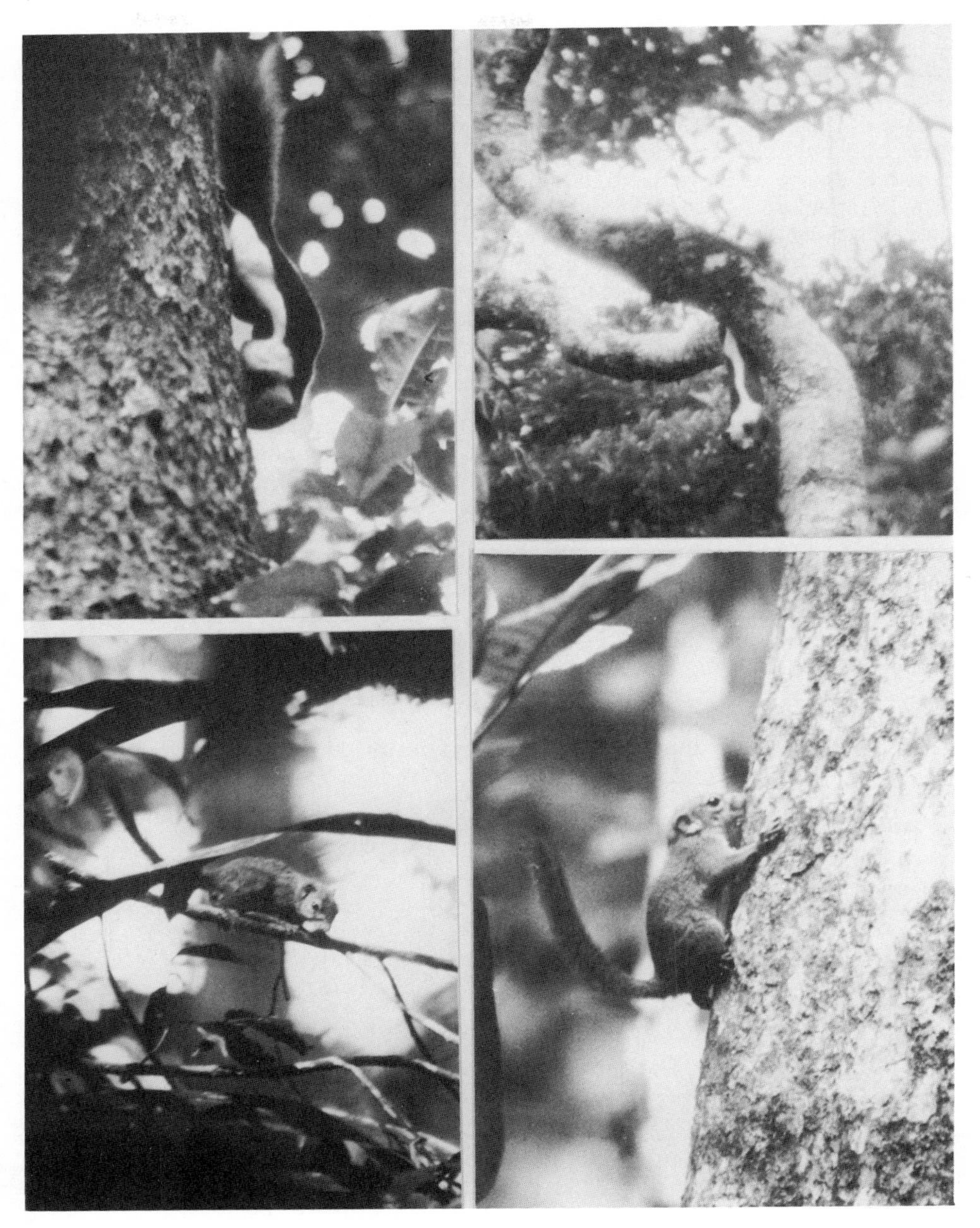

Callosciurus prevostii feeding on sap of a *Melanorrhoea malayana* tree (above left) and on fruit in a *Maranthes corymbosa* tree (above right). *Sundasciurus tenuis* travelling in a small tree (typical glimpse, lower left) and searching on the trunk of a *Baccaurea brevipes* tree (lower right).

laticaudatus, feed on the ground, on tree buttresses and among fallen trees and branches. A third species, *Sundasciurus lowii*, shares these habits at Kuala Lompat, but elsewhere it appears to be mainly arboreal, e.g. Kuala Tahan (pers. obs.) and Siberut Island (Whitten, 1979). The two most abundant species at Kuala Lompat, *Callosciurus notatus* and *S. tenuis*, live predominantly in in the lower canopy levels, the latter eating much sap and bark. Three species - *S. hippurus*, *C. nigrovittatus* and *C. caniceps* - were rarely seen in 1976-77 (Payne, 1979b), although the first was seen more often in 1973 (MacKinnon, 1978) and the latter two were not seen at all. *S. hippurus* spent most time in the lower canopy and on the ground. Only three species - the largest ones: *C. prevostii*, *Ratufa bicolor* and *R. affinis* - fed mainly in the middle and upper canopy levels (Plate XXI).

Comparing the use of canopy levels by primates and squirrels (fig. 9.2, cf. fig. 6.3), the primates are concentrated more into the upper layers than the squirrels. All the monkeys and apes, except *M. nemestrina*, spend more than 90% of total active and feeding times more than 8 m (26 ft) above the ground (MacKinnon and MacKinnon, 1978; Chapter 6). Although *M. nemestrina* is a terrestrial quadruped, travelling mainly on the ground, it usually climbs into the canopy to feed. If there is any competition for food resources between primates and squirrels, this will occur mainly in the upper canopy levels. Of the squirrels, only the two *Ratufa* species and *C. prevostii* habitually feed at the same levels as primates.

The composition of the diets of four diurnal tree squirrel

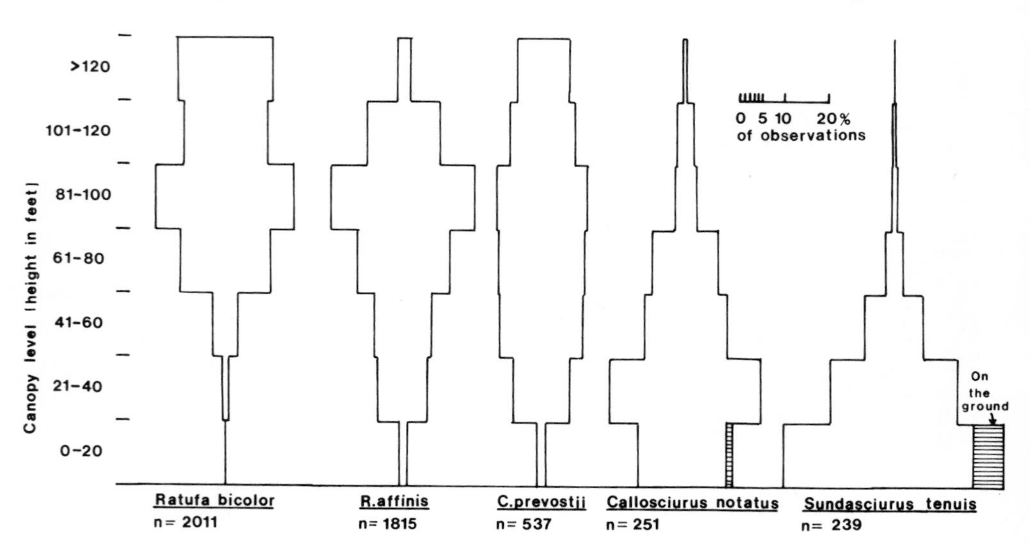

Fig. 9.2. Use of canopy levels by five species of tree squirrels at Kuala Lompat (spot and 5-min observations combined).

species - the three largest, middle-canopy dwellers and the commonest species, *C. notatus* (whose vertical range also overlaps with that of primates) are shown in fig. 9.3. Both *Callosciurus* species have a wider range of food types than either *Ratufa* or *Sundasciurus*; they eat the seeds of some fruit, the pulp of others, and both seed and pulp of yet others; they take both ripe and unripe fruit. Animal food seems to be equally important to *Callosciurus* because, although they consume less animal matter than fruit, most of their searching behaviour seems directed at finding arthropods rather than fruit. Flowers and the vegetative parts of trees and lianas (young leaves, bark and sap) are eaten as a minor part of the food intake, according to species and availability.

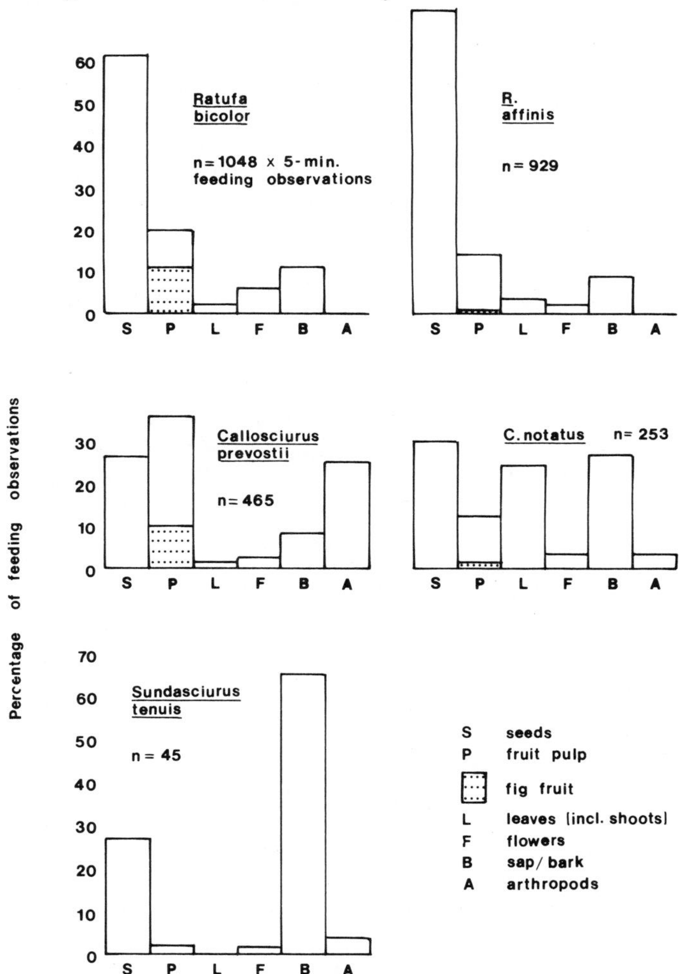

Fig. 9.3. Diet of Malayan tree squirrels: percentages of feeding observations for seven types of food.

Most of the food intake of *Ratufa* and *C. prevostii* is fruit. *Ratufa*, and probably *S. hippurus*, are essentially seed destroyers. As estimated by independent "spot observations" and 5-min. samples of continuous feeding bouts:

	% feeding time on fruit, composed of	seeds and	pulp
C. prevostii	62	26½	35½
R. bicolor	81	61	20
R. affinis	86	72	14.

The percentages given for seeds (see also fig. 9.3) include cases where both seeds and pulps were eaten, but in these cases it is believed that the seed was the main objective of the squirrel, since it is more difficult to extract than pulp.

Of a sample of 160 plant species which bore fruit during the study period 1976-77, and whose fruit were examined when ripe, 75 bore parts judged as edible by hylobatids and macaques; 85 bore seeds judged as edible by squirrels, but with no part edible to pulp-eaters. Wind dispersed seeds, including those of emergent dipterocarps and legumes, and figs were not included in these figures. Thus, since squirrels, especially *Ratufa* species, are seed-eaters, they have a wider range of fruit species available to them than do primates if they also eat pulp.

Unlike gibbons and macaques, leaf monkeys eat and digest seeds of both ripe and unripe fruit. Neither species of leaf monkey at Kuala Lompat, however, have been seen eating the seeds of certain species used heavily by squirrels, e.g. *Triomma malaccensis*, Burseraceae; *Combretum* sp., Combretaceae; *Mallotus leucodermis*, Euphorbiaceae; *Aquilaria malaccensis*, Thymelaceae; *Buchanania sessilifolia*, Anacardiaceae; *Lithocarpus sundaicus*, Fagaceae.

If fruit-eating records for *Ratufa* and *C. prevostii* are combined, the five most important plant families are:

Burseraceae	14.2%	of feeding records for fruit
Annonaceae	11.8	
Anacardiaceae	10.1	
Euphorbiaceae	9.9	
Moraceae	7.3	(6.8% for figs), a total of 53.3% for 18 months of observation.

Trees of the families Burseraceae and Euphorbiaceae are hardly used at all by primates for food. Both *Ratufa* species show a preference for small, oily seeds in large sources, whereas records of seed-eating by leaf monkeys reveal no such preference. Neither squirrels nor leaf monkeys take much mast fruit from trees of the families Dipterocarpaceae and Sterculiaceae. The predator satiation mechanis (Chapter 2), seems to be successful against arboreal mammalian seed destroyers, although other factors, such as fruit chemistry, may be involved.

The most obvious difference between the diets of squirrels and primates is the intake of significant quantities of leaves by the latter. MacKinnon and MacKinnon (1978) record the following percentages of feeding time on leaves, from their 6-month study (Chapter 6):

H. syndactylus 45.3% *P. obscura* 37.9% *M. fascicularis* 25.8%
H. lar 35.7% *P. melalophos* 38.2%

Both *H. syndactylus* and *P. obscura* may have annual scores around 60% (Chivers, 1974; Curtin, 1976a). Comparable scores for squirrels are:

R. bicolor 0.5% *R. affinis* 3.6% *C. prevostii* 1.7% (Payne, 1979b).

There is no evidence that Malayan tree squirrels ever cache seeds where they are likely to germinate. *Ratufa* do not store food, but there are observations to indicate that *Callosciurus* and *Sundasciurus* sometimes make short-term stores in trees. The co-evolutionary relationship between squirrels and trees in temperate forests of the northern hemisphere appears to have no equivalent among the squirrels of South-east Asia (Chapter 2).

Many tropical plants have evolved the characteristic of producing seeds coated with pulp attractive to animals. Because of the equable climate, ripe fruit are produced at all times of the year, and pulp-eating specialists have been able to evolve. Plants can increase their reproductive success by bearing fruit out of synchrony with similar plants. It is important also for tropical plants to supply the seed with food to start germination and growth, since light conditions are likely to be poor where the seed is deposited.

The availability of nutritious seeds throughout the year creates a niche for specialists in seed-eating. It might be expected that the kinds of plants which produce seeds synchronously for cache dispersal by squirrels in temperate climatic regions would not survive in tropical rain-forest. The fact that trees of the oak family, which produce hard fruit such as the acorns and nuts of the genera *Lithocarpus* and *Castanopsis*, are common throughout South-east Asia, suggests an alternative mode of dispersal. Evidence from African and American forests (Emmons, 1975; Smythe, 1970) indicate terrestrial rodents as the most likely dispersers of hard fruits. The availability of fruit is perhaps less certain on the ground than in the trees, and caches would be important to terrestrial seed destroyers.

The density of tree squirrels (table 9.3) is somewhat lower

Table 9.3. Estimates of the relative and absolute density of tree squirrels at Kuala Lompat, 1977.
Estimated density is of adult individuals in 100 ha.
Spot observations - collected April 1976 to September 1977.
Calls - (a) all calls heard, January-September 1977.
(b) calls heard before 1000 hrs and after 1500 hrs, same period.
Estimated biomass - average adult body weights, multiplied by estimated density.

Species	Estimated density (/100 ha)	Spot observations	Calls (a)	Calls (b)	Estimated biomass (kg/100 ha)
Ratufa bicolor	11	95	26	13	15.9
R. affinis	26	85	193	106	28.0
Callosciurus prevostii	38	303	195	164	17.3
C. notatus	244	263	69	55	55.4
C. nigrovittatus	5	5	3	2	1.2
C. caniceps	0	0	0	0	0
Sundasciurus tenuis	160	172	25	17	11.2
S. hippurus	2	9	8	6	0.8
Estimated density of tree squirrels, all species in 100 ha	486	Estimated total biomass of tree squirrels in 100 ha			130 kg

than that of monkeys and apes, and the biomass of all squirrels is considerably lower (cf. table 10.1). The low biomass of squirrels would seem to relate to their small size and simple digestive tract, which is indicative of their inability to detoxify plant defences and thus to subsist on significant proportions of leaves in their diets. Their gnawing dentition enables them to overcome the mechanical defences, and to extract beetle larvae, termites, sap and inner bark from 'wood'; such sources are not exploited by Malayan primates.

Individual squirrels have not been observed over periods greater than about two weeks. Within such a period, *Ratufa* may range within about 9 ha (fig. 9.4); nothing is known of the ranging habits of the other squirrels. *Callosciurus* tends to be most active at two times each day, at or soon after dawn, and late in the afternoon. *Ratufa* shows no such bimodality in activity pattern; they tend to become active later and to cease activity earlier than other arboreal, diurnal frugivores.

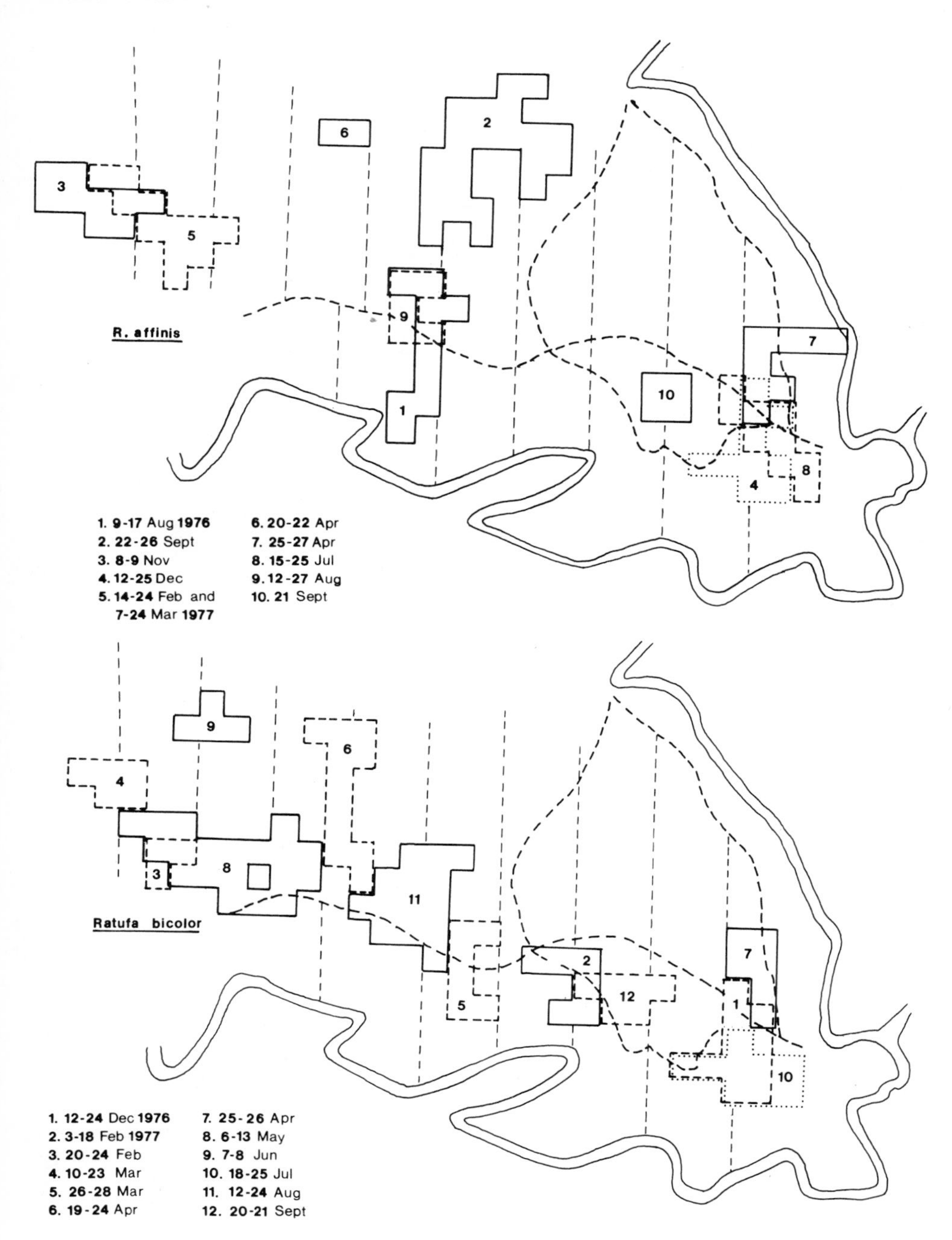

Fig. 9.4. Ranging of *Ratufa* species from distribution of feeding localities (in terms of $\frac{1}{4}$-ha quadrats) during prolonged observations.

SOCIO-ECOLOGY OF MALAYAN FOREST PRIMATES

J.J. Raemaekers and D.J. Chivers

Department of Physical Anthropology and
Sub-department of Veterinary Anatomy
University of Cambridge

INTRODUCTION

In the preceding chapters we have described the Malay Peninsula and its forests (against a background of previous primate field studies) and the natural history of the three kinds of diurnal primate found there - gibbons, leaf monkeys (or langurs) and macaques. Then we have analysed various features of primate socio-ecology - niche differentiation, positional behaviour and habitat use, long-term changes in behaviour in relation to the environment and competition with other arboreal animals.

The significance of many of the results have been discussed in passing, but we can now draw together and discuss more fully certain aspects of socio-ecology attracting wide interest. In this chapter we first compare the primate species at Kuala Lompat with each other and, where relevant, with results from elsewhere in South-east Asia and beyond; we also consider the extent to which differences in results between observers reflect differences in methods or sample size rather than the different timings of their respective studies.

Next we deal with very broad relationships between diet and body size, and the way species differing in these respects fit together into an ecological community. This is followed by a finer analysis of the role of ecological factors in shaping differences between species in various aspects of social organisation - group size sex ratio, inter-group relations. We then consider why the species differ in one aspect of behaviour, which, while often documented, has proved difficult to explain - the distribution of feeding through the day.

Table 10.1. Summary of grouping, ranging and feeding for the primates of Kuala Lompat, compounded from the study periods of Aldrich-Blake, Chivers, Curtin, Raemaekers and the MacKinnons ('weighted means' and ranges).

	Siamang *Hylobates syndactylus*	Lar gibbon *H. lar*	Dusky leaf monkey *Presbytis obscura*	Banded leaf monkey *P. melalophos*	Long-tail macaque *Macaca fascicularis*
GROUP SIZE	4 2-5	4 2-5	14 10-17	12 9-13	23 17-27
GROUP WEIGHT kg	31 26-33	16 15-17	72 57-82	60 51-76	73 39-101
BIOMASS kg/km^2	97 39-118	29 30-34	240 172-574	286 406-490	182 89-163
HOME RANGE km^2	0.32	0.55	0.30	0.21	0.40
ha	23-48	53-59	28-33	21-21	35-46
DAY RANGE km	0.87 0.64-1.15	1.67 1.49-1.85	0.76 0.56-0.95	0.95 0.75-1.15	1.08 0.76-1.40
ACTIVITY active, h	10.4	8.6	*c.*12	*c.*12	*c.*12
REST min/day	184 168-214	122	334	334	305
FEED	342 310-362	217 204-217	242	238	270
TRAVEL	101 67-136	166 166-186	144	148	145 144-146
% daytime, 12h (% active period)					
REST	39 (29)	45 (24)	46	46	42
FEED	48 (55)	30 (42)	34	33	38
TRAVEL	14 (16)	23 (32)	20	21	20
DIET					
LEAVES min/day	164	65	136	69	54
FRUIT + FLOWERS	150	132	106	138	170
ARTHROPODS	27	20	1	3	46

	Siamang *Hylobates syndactylus*		Lar gibbon *H. lar*		Dusky leaf monkey *Presbytis obscura*		Banded leaf monkey *P. melalophos*		Long-tail macaque *Macaca fascicularis*	
% feed time										
LEAVES	48	35-58	30	29-31	56	48-66	39	35-46	20	19-25
FRUIT + FLOWERS	44	41-56	61	57-71	44	35-52	58	48-62	63	58-73
ARTHROPODS	8	2-15	8	0-13	$\frac{1}{2}$	0-$\frac{1}{2}$	3	0-3	17	3-23

Finally, we outline the alarming threats to primates in South-east Asia, the forces which create them and some of the approaches to counter them, so as to increase productivity and economic viability without further upset to the very delicate and essential balance of vegetation, soil, water and atmospheric gases.

THE PRIMATES OF KUALA LOMPAT: RESULTS AND METHODS

The studies of Chivers on siamang, Curtin on leaf monkeys, Raemaekers on lar gibbon and siamang, and Aldrich-Blake on long-tail macaques (all lasting a year or more), along with the intensive 6-month study of all species by the MacKinnons, provide us with the opportunity to collate the results and assess the merits of different methods. Modal scores for each parameter of grouping, ranging and feeding (table 10.1) have been derived by weighting the averages derived for each primate species from the available studies in favour of the longer ones (where necessary).

The modal group size of gibbons must be taken as four - two adults and two offspring; Curtin's figures are preferred for leaf monkeys, although groups were clearly smaller, with a higher proportion of adults, during the MacKinnons' study (see Chapter 8); the sample size for long-tail macaques is small, and the modal size chosen may be a little too high. Group weights have been recalculated since Table 6.1 and Curtin and Chivers, again with the immature animals assumed to be half the weight of the adult female on average. While biomass calculations with regard to available space are useful for conservation purposes, for ecological analyses it is more important to relate group weight to the area of habitat used. Thus the biomass figures represent the number of kilograms of each species in relation to the size of home range (HR), and they are not necessarily within the range of figures reported previously.

Home ranges vary between 20 and 60 hectares; *H. lar* have the largest, *P. melalophos* the smallest, but otherwise they are surprisingly similar considering the differing group sizes of the various species. Mean day ranges vary between 640 and 1850 m; *H. lar* have the longest, *P. obscura* the shortest. *H. syndactylus* had a smaller range and shorter day range, and *M. fascicularis* a larger range and longer day range during the MacKinnons' study than appears usual, even compared with other observations at the same time of year; the smaller groups of *Presbytis* species had longer day ranges, but range sizes were close to those of Curtin.

There is good correspondence between different measures of the activity budget of each species, but the length of the activity period has been measured accurately only for *Hylobates* species; the monkeys are active for several hours longer each day - for about 12 hours, it is assumed, but with a conspicuous mid-day rest period. Compared to day length there is remarkable similarity between species in the proportion of time devoted to each major activity, except for the *H. syndactylus* which feeds more and travels less than the others (table 10.1). *H. lar* spends less time feeding and more time travelling than the *Presbytis* species, and *M. fascicularis* feeds more, but the gibbon's travel time and the macaque's feeding time include a proportion of time spent searching for food (small fruit sources and/or arthropods).

With regard to diet, in terms of the modal proportions of time spent eating the vegetative and reproductive parts of plants and animal matter, there is again good correspondence between studies of dawn-to-dusk observation in particular (table 10.1, cf. fig. 5.12a). *H. lar*, *P. melalophos* and *M. fascicularis* emerge as fruit-eaters, and *M. fascicularis* as the most faunivorous; *P. obscura* is the most folivorous, and *H. syndactylus* maintains a closer balance between fruit- and leaf-eating. In putting flowers with animal matter in their triangular projections of diet (fig. 6.4), the MacKinnons introduce a variability to the diet of each species between months that is not very meaningful in terms of harvesting and digestion; that is, fruit and flowers are comparable in abundance, structure and digestibility. It also seems that they complement each other in their availability, since repartitioning of the data reveals less variation between months, most obviously for *M. fascicularis* and to a lesser extent for the *Hylobates* species; variation is still marked for the *Presbytis* species, but the monthly balance between the reproductive and vegetative parts of plants is thereby clarified.

We can be confident of the comparability of these data (table 10.1), because they are derived mostly from dawn-to-dusk observations on several consecutive days, with 10-min samples of behaviours that are sufficiently distinctive (gross) not to be confused by different observers. This is the key factor in quantifying behavioural observations for comparative purposes - the finer the sampling, e.g. 1-min samples of the components of feeding behaviour, the greater the problems of comparisons between observers. Apart from the comparison of 1-, 5- and 10-min samples of *H. syndactylus* behaviour (Chivers, 1974) revealing the adequacy of 10-min samples for depicting the patterns of major activities, such as rest, feed and travel, our assessment of quantitative methods is mainly qualitative; comparisons between the authors in 1972 revealed little discrepancy, but in 1974 J.J.R. found that Kalang consistently recorded about 20 min more feeding by *H. syndactylus* each day, and, although correspondence increased, he treated the two sets of data from his study period with caution. Nevertheless, considering the gross level of analysis, the problems of refining comparability and other priorities, more elaborate efforts (in the field or in subsequent analysis) do not seem appropriate.

Similarly, we could calculate correcting factors for behaviour patterns deduced from spot observations on survey - by separating out the two kinds of data collected by the MacKinnons, and by comparing Aldrich-Blake's data with those presented in Chapters 3 and 4. It makes more sense, however, to restrict the use of survey data for defining the distribution of groups, when continuous observations are available to deduce group size and to determine activity patterns. Such spot observations are biased towards more conspicuous activities - feeding and travelling rather than resting, and eating fruit rather than leaves; group size is likely to be under-estimated. Studies of less than a year are likely to under-estimate home range size, especially in species lacking distinct boundaries and boundary-orientated behaviour, and to give a distorted picture of day range length and diet over the year (table 10.1, cf. table 6.1).

There is no doubt that, in the very variable and complex forest environment, the behaviour of a primate social group can only be depicted accurately by continual monitoring and frequent sampling of as many individuals as possible from dawn to dusk on at least 5, preferably 8-10, consecutive days. Such observations need to be sustained for at least two years, or repeated in later years; successful interpretation of the data depends on good records of changes in the weather and vegetation. Because of the restricting demands of such longitudinal studies of single species, short intensive studies of all species are essential to give perspective in a particular locality (see Chapter 6). The main advantage of Kuala Lompat, leading to sufficient material for a book, has been the sequential studies in an area of forest with an increasingly detailed system of trails and marked trees. The picture is completed by the wider geographical perspective afforded by surveys, the results of which can be interpreted more fully in the light of detailed community studies.

Surveys through the Krau Game Reserve (fig. 10.1) revealed Kuala Lompat to be the richest part with regard to the abundance of primates, with the hill forest around Gunung Benom in the north of the Reserve a close second (Chivers and Davies, 1979). In contrast,

Plate XXII. Bukit Patong, on the west edge of the Krau Game Reserve at dawn (above). The Rest House at the Kuala Lompat Ranger Post of the Krau Game Reserve (below), our home for the 10 years courtesy of the Pahang Game Department.

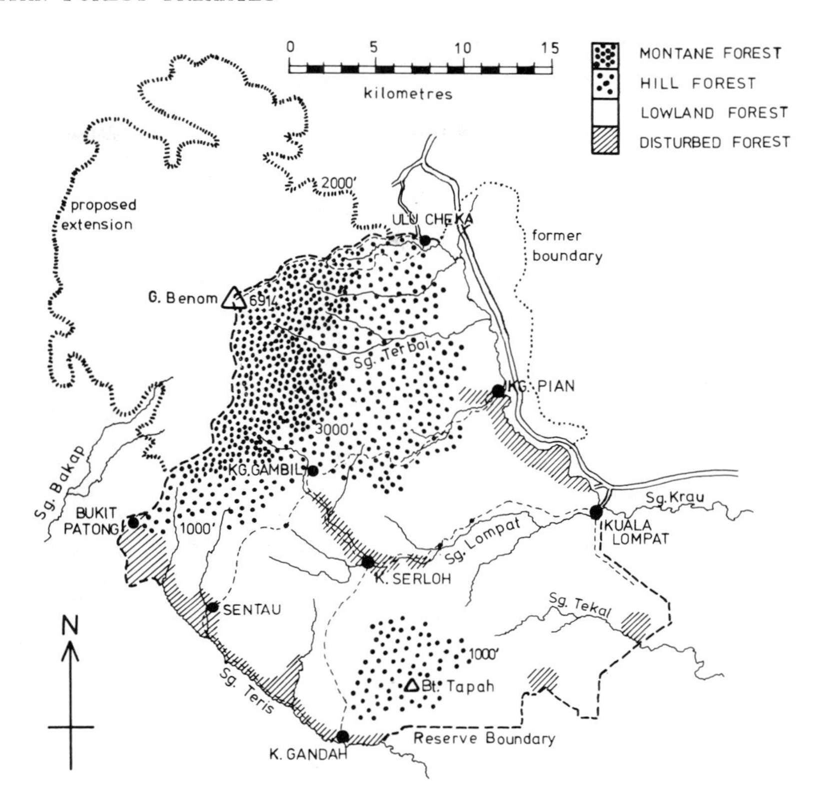

Fig. 10.1. Krau Game Reserve, Pahang, Peninsular Malaysia (from Chivers and Davies, 1979).

the fauna in the west and south of the Reserve was more impoverished, with fewer, smaller groups of primates; these are the parts inhabited by Orang Asli (indigenous people), especially around Kuala Gandah in the south (Plate XXII).

Comparison with other sites in South-east Asia (fig. 10.2), whether they have been subjected to detailed socio-ecological study or to briefer faunal surveys, reveal consistent patterns, which reduce worries that results from these contrasting approaches are not truly comparable (Chivers and Davies, 1979, table IX). Impoverished areas, such as some of the sites in Malaysia surveyed by Southwick and Cadigan (1972), may have only 2 primate groups/km^2 (or 100 kg/km^2), whereas the richest, such as Ketambe in Sumatra and Kuala Lompat, have 14-19 groups/km^2 (or 700-1000 kg/km^2). A general picture emerges from South-east Asian forests of 5-6 groups of 6 species (2 apes, 2 macaques and 2 leaf monkeys) with 300-400 kg/km^2 (a lower value than kg/km^2 of home range). Current studies in East Malaysia by J.B. Payne and A.G. Davies, and in West Malaysia by C.W. Marsh, W.L. Wilson and A.D. Johns (with some help from

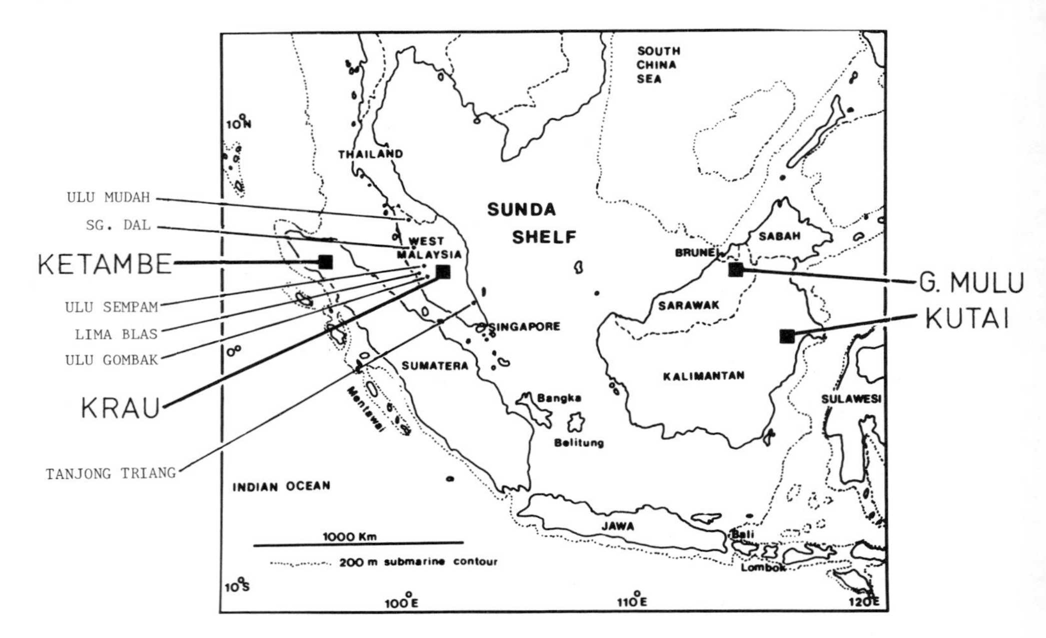

Fig. 10.2. South-east Asia (Sundaland) showing main sites of primate field study mentioned herein (based on drawing by J.O. Caldecott).

J.O. Caldecott and E.L. Bennett and, by night, E.B.M. Barrett), are evaluating more precisely the relative merits of different survey techniques and of their use in determining behaviour patterns in each species, by reference to intensive study at selected sites.

Having summarised the results from Malaysia, and commented on methodological problems, we must now seek to set them in wider conceptual and geographical perspectives.

DIET AND COMMUNITY STRUCTURE

The escalation of studies of primate ecology over the past 20 years (cf. Altmann, 1967) has brought primatologists to the point where they can begin to make some meaningful generalisations about the ways in which primate communities are structured. In this section we will outline some of these generalisations and compare them with data from the community at Kuala Lompat.

Firstly, there is general correspondence between diet and body size among animals of similar body design: leaf-eating predominates among the larger primate species, fruit-eating among intermediate ones, and insect-eating amont the small ones (Milton and May, 1976; Hladik, 1975; Clutton-Brock and Harvey, 1977; Hladik and Chivers, 1978). The reasons are (1) that a large body loses heat more slowly than a small one of the same shape, so that large animals have relatively lower metabolic rates than do small ones (although the rate scales according to $^3/_4$ power of body weight, and not $^2/_3$ as expected, Kleiber, 1961) and can eat diets lower in energy, and (2) that, very broadly, insects yield more net energy/unit weight than do fruit, which yield more than do leaves, especially mature ones (e.g. Hladik *et al.*, 1971; A. Hladik, 1978). The same relationship between body size and diet quality has been demonstrated empirically in ungulates, although in that case it involves only more or less nutritious foliage (Bell, 1971; Jarman, 1974). The relationship is further reinforced in insectivorous primates by a closely-related effect: success in hunting insects does not increase with body size, so that an hour's hunting will provide a smaller proportion of the needs of a large primate than of a small one (Hladik and Hladik, 1969).

The relationship between body weight and diet is confounded somewhat by specialisations such as the harbouring of large quantities of symbiotic fermenting bacteria in the gastro-intestinal tract (colobine monkeys: Kuhn, 1964; Bauchop and Martucci, 1968), or by varying metabolic rate (some small prosimians: Petter, 1978), but it remains strong enough to show through. Since leaves are an order of magnitude more abundant than fruit, and fruit several orders of magnitude more abundant than insects (fig. 10.3), diet also predicts the relative biomasses of primates depending on each (Hladik and Chivers, 1978). We can also predict (table 10.2) that the more abundant the modal food type of a species, the less far it will travel daily to obtain its food, and the smaller will be its home range relative to body weight (Clutton-Brock and Harvey, 1977). The ecological grades 1, 2 and 3 from Hladik and Chivers (1978) are a convenient short-hand for denoting emphasis on one of the three main food types - animal matter, fruit, leaves.

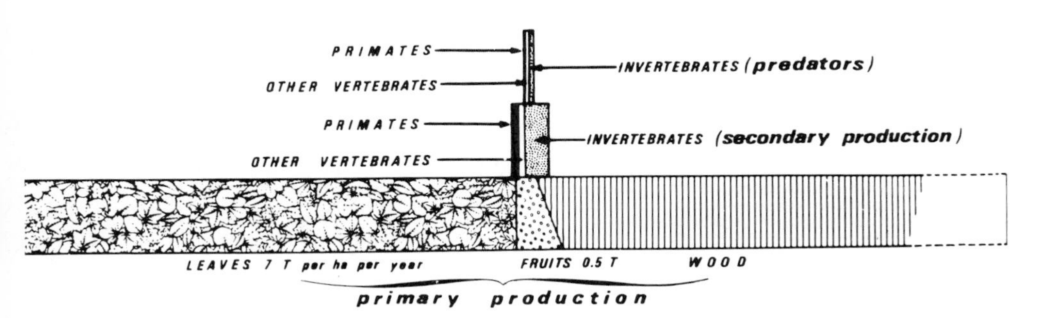

Fig. 10.3. Simplified pyramid of production in a rain-forest (from Hladik and Chivers, 1978).

Table 10.2. Correlated features of primate ecological niches.

ECOLOGICAL GRADE	1	2	3
MODAL FOOD TYPE	insects	fruit	leaves
BODY WEIGHT	low	medium	high
BIOMASS	low	medium	high
DAY RANGE / BODY WEIGHT	long	long	short
HOME RANGE / BODY WEIGHT	large	large	small

How well does Kuala Lompat community fit this picture? Data from table 10.1 (summarising the findings in Chapters 3-6), together with data on the slender loris, *Loris tardigradus*, from Sri Lanka (C.M. Hladik, 1978), a close relative of the Malayan slow loris, are summarised in table 10.3. The fit is clearly far from perfect; for example, the biomass of long-tailed macaques in grade 2 exceeds that of the siamangs at the grade 2/3 interface, reflecting the greater selectivity of siamangs in their choice of species of fruit. Nevertheless, as predicted, the highest biomasses lie at the grade 3 end of the spectrum, and the lowest body weights, highest day range/body weight ratios and highest home range/body weight ratios all lie at the grade 1 end. The data from Kuala Lompat thus indicate that the crude relationships outlined above persist despite the complications engendered by other factors, including such basic ones as the sizes of groups and of home range, and the kind of fruit selected.

Table 10.3. Distribution of Kuala Lompat primates across ecological grades.

SPECIES	Slow loris	Long-tail macaque	Pig-tail macaque	Lar gibbon	Siamang	Banded leaf monkey	Dusky leaf monkey
ECOLOGICAL GRADE	1	2	2	2	2/3	3	3
MODAL FOOD TYPE	insects	fruit	fruit	fruit	f/l	f/l	leaves
BODY WEIGHT, kg (adult female)	1.5	3.5	7	5	10	6.5	6.5
BIOMASS, kg/km^2 HR	<25	182	?86	29	97	286	240
DAY RANGE / BODY WEIGHT	?	0.31	0.29	0.33	0.09	0.15	0.12
HOME RANGE / BODY WEIGHT	?	0.11	0.29	0.11	0.03	0.03	0.05

The few communities of forest primates which have been analysed in detail differ in which grades are emphasised (Hladik and Chivers, 1978). The Kuala Lompat community of 7 species has 3 species at the grade 3 end and only 1 in grade 1, whereas the community of 5 species on Barro Colorado Island, Panama, has only 1 species at the grade 3 end and 2 at the grade 1 end, and of the 12 primate species at Makokou, Gabon, there is again only 1 folivore (grade 3) but 4 insectivores (grade 1). These differences, which are reflected at a continental level, might be due to differences in forest type, but are more probably due to historical accidents determining the degree of competition for niches with non-primates. The presence of several kinds of large, arboreal folivorous non-primate mammals in the neotropics, including the highly specialised sloths, might account for the absence there of a primate with digestive specialisations equivalent to those of the colobine monkeys in Africa and Asia (Montgomery and Sunquist, 1978). Likewise, the relative lack of insectivorous primates in Asia, compared with Africa and the New World, could be explained partly by the presence of tree shrews only in Asia.

Another principle of community structure, which emerges clearly from Chapters 6 and 7, is that the similarities in ecological niche between species within genera are greater than those between species of different genera. This indicates that phylogenetic inheritance is a relatively strong force, which will tend to channel diversification within the community. A force acting in the same direction is the tendency for species inhabiting the same environment to resemble each other ecologically. This was a premise of the original attempt to correlate habitat and social organisation (Crook and Gartlan, 1966).

That premise is only broadly true, however, for it has also become clear that competition between species acts in the opposite direction to create ecological differences between co-existing species (see Lack, 1971, for a history of their principle of competitive exclusion). Thus, while there is a broad similarity between species occupying the same habitat, competition will create finer differences between them. These differences may be quite subtle and indetectable by the crude kind of correlation approach outlined above (Clutton-Brock, 1974).

Among the primates this is well illustrated by colobine monkeys in Africa and Asia. Pairs of sympatric species occur throughout their range; these have been studied in detail at three sites, of which Kuala Lompat is one (Chapters 4, 6 and 7). The other sites are the overgrown temple at Polonnaruwa in Sri Lanka (Hladik, 1977) and the hill forest of Kibale in Uganda (Struhsaker and Oates, 1975; Clutton-Brock, 1975b). Each site contains two species belonging to the same genus: *Presbytis melalophos* and *P. obscura* at Kuala Lompat, *P. entellus* and *P. senex* at Polonnaruwa, and *Colobus badius* and *C. guereza* at Kibale. At each site the two species seem to be segregated ecologically in the same way: one species (*P. melalophos*, *P. entellus* and *C. badius*) eats relatively less leaves, especially mature ones, is more mobile, tends to be active lower in the canopy, lives in larger social groups, shows more overlap in range between groups, and conflicts more with neighbouring groups than does the other species (*P. obscura*, *P. senex* and *C. guereza*). The correspondence breaks down in the case of biomass, but that may be due to variation between sites in habitat type, number of other primate species present and presence of predators.

In each pair, therefore, one species appears to pursue a higher energy budget (moving more and 'fighting' more for richer food), while the other follows a lower energy budget (putting in less effort, needing less and getting less in return). In these colobine monkeys, the species adopt high or low energy budgets without a difference in body size (except perhaps in Sri Lanka, where *P. entellus* is more terrestrial and larger than *P. senex*). In other cases, adoption of one or other approach may be related to body size, because of the relationship between body weight and diet described above; for example, the siamang and lar gibbon at Kuala Lompat (Chapters 3 and 6). Such ecological segregation of closely-related species, by a difference in body weight, is probably quite easily evolved, since body size is essentially determined by the cessation of growth, which is probably a very plastic trait.

SOCIAL ORGANISATION

Having briefly considered ecological relationships at the community level, in this section we shall try to account for some of the differences between species at Kuala Lompat in the main elements of social organisation: group size, socionomic sex ratio, sex-linked differences in behaviour and inter-group relations. These features are determined by a complex inter-play of forces including parental investment, sexual selection, selectivity of feeding, food source size, phylogenetic inertia in behaviour and predator pressure. We tried initially a sequential consideration of the effects of each determinant on each feature of social organisation, but this approach failed because a much larger sample is needed before any one effect can show through the complications produced by others.

The starting points chosen are (1) selectivity of feeding and (2) the polygynous tendency of mammals. The feeding niche appears to be a relatively stable feature constraining evolutionary change in other features, rather than being constrained by them (Clutton-Brock and Harvey, 1977:574-575, discuss this point in depth). The polygynous tendency in mammals arises from the different parental investment in offspring shown by the two sexes (Trivers, 1972). It is males who maximise the number of matings in mammals, because only females directly feed the young. Males are, therefore, typically subject to strong intra-sexual selection through competition for mates. Some males are bound to compete better than others and will mate with a disproportionate number of females to the exclusion of other males, resulting in polygyny. The situation may only be altered if for some reason the male can increase his fitness by increasing his parental investment. In birds monogamy is the rule, because males can do everything as well as females except lay eggs (Orians, 1969); 91% of bird species are monogamous (Lack, 1968). In mammals, the factors favouring paternal investment seldom arise, and only 3% of species are monogamous (Eisenberg, 1966; Kleiman, 1977).

Apes

Gibbons (Plate XXIII) number among those 3%, and we must ask why the polygynous tendency is suppressed in these cases (see also Chapter 3). The gibbons are among the most selective feeders of the six large primate species at Kuala Lompat (table 6.6). They depend to a large degree on rare, scattered, and often small, sources of food. While the individual's feeding efficiency might be promoted by grouping for reasons such as (1) improved location of food (e.g. chimpanzees: Menzel, 1971) and (2) saving time by not visiting food sources cropped recently by others (e.g. finch flocks: Cody, 1974), small scattered sources will select for small size of feeding group through inter-individual competition (e.g. mangabeys: Chalmers, 1968). Such competition is apparent in gibbons in the exclusion of sub-adults from food sources by parents; the smaller the tree, and the more selected the food, the more likely is the exclusion (Raemaekers, 1978c).

Selective feeding may also be expected to promote defence of the food supply, because site attachment should develop to aid efficient harvesting of scattered sources, and the loss of rare foods from the known area would be costly in proportion to their rarity. Since it would clearly be impossible to defend scattered and unpredictable food sources as and when they appear, it is necessary to defend an area which will contain enough of them over a long period. The relative stability of production of the tropical forest helps to make that area predictable, and predictability favours the evolution of territoriality (review in Davies, 1978). Selective feeding means, however, that the area defendable by a gibbon can only support a small number of animals, since defendability is essentially fixed by the ability to reach the boundary economically (Brown, 1964; Mitani and Rodman, 1979).

Small feeding group size and small territorial group size are thus selected. The two largely coincide, especially in the siamang, in which the territorial group seldom breaks up to forage. It is presumably the combination of territory and small group size which force monogamy on the male gibbon. The female cannot defend a territory and raise young alone, so the male must remain with the

Plate XXIII. Siamang swinging through canopy (above left) and a lar gibbon leaping spectacularly from one tree to another (above right); young adult female siamang playing among the foliage (below).

female and assume the bulk of territorial behaviour so as to ensure the survival of his offspring. Because of the selectivity of feeding, the area he can defend will only support one female and her offspring, and monogamy is added to male residence. This hypothesis is supported by the observations that monogamy in mammals is typically associated with high paternal investment (Kleiman, 1977), and that all monogamous primates are territorial (Chivers, 1974; Clutton-Brock and Harvey, 1977).

Monogamous territoriality means that both sexes of young necessarily leave the natal group, except in those cases where a late offspring may replace a parent (Chapter 8), but see Tilson (in press b). This contrasts with the majority of social mammals, in which only one sex leaves (Packer, 1979a). The transference of both sexes, combined with probably low dispersal, means that a moderate degree of inbreeding is likely, although sib matings are unlikely because of the age difference (Tilson, in press b).

Kleiman (1977) lists a number of other features associated with monogamy in mammals, which can be compared with our data here:

(1) Where older offspring are retained in the natal group, they assist with raising their siblings. The one way in which a gibbon could do this is by helping with territory defence, which the sub-adult lar George did for several years, and to which immature male and female offspring contribute by calling with their parents (immature males only in siamang) (Chapter 8). Sib-helping behaviour has been shown in cooperatively breeding birds to fit the hypothesis that parents allow their older offspring to remain and help when this will help the parent's fitness more than excluding it, because the chances of the older offspring establishing a breeding territory at that time are low (Emlen, 1978).

(2) Sexual behaviour and intense social activity are low in established pairs, although higher in newly formed ones. Copulation and sex play are indeed rare in established gibbon groups (table 8.5), and grooming occupies only 1-6% of the day compared with 12% in macaques (no comparable figures for leaf monkeys). Much more play was seen between two siamangs consorting in 1972 (TS1a) than is usual in adults.

(3) Monogamous mammals are socially inflexible. This is a striking feature of gibbons emphasised by Ellefson (1968, 1974). It is remarkable that the monogamous pair bond is universal throughout the geographical range of gibbons, from Assam to Java (with the exception of some hybrid groups, Brockelman, 1978, and rehabilitants, Proud, pers. comm., Brockelman *et al.*, 1974), considering the differences which must exist in ecology and population dynamics. This inflexibility is echoed in another monogamous primate, the common marmoset (*Callithrix jacchus*), in which only one female will

breed in captive artificial groups containing more than one female (Rothe, 1975). It contrasts with the flexibility of what Kleiman terms 'facultatively monogamous' species of mammals, which may be polygynous when the conditions are favourable. A similar dichotomy may exist among birds, between obligate and facultative monogamous species; as an example of the latter Verner (1964) described variation in the long-billed marsh wren, in which bachelor males occupy small territories, monogamous ones larger territories, and bigamous ones still larger territories.

(4) Young show delayed maturation if retained in the natal group. We have no hard data on this, but the female lar gibbon Gail was still 11 years old in 1979, differing little in size or demeanour from 1976. Also, Tilson (in press b) describes the onset of singing in a sub-adult male after his father died, in a way which suggests that it was suppressed previously. These are cases of delay in social rather than sexual maturity.

(5) Sexual dimorphism in body size and role in territorial defence are less than in related polygynous species. Gibbons have no close relatives that are polygynous, but as a group they show much less sexual dimorphism in body size than the polygynous monkeys at Kuala Lompat (other than *P. melalophos*), and many other polygynous primates (e.g. Brown, 1975, table 12.3). Moreover, the gibbon sexes differ less in territorial defence roles than do the more territorial monkeys at Kuala Lompat, since only male monkeys give loud calls and take the leading role in inter-group conflicts.

With respect to parental investment, Kleiman notes that, among marmoset species, the weight of the neonate relative to that of the mother corresponds roughly with the time at which the father begins to carry the young twins. As she points out, the relationship does not hold in gibbons: the weight ratio in the lar is higher than in siamang, yet the male lar never carries the young, and the male siamang carries it from about 10-12 months of age (Chapter 3). Kleiman quotes Tenaza's suggestion that this is because predation on the lar is heavier and the lar male therefore spends more time as a look-out. There is no evidence to support this; indeed, if man is the main predator on gibbons, the slower siamang is the more susceptible species. A more plausible explanation is that the lar gibbon's higher rate of inter-group conflicts (see below) precludes him from carrying the young and/or the greater role of the female siamang in food-finding (Chivers, 1974) precludes her from carrying the heavier infant. Furthermore, if the female has stopped lactating, she has to build up her reserves to produce another offspring. Chivers also argues that the close interaction between the growing infant and adult male (and later with other group members) contributes to the greater cohesiveness of the siamang family group. It may still be true that the male gibbon increases his parental investment by taking the major role in protection from predators;

Tenaza (1975) states that the male Kloss gibbons placed themselves between him and the rest of the family when frightened by him, but D.J.C. observed the female siamang to be the usual detector of potential non-human predators.

The model of forces shaping the social organisation of the gibbons, and some of its associated features are summarised diagrammatically in fig. 10.4.

Monkeys

In the cercopithecoid monkeys the polygynous tendency is not suppressed (table 10.4). This may in part be because the constraint of feeding efficiency on feeding group size is weaker: all the monkey species are less selective feeders than are the gibbons (Chapter 6). This allows more animals to remain together without

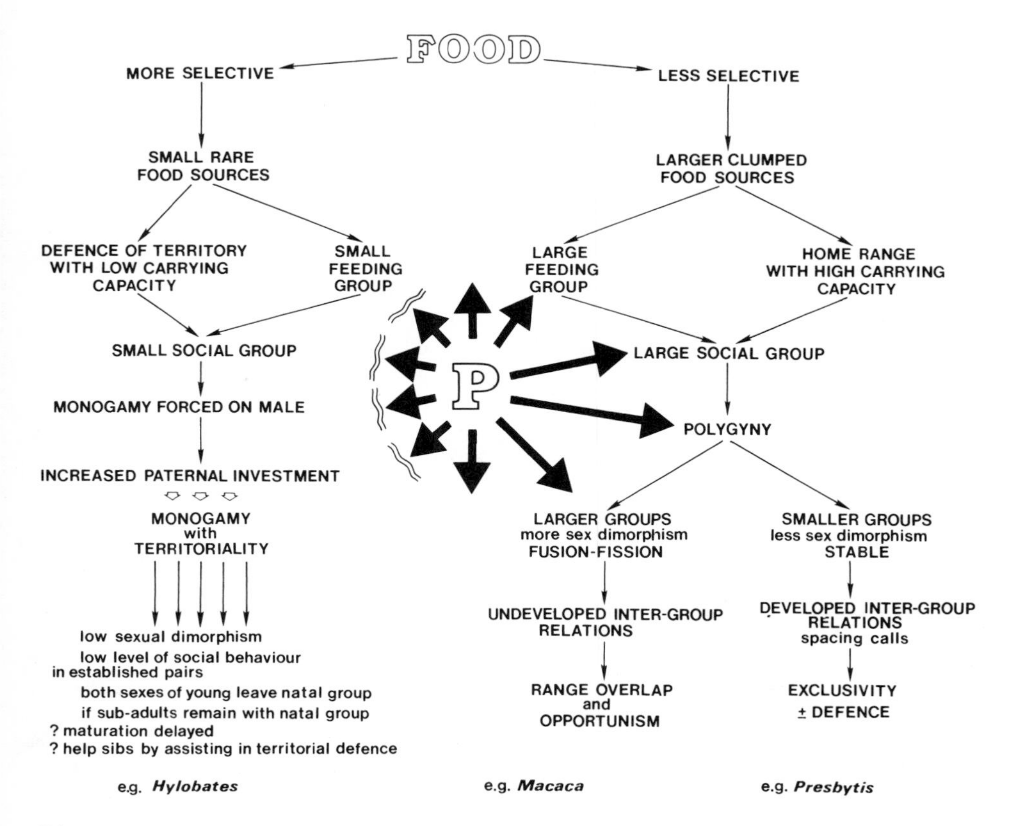

Fig. 10.4. Model of forces shaping social organisation of Malayan forest primates (P = basic polygynous tendency of mammals).

Table 10.4. Sexual dimorphism in primates at Kuala Lompat (body weights from Napier and Napier, 1967, and Burton, in prep.; other data from Chapters 3-6).

SPECIES	Siamang	Lar gibbon	Pig-tail macaque	Long-tail macaque	Dusky leaf monkey	Banded leaf monkey
BODY WEIGHT female as % of male	92	94	74	70	88	101
SEX RATIO adult females: adult males in mixed groups	1	1	?3	3	3.5*	5-6* 4 1.5+ 1.5-2
NO. OF ADULT MALES IN MIXED GROUPS	1	1	several	2-3	2 3	1 2 4 2

* number for each group studied

inter-individual feeding competition, because (1) food sources tend to be larger, e.g. the big deciduous trees which produce massive bean crops the gibbons cannot eat, and (2) even when food sources are small, a more varied diet allows different individuals to eat at the same time different foods growing close together. Less selective feeding also increases the number of females and young supported by the area defended by a male, favouring polygyny. The male may not even defend a territory, because less selective feeding reduces the pressure to defend a food supply, and he may be able to obtain exclusive access to females without recourse to defending the area in which they feed. It is not surprising to find that all four monkey species invest far less than the gibbons in defence of an area (table 10.5).

This very simplistic approach suggests that one can account for the wide differences between gibbons and monkeys in social organisation by assuming that the feeding niche is the constraining variable (fig. 10.4). It is not so easy to account for differences between macaques and leaf monkeys in the same way. The two macaques live in large multi-male groups (as probably do all macaques, e.g. Eisenberg *et al.*, 1972), whereas the leaf monkeys at Kuala Lompat live in groups of varying size with one to several males (as do some other leaf monkey species e.g. the grey langur, *Presbytis entellus*: Yoshiba, 1968; Hrdy, 1977). We would argue, however, following Eisenberg *et al.*, that multi-male groups in leaf monkeys are not truly comparable with the macaques as the males are age-graded and there is one distinct leader. Can such differences be accounted for

Table 10.5. Inter-group relations and index of defendability, D (mean day range/diameter of circle of area equal to home range, Mitani and Rodman, 1979) in the primates at Kuala Lompat.

SPECIES	% OF TOTAL AREA USED BY MORE THAN ONE GROUP	LONG-RANGE ADVERTISEMENT	INTER-GROUP CONFLICTS	D
Lar gibbon *Hylobates lar*	15	frequent rounds of morning duets lasting 15 min; large evolutionary investment in elaborate song with sex-specific parts; males may also sing long solos	frequent and long, lasting up to 2 h; special calls in males	2.3
Siamang *H. syndactylus*	little	as *H. lar*, except that duets less frequent and more complex, no male solos; throat sac evolved to amplify sound	infrequent	1.2
Banded leaf monkeys *Presbytis melalophos*	20	brief male calls by day; frequent rounds of male calls at night, perhaps because of disturbances unrelated to inter-group relations	few	2.2
Dusky leaf monkey *P. obscura*	3	little reciprocal calling; male calls by day	few	1.6
Long-tail macaque *Macaca fascicularis*	?10	none	few	1.8
Pig-tail macaque *M. nemestrina*	?	none	?	1.8

in terms of feeding and ranging strategies?

The great difference between macaques and leaf monkeys in feeding strategies (Chapter 6) may be partly accounted for in terms of a phylogenetic difference in the flexibility of grouping. The key adaptation of the macaques may be the ability to split and regroup with ease, matching the size of feeding group to the prevalent distribution of food. A similar ability has been described quantitatively for Colombian monkeys (Klein and Klein, 1975). This flexibility may mean that the size of social group (effectively the largest sleeping group) may not depend on the size of the modal food source, if such a concept is even applicable to macaques. Groups of macaques in forest can grow large (a count of 79 for a pig-tail macaque group in Thai forest seems average, Srikosamatara, pers. comm.); the upper limit of social group size may be set, therefore, more by constraints of maintaining contact during foraging (as much as 500 m apart) and by the maintenance of social ties, than by the size of food sources.

In the pig-tail macaque, the adoption of partial terrestriality may also allow larger groups, because the lower costs of travel on the ground would allow the exploitation of a larger home range, and the more economic fission and fusion of 'groups'. The larger body size of this species, compared to the long-tail macaque, might also be an adaptation to reduce travel costs, since large animals travel relatively more cheaply than small ones (Schmidt-Nielsen, 1972; Dawson, 1977, McNab, 1963). Large group size could also be selected by competition of resources such as sleeping sites and feeding areas. Angst (1975) reports that on Bali the ability of long-tail macaque groups to displace each other corresponds with group size. The MacKinnons (Chapter 6) suggest that large group size might also help this species in competition with banded leaf monkeys. The advantage of large group size in inter-specific competition for food has been demonstrated in aggressive bees, which owe their ability to displace other species from food sources partly to recruitment of the number required to overpower the opposition (Heinrich, 1978).

The macaques' social flexibility allows great variation from day to day in travel distance and diet (Chapter 5), so that large group size does not preclude the harvesting of high-quality foods on an opportunistic basis. Their diet consists mainly of high-energy fruit, as well as more insects than eaten by leaf monkeys (fig. 5.12). Gibbons certainly lack such flexibility (see above); perhaps this is also true of the leaf monkeys, although Curtin reports frequent group splitting over moderate distances in the dusky leaf monkey (Chapter 4), and the MacKinnons report some splitting of groups in the banded leaf monkey.

In the dense rain forest the macaques' combination of large group size, large home range, frequent sub-grouping and wide group spread may prevent males from establishing the one-male groups achieved by some male leaf monkeys. This is supported by the observation that the small bisexual groups of banded leaf monkeys studied by the MacKinnons contained only one adult male, whereas the larger groups contained several males (Chapter 6). Thus, the differences in group size and composition between leaf monkeys and macaques may be a compromise between the interests of females grouping to maximise feeding efficiency, and those of the males trying to control access to as many females as possible to maximise breeding efficiency, which would appear to be easier at some times than others (cf. chimpanzees: Wrangham, 1979; Tana mangabeys: Waser and Homewood, 1979; Dawkins, 1976, chapter 9, for a review of this 'battle of the sexes').

Territoriality

The monkey species at Kuala Lompat also differ from one another in inter-group relations, although not nearly so much as they all differ from the gibbons; they all show relatively little range overlap compared with some forest primates, such as gorillas (Schaller, 1963) or red colobus (Struhsaker, 1975). The macaques lack a loud call which might function in the remote spacing of groups, whereas the two leaf monkey species both have one; the two leaf monkeys differ from each other in the frequency of inter-group conflicts and in the degree of range overlap (table 10.5). There appears to be no consistent relationship between loud calls, conflicts and range overlap.

The difference between macaques and leaf monkeys in the possession of a loud call seems to be deeply rooted in phylogeny: no macaque species has one (apart from the loud moaning intra-group contact call of pig-tail and lion-tail macaques), and all Colobine monkeys, in both Asia and Africa, have a loud call, as far as we know. We cannot expect, therefore, a relationship between loud calls and contemporary ecological conditions. The differences between the two leaf monkey species in calling, conflict and range overlap seems to be analogous to those between the two gibbon species. In either genus, one species achieves virtually complete range separation by means of loud calls and few conflicts, whereas the other species, the more frugivorous, achieves somewhat less separation with more calls and conflicts. This emphasises the distinction to be drawn between the two aspects of territory, defence of an area (Burt, 1943) and exclusive use of it (Nice, 1941), a distinction which Struhsaker and Oates (1975) were also forced to draw in describing the behaviour of the black-and-white colobus.

The extent to which the exclusive area is defended may relate to the degree of mobility relative to range area. Mitani and Rodman (1979) derived a simple index of the defendability of an area

$$D = \frac{\text{mean daily route length}}{\text{diameter of a circle with area equal to home range}}$$

based on the assumption that defence of a territory requires patrolling, and that this is only economical if D is greater than or equal to 1. They found that, of the primate groups they examined, the 33 reported as territorial conformed to this expectation, while 14 of

20 reported as not territorial had D values of less than one. If D values are calculated for the Kuala Lompat primates, we find that all are greater than 1 (table 10.5, data from table 6.1), despite wide variation in the degree of area defence. Nevertheless, dusky leaf monkeys and siamang have lower D values than banded leaf monkeys and lar gibbon; thus, higher D values correspond with more range overlap, more conflicts and more calling .... and more fruit-eating.

Sexual dimorphism

The presence of sexual dimorphism in body size in the dusky, but not the banded, leaf monkey is intriguing. In this respect they correspond with the conditions prevailing in the two *Presbytis* groups to which they belong (female weight is 99% of male weight in the (*Presbytis*) group, and 93% in the (*Trachypithecus*) group - aygula and cristata, respectively, in Napier and Napier, 1967). Sexual dimorphism in body size in primates is generally associated with polygyny (Brown, 1975; Leutenegger, 1978), presumably as a result of selection between males for fighting ability, which is correlated with body size in many animals (e.g. toads: Davies and Halliday, 1979; baboons: Packer, 1979b). An extreme manifestation of inter-male competition - infanticide following take-over of a group - has been reported in three species groups of *Presbytis* (Hrdy, 1974; Rudran, 1973; Wolf and Fleagle, 1977) and in black-and-white colobus, *Colobus guereza* (Oates in Hrdy, 1977), as well as in several non-colobine species (Angst and Thommen, 1977). While it has not been reported in either species at Kuala Lompat, there is strong evidence of a case in the silver leaf monkey, *Presbytis cristata* (Wolf and Fleagle, 1977), in the same species group as the dusky leaf monkey. The expression of sexual selection through infanticide and sexual dimorphism in body size might then perhaps be correlated within, and differ between, species groups of *Presbytis*: the species group containing the dusky leaf monkey shows less dimorphism and some infanticide, whereas that of the banded leaf monkey shows virtually no dimorphism and no infanticide. Most infanticide has been reported in the grey or Hanuman langur, *Presbytis entellus*, which is the most dimorphic, with female weight only 68% of male weight (Napier and Napier, 1967).

Predation

We have so far largely ignored predator pressure in our analysis of species differences in organisation. It is difficult to assess how strong this pressure is. We have no direct evidence of predation at Kuala Lompat, but there is abundant indirect evidence in the primates' fear of man, and in the alarm they display at raptorial birds and predatory arboreal mammals. Since the different monkey and ape species are all arboreal, diurnal and of much the same size, they would seem to be subject to similar pressures. Predation might be more of a threat on the ground, however, where cats are at an advantage. This would favour stronger measures for avoidance in terrestrial species such as the pig-tail macaque. Large group size and large male body size might help in this respect, although they are explained more readily as adaptations to foraging and competition between males (Clutton-Brock and Harvey, 1977). In the trees escape is easy and extant arboreal primates are not exposed to serious predation.

In all monkey species at Kuala Lompat, the adult male definitely takes the lead role in defence of the group. They are more likely to be seen, which could be a product of vigilance for rival males rather than for predators, and they generally face any threats while the rest of the group flees or hides. In the macaques there is some tendency towards mobbing (cooperative defence) by all age-sex classes (cf. the macaques on Bali, Angst, 1975), but in the leaf monkeys there is a clearer distinction between the behaviour of the leader male and of other group members. The male banded leaf monkey often makes an elaborate distraction display, calling and leading the observer away from the group, which hides and may stay still for hours (cf. patas monkeys, Hall, 1967, and capuchin monkeys, D.J.C., pers. observ.). Assuming that the male incurs some danger to himself, his benefits are best explained in terms of kin selection, insofar as his group contains his offspring and maybe other relatives.

Thus in the larger polygynous monkey groups, it is usual to argue that the male's contribution in return for acceptance by a feeding and rearing group of females may be mainly in defence against predators, whereas in the monogamous gibbons it is at least as much the defence of food against conspecifics. Our observations suggest that this interaction with conspecifics may also be a significant feature of male behaviour in polygynous monkey groups. Note that the benefit of having a leader male as protector against predators may not only be in reducing one's chance of falling prey; by off-loading vigilance onto the leader, the other group members may also increase the time available for other activities. There is strong evidence that flocking in birds in response to predator pressure is partly a way of reducing vigilance and thus increasing feeding time, which may be critical to survival (Powell, 1974; Caraco, 1980). If take-over with infanticide does occur in the monkeys (cases have been reported in the long-tail macaques under artificial conditions, Angst and Thommen, 1977), then the females may also be obtaining protection for their infants from the leader male against other males (Hrdy, 1977).

In conclusion, the differences in social organisation between gibbons and monkeys at Kuala Lompat seem to be explained best by viewing diet as constraining the effects of other variables such as sexual selection and predator pressure (fig. 10.4). The inherited ability to aggregate and disperse appears to differ more between species than one might have supposed, being rather inflexible in some species. This may seem surprising in view of the primates' renowned adaptability in both diet and social grouping tendencies. We must remember, however, that we are analysing a community in a relatively benign and stable environment, in which inter-specific competition is bound to be high (Chapter 2), with species tightly packed, so that each species occupies a narrower niche than it could under other circumstances. Ironically, one of the best ways of finding out the relative plasticity of species traits is to compare the response of species to severe habitat disturbance (A.D. Johns, in progress). Of the Asian primates, the long-tail macaque adapts best to disturbance (see below under Conservation), but it is not clear just what traits permit this. Some light may also be shed on this by Mah's (in prep.) comparison of long-tail macaques at Kuala Lompat and in a 'disturbed' environment on the outskirts of Kuala Lumpur.

DAILY PATTERN OF FEEDING

In this section we shall discuss the distribution through the day of just one activity - feeding - which shows more striking and consistent patterns than most other activities, except calling (Chapter 8), and which is most frequently reported in the literature.

Different studies have reported different daily feeding patterns for the same species at Kuala Lompat (table 10.6). It is not easy to say whether such differences are genuine, or due to different methods of data collection (see above), or different sample sizes. They are in any case insufficient to mask certain differences between the species. Gibbons consistently cease feeding (and all other activities) in the middle of the afternoon, whereas the monkeys cease around dusk. Early retirement seems to be universal among gibbons (Ellefson, 1974; Whitten, 1980; D. Leighton, in prep.; S. Srikosamatara, in prep.). Three of the four graphs of leaf monkey feeding patterns (figs. 4.11 and 6.6) are strongly bimodal with peaks in the early morning and late afternoon, as reported in all other studies of *Presbytis* species (see below), whereas the gibbons show at most a weak bimodal pattern (fig. 6.6), or usually a morning peak followed by gradual decline through the day (fig. 3.14).

What factors might account for such differences? There is evidence suggesting that weather can directly affect the distribution of activities through the day; for example, cool, dry weather seems to delay rising and morning feeding in sifakas, *Propithecus verreauxi* (Richard, 1974), whereas cool, wet weather delays rising in siamang (Chivers, 1974), and rain-storms might explain shifts in an habitual feeding period in the red colobus, *Colobus badius*

Table 10.6. Daily patterns of feeding in Malayan forest primates - at Kuala Lompat and Sungai Dal.

SPECIES	DAILY FEEDING PATTERN	SOURCE
Siamang *Hylobates syndactylus*	peaks morning, falls through day, ends early	DJC & JJR
Lar gibbon *H. lar*	peaks morning, falls through day, ends early	JJR
	weakly bimodal, peaks morning and afternoon	JRM & KSM
Agile gibbon *H. agilis*	peaks morning, falls through day, ends early	SPG
Long-tail macaque *Macaca fascicularis*	bimodal	FPGA-B
	even through day	JRM & KSM
Banded leaf monkey *Presbytis melalophos*	bimodal	SHC
	even	JRM & KSM
Dusky leaf monkey *P. obscura*	bimodal	SHC, JRM & KSM

(Clutton-Brock, 1974). Chalmers (1968) demonstrated a loss of the bimodal activity pattern in mangabeys, *Cercocebus albigena*, during wet, cloudy times of the year; a similar change in behaviour was observed in howling monkeys, *Alouatta villosa* (Chivers, 1969), and in the agile gibbon, *Hylobates agilis* (Gittins, 1979). It is often implied that primates follow a bimodal pattern of activity with a rest period in the middle of the day to avoid over-heating (e.g. Chivers, 1969; Mukherjee and Saha, 1974), and this is supported by the adoption of a trimodal activity pattern by mangabeys in the cloudy season (Chalmers, 1968).

Temperature, however, does not readily account for those cases where the period of activity falls before the hottest part of the day, nor for the presence of the same bimodal pattern of feeding in some nocturnal prosimians (Charles-Dominique, 1977; Clutton-Brock, 1977a). Temperature declines steadily through the active period of nocturnal prosimians (Pariente, 1974), and so cannot account for a fall in activity in the middle of the night (for example, to conserve heat). To explain the differences in daily feeding pattern between species at Kuala Lompat by direct weather effects, it is necessary to invoke differences between species in their responses to them; yet it is not obvious that an active animal like a lar gibbon should be less sensitive to heat than a less active leaf monkey, and thus not have to rest during the middle of the day. Nevertheless, the gibbons do spend most time in the under-storey in the middle of the day.

A quite different factor affecting the daily feeding pattern might be competition between species. Moynihan (1970) suggests that the timing of feeding by insectivorous tamarins, *Callithrix* spp., is determined partly by the avoidance of competition with flycatchers. A similar argument might be applied to gibbons. Perhaps they feed more continuously through the day and finish earlier than monkeys, so as to avoid losing food to them in the evening (Chivers, 1973). The larger groups of monkey would affect the gibbons' food supply more than the other way about. Gibbons in Khao Yai National Park in Thailand, however, also cease activity early, even though there is only one other primate present at low density (pig-tail macaque, J.J.R., pers. observ.).

As suggested by the MacKinnons (Chapter 6), one would expect that some sort of relationship between diet, feeding rate and digestive through-put would be the most widespread determinant of feeding pattern. A breed of cattle with small stomach capacity has been shown to feed for longer and travel further in the same environment than another breed with a larger stomach, presumably because it cannot eat so much at a time before becoming satiated (Hafez *et al.*, 1969). Clutton-Brock (1974) suggests that similar principles might apply within the primates. One would expect a leaf-eater, whose food is abundant and clumped, but bulky and slow to digest, to have a few long feeding bouts daily - each one filling its stomach - followed by long digestive pauses. A fruit-eater, by contrast, seeking higher-energy, more scattered food, might be expected to break up its day into shorter bouts of feeding and travel, with less resting. With more feeding bouts daily, and more variation in the distance moved to find food, small differences between days would tend to smooth out the feeding curve in comparison with the leaf-eater. This is the sort of difference observed between the bimodal pattern towards which the leaf monkeys at Kuala Lompat tend, and the more even pattern reported for the three gibbon species studied (Chapter 3).

Another aspect that has become topical, is the problem of toxins in food.* As the primate feeds at the start of the day, food will

* Future collaboration between primatologist and plant biochemist should resolve the question as to whether there is chemical defence in plants against mammalian folivores, including primates, as well as against leaf-eating insects (Janzen, 1978; McKey, 1978). The probability that plant secondary compounds influence primate diets negatively has been raised repeatedly in this book, but no tests have yet been carried out in South-east Asia. Certain principles are emerging; more chemical defence is expected in (1) mature than in young leaves, (2) in climax than in colonising species, (3) in ever-green than in deciduous species and (4) in trees than in vines (these features often being inter-correlated as climax and ever-green species, Opler, 1978). We need tests of more specific cases, however, before such principles should be allowed wide acceptance.

accumulate in the gut as feeding rate exceeds digestive through-put. This is particularly true of species eating more coarse and toxic foods, such as mature leaves, because the fermentation associated with folivory is more complete the longer it is allowed to continue (which will select for as long a fermentation period as metabolic rate allows), and because the effect of many toxins is to slow the rate of assimilation (Bauchop, 1978; Parra, 1978; Freeland and Janzen, 1974). This might be a contributory factor to the long mid-day rest period that distinguishes leaf monkeys from gibbons, although it cannot be the whole story since one of the leaf monkeys eats many fruit, and one gibbon eats many leaves, albeit young ones.

The hypothesis that diet is related to daily feeding pattern can be tested between species of primate and, in a more oblique way, within species. Within species we should expect a correspondence between the amount of leaf eaten and the amount of resting in different sample periods (Struhsaker, 1978). This does not, however, appear to happen. Struhsaker found no correlation across monthly samples between the proportion of mature leaves eaten and the amount of resting in red colobus, *Colobus badius*; J.J.R. (unpubl.) also found no such correlation in siamang and lar gibbon. A more precise test is to determine whether leaf-eating is more likely to be followed by resting than is the eating of other foods. Looking at the sequence of activity in individual siamang and lar gibbons, J.J.R. found that this is not so, and, moreover, that the length of a feeding bout on any kind of food is not related to the probability that resting will follow, nor to the length of resting bout if resting does follow.

To make a quick and crude test between species for a relationship between diet and daily feeding pattern, J.J.R. searched the primate literature for reports of daily feeding pattern, main food type, body weights (which ought to be related to diet) and habitat type (which need not be). The results summarised here are being prepared for future publication. Feeding patterns are characterised as:

(1) unimodal, with one peak in the middle of the day (or night),
(2) unimodal, rising to a peak at the end of the day,
(3) unimodal, falling through the day from an early morning peak,
(4) spread evenly through the day,
(5) bimodal, with peaks in the morning and afternoon or evening, and
(6) trimodal, with a third peak between these two.

Diet is represented by the modal food type; body weight is that of the female to avoid the effects of sexual selection on male body size. Habitat is categorised, according to the continuity of the canopy as:-

(1) forest, including gallery forest,
(2) parkland, including woodland and scrub, and

(3) open country, including fields.

The quality of the data vary enormously, and particular problems concern seasonal variation within a population and differences within species.

The bimodal pattern is by far the commonest, being present in 71% of 77 populations and 77% of 48 species considered. There is a slight tendency for primates inhabiting open country to deviate from the commonest bimodal and trimodal patterns. While the overall lack of clear relationships between feeding pattern and diet, body weight and habitat may be due to the coarseness of the analysis, it also appears to be in part due to a closer relationship between taxon and feeding pattern, than between ecology and feeding pattern. The bimodal feeding pattern predominates in most of the major taxa, despite wide variation in ecology between species. The marmosets, gibbons and macaque-baboon tribe depart most from bimodality; the latter spans the whole range of habitats, which should provide a good test of the relative influences of phylogeny and ecology, but baboons show bimodal patterns in both open and closed habitats.

The general tendency towards similarity among related species, however, is 'softened' by different patterns in the same species, particularly between seasons (or even months) in the same group (e.g. *Colobus badius*: Clutton-Brock, 1974; *Cercocebus albigena*: Chalmers, 1968; *Alouatta villosa*: Chivers, 1969). Such differences are best explained in terms of changes in food distribution and abundance; they are not consistent across species, which is to be expected from the differing conditions between sites.

In summary, while it appears that factors such as changes in temperature and food distribution may affect the daily feeding patterns within species, the major determinants of differences between species appears to be phylogenetic inheritance rather than behavioural response to contemporary conditions. There is little evidence that differences between species in daily feeding pattern are related to gross dietary differences. The observation that primates in captivity may show activity patterns found in their wild counterparts also suggests species-specificity and lack of plasticity (Clutton-Brock, 1977b). Unsatisfactory as it may be, the differences between primate species at Kuala Lompat are perhaps best explained in such terms.

It should be noted that the bimodal pattern of activity predominant in primates is by no means confined to them. Spot observations of tree squirrels at Kuala Lompat indicate a bimodal pattern in all five species investigated, although continuous following of two species does not substantiate this (Payne, 1979b). A bimodal pattern may also be general among the larger African forest squirrels and Holarctic tree squirrels (Emmons, 1975; Layne, 1954; Corbet and Southern, 1977). In the latter, however, the pattern may change in winter (Layne, 1954; Tittensor, 1970). Bimodal patterns have also been reported in impala, zebra, sitatunga and elephant on the African plains (see Harcourt, 1978); and the number of species of birds and mammals seen and heard from a tower in Thai rain forest by J.J.R. peaked in the morning and evening. A unitary explanation of this phenomenon, if such is appropriate, is still lacking. The idea that an animal rises feeling hungry, feeds to satisfy this hunger, remains relatively inactive until it is hungry again in the evening, when it eats as much as it can before a long fast of some 12 hours, is still the most attractive; yet, as shown above, it is difficult to relate variation on this pattern to diet, which argues against the whole relationship unless the answer can still be found at a physiological or biochemical level.

CONSERVATION IN SOUTH-EAST ASIA

Since the primates of South-east Asia mostly inhabit tropical rain-forests, their fate depends essentially upon the pressures to fell these forests and the reasons for preserving them. Thus, in this section we shall outline these pressures, their effects and the main counter-measures for conservation; conservation, it should be stressed, involves management as well as preservation (or protection). First, however, some comments about other threats to primates in the region - hunting for food and sport, and trade.

The pressures

Statistics on hunting by local people are hard to obtain. Tilson (1974) and Whitten (1980) have recorded the numbers of skulls of each primate species in the houses of the people of Siberut, but no time scale can be assigned accurately; the Orang Asli (indigenous people) around the Krau Game Reserve in Malaysia kill monkeys and other forest animals quite regularly for food, but say that they can only catch about one gibbon a year, because they are so elusive. Generally speaking, it seems that, so long as a population depends on hunting and gathering for the bulk of its food and uses primitive weapons, it will not exceed the sustained yield carrying capacity of the natural environment, and the wildlife will not therefore be endangered.

Paradoxically, depletion of wildlife begins with farming, which permits a leap in population density from around 1 person/km^2 to as many as 700/km^2 in the rice-growing countryside of South-east Asia. The much larger population still derives part of its protein intake from wild game, yet, being less dependent on it, no longer feels the need to conserve. Moreover, as economies grow more sophisticated and people become more removed from wildlife, so hunting for sport, seemingly an unfortunate attempt to recreate the hunting life of the past, accelerates the drain on wildlife; the introduction of firearms completes the rout. Salvation only comes for wildlife, perhaps, when education and well-being of the people lead to a renewed understanding of the importance of the natural environment to them, both economically and aesthetically.

Primates are traded within South-east Asia as pets (e.g. gibbons in Thailand, Brockelman, 1975; silver leaf monkeys and lorises in Java, Brotoisworo, 1978), to make local medicine (silver leaf monkeys in Bali, Brotoisworo, 1978), for food (macaque brains are a delicacy in certain sectors of the urban population in Malaysia, Taiwan and Japan), and to a small extent in biomedical research. Primates are also exported from the region, mainly to Japan, Europe and the U.S.A. for biomedical research and stocking zoos, but also for pets. Most of the exports are of macaques. The pressure on South-east Asian macaques has increased recently, following the ban on the export of rhesus macaques, *Macaca mulatta*, from India - previously a major source of laboratory monkeys (Mohnot, 1978). The export of long-tail macaques from West Malaysia has increased from about 8,000 to 14,000 in the last two years. Stump-tail macaques, *M. arctoides*, in peninsular Thailand have been badly hit by exports of 2-3,000 annually (Eudey, 1978). Indonesia contains most of the region's forest and primate species, and events there over the next few years will be critical. It should be borne in mind that export figures represent the number of animals leaving the country; many more die before that point, and some after. Harrisson (1973) estimates that for every monkey reaching its destination in the user country, 5-8 die; many more are disrupted socially.

Although these figures are disturbing, the most important threat to primates (and other wildlife) in South-east Asia, as elsewhere, is habitat destruction, causing death and suffering to tens of thousands annually. Such forest disturbance can be divided into two broad categories - selective logging and clear felling. Selective logging is the extraction of the few commercially viable trees, leaving the remainder standing ... in theory. In West Malaysia only about 150 tree species are regularly exploited out of some 2500 (Whitmore, 1975). There are many rules governing selective logging, but they are often abused; Burgess (1971) calculated that in extracting 10% of the standing crop of a lowland mixed dipterocarp forest in Malaya, another 55% was destroyed by falling trees and logging tracks.

The effect on the fauna is proportional; Stevens (1968) estimates that 37% of non-flying Malayan mammal species are tree-living, and Medway and Wells (1971) estimate that 70% of South-east Asian mammals depend on undisturbed forest. Not surprisingly, then, Harrisson (1969) reported that a disturbed forest in Malaya supported only 40% as many mammal species as an undisturbed forest. Among the primates, the orang-utan seems to be most vulnerable (Southwick and Cadigan, 1972; Rijksen, 1978). Censuses before and after selective logging in the Gunung Leuser Reserve in Sumatra indicated a halving or even elimination of orang-utan populations (Rijksen, 1978). MacKinnon (1974) documents the disturbed behaviour and mobility of a population of orang-utans in Sabah as logging approached from the north. Gibbons may benefit slightly from very light logging of upper canopy trees, which lets in light to promote growth of food trees (Chivers, 1977), but they are seriously disrupted by moderate to heavy disturbance. The monkeys seem more resilient, especially the ever-adaptable macaques, which can thrive in areas densely settled by humans; even leaf monkey groups are found living in unusually small patches of forest (Wilson and Wilson, 1975). Forest birds may be among the most vulnerable groups (Wells, 1971); for instance, selective logging led to the disappearance of trogons from forest on Gunung Angsi in Malaya. Selective logging is practised mainly by large commercial concerns operating in concessions granted by governments. In 1974, about half of the 1.2 million km^2 of forest in Indonesia had been, or probably would be, leased as logging concessions (Brotoisworo, 1978). The commercial loggers are sometimes followed by local people operating illegally on a small scale (Rijksen, 1978); this opening up of the forest also promotes hunting. The increasing pressure to log for timber is mainly due to rocketing demand for wood from Japan and the Western block (table 10.7), especially now that tropical hardwoods can substitute for some of the uses of temperate hardwoods, thanks to new pulping and chipping techniques. Exports from Malaysia increased 9-fold between 1963 and 1975, far in excess of prediction (Stacey in Whitmore, 1975). South-east Asian governments, anxious for cash to drive the development programmes advocated as the solution to socio-economic problems attendant on population growth, readily sell off their forests. Timber is Indonesia's second highest source of foreign exchange, and one of the highest in Malaysia.

Clear-felling, in contrast with selective logging, is generally undertaken to obtain land rather than timber. Clearance for fields or tree crops such as rubber and palm oil is carried out both at government level and in an uncontrolled manner by small farmers hungry for land. Government schemes in West Malaysia, like the vast

Table 10.7. Consumption of tropical hardwood (from IUCN, 1980).

COUNTRY/REGION	YEAR 1950	1976	2000 (projected)
Japan	1.5	22.7	48
U.S.A.	0.8	5.7	20
Europe	1.9	13.8	35
total	4.2	42.2	103

million m^3 of roundwood equivalent to logs

Pahang Tenggara, are projected to clear 400-800 km^2 annually over the next 10-20 years (FAO, 1976). Population pressure and cultural change have disrupted the slash-and-burn agriculture practised widely over forested South-east Asia; this relies on rotating 'fields' in the forest, cropping each for 1-3 years, then leaving it fallow for 10-20 years. Now fields are cropped for too long and too often, the soil is quickly exhausted, and the forest is used up in the effort to find new land. Slash-and-burn farmers are reckoned to be clearing 85,000 km^2 annually in South-east Asia (IUCN, 1980), 9.4% of the 1980 estimate of forest area. Moreover, large-scale clearance schemes have engendered a new pioneer spirit in some rural areas, whereby farmers follow the chain-saw, crop the land until it is exhaused and move on to newly-felled areas (Indonesia: Rijksen, 1978; Thailand: Boulbet, 1979). The spent land is often as not colonised irreversibly by useless *Imperata* grass, which already occupies 160,000 km^2 (8.4%) of Indonesia (Kartawinata in Brotoisworo, 1978). Such clearance naturally wipes out the forest fauna and substitutes a relatively poor one.

The pressures for timber and farmland are compounded further by local need for fuel-wood. This is relatively unimportant in West Malaysia, where most people cook on paraffin stoves, but in other countries in the region it is a very important factor. In Thailand, 76% of total wood consumption is for fuel, and 97% of the population uses wood for fuel, mainly in the form of charcoal (Eckholm, 1975).

The problem

The overall effect of these pressures on the forest is rapid, perhaps catastrophic, depletion. In 1969 South-east Asia contained some 2.5 million km^2 of tropical rain forest, 30% of the world total (Pringle, 1969), by 1980 only 0.9 million km^2 (a reduction of 64% if the two sets of figures correspond), only 13% of the world total (IUCN, 1980); predictions for most countries indicate denudation by the end of the century. At present rates of consumption Thai forests, now covering only 25% of the country, should disappear by 1987 (Lekagul and McNeely, 1978). Over South-east Asia as a whole, gibbon populations were projected to fall by 84% between 1975 and 1990, on the assumption that by 1990 no viable habitat remains outside protected areas (Chivers, 1977). Statistics on land area, population size, density and increase, GNP, life expectancy, areas forested and annual forest loss are presented for a selection of South-east Asian countries, in comparison with some from tropical Africa and America in table 10.8 (from Stoel, 1980).

The nature of these pressures make it clear that arguments for conserving forests must be economically convincing. What economic disadvantages of deforestation can be advanced to counterbalance the obvious benefits of more land for food production and of foreign exchange from timber exports and cash crop exports? In essence, it

Table 10.8. Population size and increase, and forest areas and loss, in certain countries in Asia, Africa and America (from Stoel, 1980).

COUNTRY	AREA	POPULATION			GNP $	av.life	FORESTS remaining 1980	annual loss			remaining 2000	
	M km²	size M	density /km²	increase %/yr	/head	expect-ancy yr	% land area	1°	% land area cultiv.	logging	% land area forestry	wildlife
MALAYSIA	0.328	12.6	38	2.5	930	69	70					
Peninsular	0.132						54		0.8		25	14
Sarawak	0.124						78	44	++			
Sabah	0.071						86	40	1.4		40	7
INDONESIA	1.904	143.3	75	2.0	300	48	43		0.8	2.4		2
THAILAND	0.520	45.1	87	2.3	410	55	25		0.8			7
BURMA	0.697	32.2	46	2.4	140	51	52		0.2			14
PHILIPPINES	0.300	47.4	158	2.4	450	58	33		2.0			
PAPUA NEW GUINEA	0.464	3.1	7	2.9	480	48	86		++	0.3		
CAMEROON	0.475	8.3	17	2.3	340	*c.*40	32	*c.*10				2
GHANA	0.238	11.3	47	3.1	380	43	9					8
IVORY COAST	0.330	7.7	23	2.9	710	43	33	10	1.2			1
ZAIRE	2.345	28	12	2.8	130	43	41		0.9			2
MADAGASCAR	0.588	8.5	15	2.6	210	38	5		0.9			
LIBERIA	0.110	1.8	16	3.2	430	45	23		0.3			
BOLIVIA	1.074	5.2	5	2.8	540	47	7		0.4			
PANAMA	0.075	1.9	25	2.3	1220	66	54	41				
PERU	1.285	17.3	13	2.8	830	54	60		3.3		28	20
VENEZUELA	0.916	13.5	15	3.0	2820	66	69					
COLOMBIA	1.138	26.1	23	2.2	710	61	32					
ECUADOR	0.300	8.0	27	3.1	770	57	60					
HONDURAS	0.112	3.1	27	3.5	450	48	36					
MEXICO	1.963	67.7	34	3.4	1110	65	6					

is a trade-off between the short-term advantages and the long-term disadvantages. Unfortunately, economic plans rarely look more than 10 years into the future, and many countries plan only on a 5-year basis; this is partly because of the difficulties of predicting that far ahead, and partly because most governments do not hold office that long. This problem is compounded by the difficulty of expressing long-term benefits in cash terms, and thus of incorporating them into development planning.

The main economic arguments against wholesale deforestation are well known:

(1) exhaustion of timber stocks will lead to loss of future revenue and to lower profitability in the end than would careful exploitation

(2) clearance of upland forests leads to loss of soil fertility and to erosion. Most of the nutrients in tropical forest ecosystems are held in the live biomass, so that clearance removes them; heavy rain on the unprotected soil accelerates leaching of the remaining nutrients out of reach of plant roots, and washes soil downstream. This erosion silts up rivers in the lowlands, which flood and ruin crops and property; reservoirs can silt up in as little as 50 years. In the long term rainfall may be reduced.

(3) clearance leads to loss of the genetic reservoir of diversity, which is a source of insurance against the future and of many potential economic benefits. Tropical forests may contain 40-50% of the world's animal and plant species (UNEP in IUCN, 1980). Wild plants and animals provide a pool from which to extract new crops, predators on crop pests, drugs and new materials, and from which to strengthen existing domestic breeds by cross-breeding with wild relatives (Lucas and Synge, 1978; Myers, 1979). For example, little selective breeding of popular domestic fruit species has been done in Malaysia, and there is wide scope for cross-breeding with wild relatives (Jong *et al.*, 1973).

(4) loss of natural vegetation means loss of a yard-stick by which to measure the effects of exploitation on land productivity.

The solutions

These arguments indicate that the prime requirements for long-term profitability are a reasonable forest cover on watersheds and a representative sample of floral and faunal diversity. Since diversity is greatest in the lowlands, some of these must be preserved along with the watersheds. There are a number of approaches to achieving these aims, the main ones being (1) the creation of unexploited nature reserves, (2) the improved use of exploited forest, and (3) the practice of agro-forestry on land already cleared, which itself (4) should be used more efficiently.

The area of reserves required to maintain species diversity

can in theory be computed from habitat diversity (Vanzolini, 1978), from rates of species extinction predicted for given isolated areas by the theory of island biogeography (Diamond, 1975), and from the area thought to be required for a viable population of the most sparsely distributed species (Medway and Wells, 1971), and so forth. In practice, much of South-east Asia is already settled so densely as to make such plans academic, and the design for reserves is then designed by what is available. As a target, setting aside 10% of the land area as protected nature reserves under forest seems realistic; Malaysia and Indonesia are already approaching this figure (table 10.9) and Thailand will soon exceed 5% (Lekagul and McNeely, 1977). Half of the reserve area in West Malaysia consists of the magnificent and remote Taman Negara.

The formula generally accepted for reserves is that they should contain an inviolable core area, surrounded by an exploited buffer zone bordering densely-settled farmland. In this way local people obtain protection from crop-raiding and access to forest products without harming the core. Often, however, reserves have been created without this buffer zone, and one can only be created by impinging on the existing reserve because disturbance reaches or has penetrated its borders (e.g. Gunung Leuser Reserve in Sumatra, Kutai Reserve in Kalimantan, and Khao Yai National Park in Thailand).

It is difficult to make nature reserves in tropical forests pay for themselves by foreign tourism, simply because the animals are so difficult to see. For example, Khao Yai National Park in Thailand is visited by 4,000 tourists annually, yet the tourist accommodation does not pay for itself, let alone compensate for the land and timber tied up in the 2,000 km^2 park. A stronger justification, in addition to the economic one outlined earlier, such as watershed protection, is the recreation and education value to local people. Very large numbers of city people visit Thai National Parks; Khao Yai receives

Table 10.9. Percentage of total land area designated as nature reserves in South-east Asian countries (from Chivers, 1977, except for Thailand, from Lekagul and McNeely, 1977).

MALAYSIA		7.1
West	6.2	
East	0.9	
INDONESIA		7.9
Sumatra	3.9	
Java	1.9	
Kalimantan	2.1	
THAILAND		4.4

150,000 Thais annually, yet it is not the most popular park (Royal Thai Forest Department statistics). Similar events are starting to occur in Malaysia and Indonesia.

The watershed areas contained within nature reserves, however, will never be enough to ensure a safe water regime in hilly tropical country like most of South-east Asia, where at least 15% is needed (Myers, 1979), although escalating environmental problems as the area of forest approaches 40% suggest that this is the figure to which we should be working. Hence the need to improve selective logging techniques, but this is very difficult to bring about. Countries in the region need foreign capital and expertise to exploit their timber resources, and are constrained to offer favourable terms to foreign investors so as not to scare them away. The investing multi-national corporations, for their part, are under pressure to recoup their investment as quickly as possible; this is because logging in rough country in which only a small percentage of trees are saleable requires large initial capital, which requires very high interest rates to borrow. Furthermore, there are very real uncertainties associated with operating in South-east Asia, such as sudden expropriation, erosion of profits by poaching small-time loggers, and insurgency. Corporations are consequently unwilling to invest in a longer-term view which would ensure the most conservative use of the forest, e.g. lightly logging in areas of special interest, logging only where regeneration is assured, logging at sufficiently long intervals, refraining from excessive logging at one time, and planting fast-growing trees on already denuded land. This reluctance is further aggravated by short logging leases; the Indonesian government requires that 35 years elapse between cuttings in one place, but grants leases of up to only 20 years (Myers, 1979), perhaps as protection against the excessive influence that such corporations can wield.

The solution to this dilemma must be to channel more money into conservative logging practices, which in turn must result in more expensive timber for user countries. This is only fair, since one reason for importing is that it is cheaper than buying home-grown timber. The extra price could be regarded as a premium for insurance that supplies will continue. Hopefully the World Bank is now turning to this crucial problem, from its previous practice of supporting forest clearance and 'land development'.

The third main approach to conserving tropical rain-forest can be called agro-forestry - the encouragement of shifting cultivators to grow timber trees in with food crops, in such a way as to increase their income while also stabilising the land for the common good. Projects in the Philippines and elsewhere indicate that agro-forestry can yield in a few years more money for the government than clear-felling (see references in Myers, 1979, chapter 12). As a means of conserving wild animals, this will be less effective than the previous approaches, but it is better than leaving the land bare of natural vegetation. To leave islands of forest of a reasonable size in large plantations offers better prospects for wildlife (Caldecott and Bennett, in prep.). That land already cleared of forest should be used to its capacity speaks for itself, but sadly it is not practised often enough.

In conclusion, if the outlook for primates and forest in South-east Asia is not rosy, there are nevertheless reasons for hope. The rate of population increase is falling as fast as in developed countries, even though there is a long way to go to reach stabilisation (World Population Data Sheets for the past few years). Valuable data on the effects of logging on wildlife are accumulating in some countries, notably Malaysia and Indonesia. Most countries plan to increase their area of reserves, and have signed the Convention on International Trade of Endangered Species (CITES). National legislation against hunting and trading have been strengthened; representatives of the Association of South-east Asian nations (ASEAN) have met to discuss coordinated incorporation of conservation and the environment into development planning. Foreign aid has increased, and the role which consumers in Western countries play in causing environmental degradation is increasingly brought to their attention through the popular media. It remains only to develop constructive long-term management of the natural resources; it is almost too late, but the world-wide awareness is there and the necessary materials are at hand.

THE FUTURE

Two main trends can be discerned in primate socio-ecological studies in Malaysia and elsewhere in the humid tropics. Firstly, there is growing concern with the practicalities of conservation in the broadest sense, to discern the contribution that the field worker can best make. Secondly, such academic advances as have been achieved can be seen to result from increasing cooperation between previously rather disparate disciplines.

The generation of studies in Malaya documented in this book are mainly of academic bent, blending the experiences of zoologists, anthropologists, botanists, functional anatomists and evolutionary biologists. The realisation of the critical status of the subjects of those studies and the accumulation of sufficient basic data have given rise to a second generation of studies now in progress with a more conservationist bent, expanding the multi-disciplinary approach with a more applied ecology and with plant biochemists, physiologists (reproductive, digestive, respiratory and cardiovascular), molecular biologists, and clinicans.

Many questions posed by these studies, particularly those concerned with population size and dynamics, and with carrying capacity of the environment and diet, can only be answered properly by laboratory study - of reproductive physiology and of food composition and digestion, respectively. The value of such studies for conservation are as great as their academic ones. Universiti Malaya has shown interest in primate field studies throughout the decade, and now Cambridge University is collaborating with staff and students in animal science and veterinary, forestry and biological science faculties at Universiti Pertanian Malaysia and in the biology faculty at Universiti Kebangsaan Malaysia, both near Kuala Lumpur. Extensive surveys and intensive studies of banded leaf monkeys, pig-tail macaque and slow loris, and a detailed study of the effects of logging on a primate population, are being supplemented with nutritional, reproductive and other physiological studies; attention is also being given to the healthy maintenance of primates in captivity, on the rationale that loss of life (and genes) might be reduced at least a little in this way, so long as forest clearance and massive losses of wildlife continue. The simpler solution might be to move such homeless animals directly to remaining forest isolates lacking some or all primate species; it is embarrassing at this late stage in the game that we know neither how to effect such translocation nor whether it will succeed. The Department of Wildlife and National Parks has shown its interest in remedying this serious deficiency.

While collaboration between field and laboratory scientists is crucial to future progress in our understanding of primate biology, so too is the development of ecological community studies, with more interaction between botanists and invertebrate and vertebrate zoologists (e.g. Janzen, 1978:80) to tease out the details of competition for different parts of plants between, for example, insects, birds and mammals. The complexities of this are as awesome as the rapid disappearance of tropical forest; quantifying consumption of one food by one animal is difficult enough in the humid tropics (e.g. Eisenberg and Thorington, 1973; Montgomery and Sundquist, 1975; Hladik, 1977). Perhaps one has to take one tree species in one spot as the hub from which interactions radiate.

APPENDIX I

TREE SPECIES AT KUALA LOMPAT AND THEIR CONTRIBUTIONS TO THE DIETS OF PRIMATES, SQUIRRELS AND SOME BIRDS

Only those species which have been identified to family, genus or species are listed. At least 60 more unidentified species of tree growing to at least 20 feet (6.6 m) tall occur in the study area.

Climbers (omitted here) of the families Annonaceae, Celastraceae, Combretaceae, Gnetaceae, Leguminosae, Moraceae, Strychnaceae, Urticaceae and Vitaceae are important sources of food to primates.

PRIMATES	*Hs* = *Hylobates syndactylus*	*Hl* = *H. lar*
	Po = *Presbytis obscura*	*Pm* = *P. melalophos*
	Mf = *Macaca fascicularis*	
SQUIRRELS	*Rb* = *Ratufa bicolor*	*Ra* = *R. affinis*
	Cp = *Callosciurus prevostii*	*Cn* = *C. notatus*
	St = *Sundasciurus tenuis*	
	? = unidentified species, Kalang records	
BIRDS	C = Columbidae (pigeons)	
	Ps = Psittacidae (parrots)	
	B = Bucerotidae (hornbills)	
	P = Pycnonotidae (bulbuls)	
	A = Aegithinidae (leaf birds)	
	E = Eurylaimidae (broadbills)	
FOOD	F = fruit pulp	s = seed f = flower
	L = leaves (new, mature, stems and/or tips)	
	b = bark	x = fruit by unidentified squirrel

The distinction between fruit pulp and seed is only accurate for squirrels; in eating the fruit pulp, primates, especially the leaf monkeys and macaques are often also consuming the seed.

(continued on p. 331)

SCIENTIFIC NAME Family, genus, species	LOCAL NAME	PRIMATES *Hs*	*Hl*	*Po*	*Pm*	*Mf*	SQUIRRELS *Rb*	*Ra*	*Cp*	*Cn*	*St*	?	BIRDS C	Ps	B	P	A	E
Alangiaceae																		
Alangium ebenaceum	kerentang	F			FL							x						
Anacardiaceae																		
Bouea macrophylla	kondang				L	F			Fs									
B. oppositifolia	banderai	F	F			FL		s										
Buchania sessilifolia	rengas ayam							s		s								
Campnosperma auriculata	terentang	F	F									x						
Dracontomelum mangiferum	sengkuang	F	F	FL	F	F	f	s	F									
Mangifera gracilipes	macang hutan daun halus											x						
M. indica	macang kampong	F	F		L							x						
M. longipes	macang hutan		L									x						
M. (?) *macrocarpa*	macang temuor						Fs											
M. microphylla	macang daun halus											x						
M. quadrifida	lanjut							s				x						
Mangifera sp.	kertang																	
Mangifera sp.	macang rawa						Fs	Fs	s	F	F							
Melanochyla fulvinervis	jenis rengas							F										
Melanorrhoea malayana	rengas (1)			F			b		b	b	b							
Pentaspadon velutinum	pelong, pelom	FfL		FfL		FfL	Ff											
	menjai (1)																	
	menjai (2)																	
	pauh kijang																	
? *Gluta elegans*	rengas (2)					F												
	rengas (3)																	
Annonaceae																		
Alphonsea elliptica	chaget				F	F			F	F								
Cananga odorata	lenggian				F	F						x						
Cyathocalyx pruniferus	antoi daun besar bulu			F								x						
C. scortechinii	antoi daun besar lichin																	
Cyathocalyx carinatus	antoi daun kechil	L				F												
Mezzetia leptopoda	bujung semalam			F	F					F		x						
Monocarpia marginalis	menjarum										b	x						
Polyalthia cinnamonea	melilin																	
P. clavigera	chagau			F	F							x						
P. glauca	pennyilab daun kasar											x						

SCIENTIFIC NAME Family, genus, species	LOCAL NAME	PRIMATES *Hs*	 *Hl*	 *Po*	 *Pm*	 *Mf*	SQUIRRELS *Rb*	 *Ra*	 *Cp*	 *Cn*	 *St*	 ?	BIRDS C	 Ps	 B	 P	 A	 E
Annonaceae (cont.)																		
P. hypoleuca	pennyilab daun halus	FL	F							s		x						
P. jenkensii	mempisang		F			F						x						
P. laterifolia	lebuor											x						
P. obliqua	gaboi				F							x						
P. sumatrana	pennyilab											x						
Xylopia ferruginea	semeliang			L														
X. fusca	tempunai lempung	L	L															
X. magna	tempunai daun bulu	f	f	f	f	f						x						
X. malayana	tempunai (1)	L		FL	F	FL	s	s										
Xylopia sp.	tempunai (2)	L	L			F	s											
	mengkupas					F						x						
Apocynaceae																		
Alstonia augustiloba	pulai	L	L	fL	L													
A. pneumatophora	rejung			L														
Dyera costulata	jelutong			L														
Ervatamia corymbosa	darak mati betujuh																	
Kibitalia maingayi	jeleti			f		F						x				F		F
Bignonaceae																		
Stereospermum fimbriatum	chichor			fL	L	Ff												
Blechnaceae																		
Stenochlaena palustris					L													
Bombacaceae																		
Durio griffithii	durian linggit	L										x						
D. oxleyanus	durian tuang		L	f		F						x						
D. singaporensis	durian burung daun panjang	L			L	L			f									
Durio sp.	durian burung daun pendek																	
Kostermansia malayana	barah batuk																	
Neesia synandra	bebeapa									f	f							
Burseraceae																		
Canarium grandifolium	kedondong kenderap	f																

SCIENTIFIC NAME	LOCAL NAME	PRIMATES					SQUIRRELS						BIRDS					
Family, genus, species		*Hs*	*Hl*	*Po*	*Pm*	*Mf*	*Rb*	*Ra*	*Cp*	*Cn*	*St*	?	C	Ps	B	P	A	E
Burseraceae (cont.)																		
C. littorale	kedondong		f	F	F													
C. megalanthum	kejijak	f		L	F		F											
Canarium sp.	rawa											x						
Dacryodes rostrata	yak kijai						F	F		F					F			
D. rugosa	janging								F							F		
Santiria laevigata	kerem						Fs	Fs	Fs									
S. rubiginosa	jenis janging										s							
S. tomentosa	kerem						F	F										
Triomma malaccensis	kijai						s	s										
Celastraceae																		
Lophopetalum floribundum	kelongko	f																
Combretaceae																		
Terminalia bellirica	deher	L			FL		F											
T. citrina	belang limau																	
Datiscaceae																		
Tetrameles nudiflora	kejor																	
Dilleniaceae																		
Dillenia (= *Wormia*) *pulchella*	simpoh paya					F	F											
D. reticulata	simpoh	f		FfL	FfL	fL			F									
D. sumatrana	simpoh kola			FL	L													
Dipterocarpaceae																		
Anisoptera laevis	loh				L													
Dipterocarpus baudii	keruing mengkemas			L														
D. cornutus	keruing betul																	
D. kunstleri	keruing minyak																	
Hopea sangal	pengawan																	
Hopea sp.	mata kuching	L																
Shorea acuminata	pelir kambing																	
S. bracteolata	rambai daun			L			F											

SCIENTIFIC NAME	LOCAL NAME	PRIMATES					SQUIRRELS						BIRDS					
Family, genus, species		*Hs*	*Hl*	*Po*	*Pm*	*Mf*	*Rb*	*Ra*	*Cp*	*Cn*	*St*	?	C	Ps	B	P	A	E
Dipterocarpaceae (cont.)																		
S. hopeifolia	mesipot	L																
S. lepidota	melanggong		L									x						
S. leprosula	meranti	L	L															
S. macroptera	melantai																	
S. maxwelliana	resak																	
S. ochrophloia	remesu daun kasar																	
S. ovalis	kepong																	
S. pauciflora	remesu daun halus																	
Vatica bella	jambu hantu				F							x						
Vatica sp.	resak paya																	
Ebenaceae																		
Diospyros spp.	kayu hitam	F		F	F	f						x						
Diospyros sp.	pulot	L	L			F						x						
Diospyros	telinga kelawar	F	F			F						x						
Elaeocarpaceae																		
Elaeocarpus	pinang peregam			f		F		F										
E. floribundus	pinang yok																	
Euphorbiaceae																		
Agrostistachys longifolia	tenglom																	
Antidesma spp.	tebasah	F				F					F							
Aporusa spp.	tebasah					F					s	x	F			F		
Austrobuxus nitidus	tebasah terang							Fs										
Baccaurea brevipes	rambai	FL	F		L	F	s	s	Fs		b							
B. griffithii	tampoi							F										
B. parvifolia	tambun											x						
B. pyriformis	taban											x						
B. racemosa	tambun											x						
Baccaurea sp.	tampoi tungau											x						
Blumeodendron calophyllum	katong kura	L		F								x						
Bridelia cinnamonea	pisit											x						
Cleidion javanicum	bayam bali											x						
Croton argyratus	kenmeso											x						
Elateriospermum petiolatus	kentet		L		F							x						

SCIENTIFIC NAME Family, genus, species	LOCAL NAME	PRIMATES Hs	Hl	Po	Pm	Mf	SQUIRRELS Rb	Ra	Cp	Cn	St	?	BIRDS C	Ps	B	P	A	E
Euphorbiaceae (cont.)																		
E. tapos	perah		L	F	FL	F				f		x						
Endospermum diadenum	menchepong	FL	L									x						
E. malaccense	menchepong											x						
Macaranga gigantes	mengkubung																	
M. hosei	berenong daun kuning kechil																	
M. hypoleuca	berenong daun putih					F	F		F	s				F				
M. pruinosa	berenong daun kuning besar																	
M. (?) recurvata	tarak			F														
Mallotus griffithianus	peredai cherang																	
M. kingii	peredai																	
M. leucodermis	peropuk			L			s	s	s									
M. macrostachys	balik angin																	
Neoscortechinii kingii		L																
Pimelodendron macrocarpum	depor				F							x						
Ptychopyxis caput-medusae																		
Sapium baccatum	remayan	F	F						F	F		x						
Sapium sp.	remayan																	
S. discolor	tulang pipit																	
	chinchang pasel																	
	panchau							F										
Fagaceae																		
Castanopsis curtisii	bereh																	
C. inermis	berangan babi					Ff		F		F								
C. megacarpa	gertak tangga							F										
C. wallichii	berangan babi																	
Lithocarpus sundaicus	berangan					F	f	F	s									
Lithocarpus sp.	berangan	F																
	mempuning																	
Flacourtaceae																		
Casearia capitellata	benglin				L							x						
Flacourtia rukam	rungkam duri																	
Homalium	pipi kelah																	
Hydnocarpus	telur buaya																	
Paropsia vareciformis	minyak puteri			FL	FL							x						
Scolopia spinosa	rungkam betul											x						

SCIENTIFIC NAME Family, genus, species	LOCAL NAME	PRIMATES *Hs*	 *Hl*	 *Po*	 *Pm*	 *Mf*	SQUIRRELS *Rb*	 *Ra*	 *Cp*	 *Cn*	 *St*	 ?	BIRDS C	 Ps	 B	 P	 A	 E
Guttiferae																		
Calophyllum curtisii	mentangau halus	f	F			F	F	F										
C. floribundum	mentangau kasar	FL	F															
C. macrocarpum	bunod	L	L															
Garcinia atroviridis	gelugur	FL	F									x						
G. nervosa	lapan taun																	
G. parvifolia	tempiles	F	F		L													
G. prainiana	menchepu																	
G. pyrifolia	kandis																	
G. rostrata	tempiles kulit lichin																	
Garcinia sp.	tempiles daun kasar																	
Mesua ferrea	penaga mesenung																	
Hypericaceae																		
Cratoxylon arborescens	mampat											x						
C. cochinchinense	kelehor, kulhoi																	
C. formosum	mampat	f		L		f						x						
Icacinaceae																		
Gomphandra sp.	chemperai				L													
Lauraceae															F			
Alseodaphne sp.	berambong daun satu				L							x						
Beilschmiedia sp.	medang				FL													
Cryptocaria sp.	gamak	L			F													
Cryptocaria sp.	menyaya																	
Dehaasia elliptica	medang or kedembe				F							x						
D. incrassata	medang or kedembe				L							x						
Dehaasia sp.	medang or kedembe			FL	F	F						x						
Endiandra sp.	medang or kedembe			F								x						
Litsea sp.	medang or kedembe				FL							x						
Nothaphoebe umbelliflora	medang or kedembe			FL	fL							x						
Lecythidaceae																		
Barringtonia macrostachya	putat																	
Barringtonia sp.	putat daun halus																	

SCIENTIFIC NAME Family, genus, species	LOCAL NAME	PRIMATES *Hs*	*Hl*	*Po*	*Pm*	*Mf*	SQUIRRELS *Rb*	*Ra*	*Cp*	*Cn*	*St*	?	BIRDS C	Ps	B	P	A	E
Leguminosae																		
Cassia nodosa	merise	L																
Cynometra malaccensis	balun	L		L	Fs	F	F	F										
Dialium laurinum	keranji											x						
D. patens	kuran	L		FL	F	F						x						
D. platysepalum	keranji	FL	FL	F	FL	F	b	F	F									
D. procerum	merebau kerak	fL	fL				b											
Intsia palembanica	merebau	FL	L	FL	FfL	FfL	F	b	F	Lb	b							
Koompassia excelsa	tualang	L		L	L		F											
K. malaccensis	kempas	L	L	FL	F		F	F										
Milletia atropurpurea	tualang daing	L	L		L	F												
M. hemsleyana	jadal				L													
Parkia javanica	kerayong			FL	Ff	F	Fsb	F			b							
P. speciosa	petai			Ff	Ff	Ff				b				F				
Peltophorum pterocarpum				L														
Pithecellobium bubalinum	keredas					FL			F									
P. clypearum	gonderik buah merah											x						
P. contortum	gonderik buah hitam											x						
P. jiringa	jering					F						x						
Saraca thaipingensis	tenglan					Ff						x						
Sindora coriacea	petil	L			L	F						x						
S. velutina	petil											x						
Linaceae																		
Ctenolophon sp.	chengal																	
Ixonanthes icosandra	pagar anak	FfL	FfL	F	Ff	F						x						
Lythraceae																		
Lagerstroemia speciosa	bungor			L														
Malvaceae																		
Hibiscus floccosus	bebaru			L	fL						F							
H. macrophyllus	totor				L	F												
Melastomataceae																		
Memecylon garcinoides						F												

SCIENTIFIC NAME Family, genus, species	LOCAL NAME	PRIMATES *Hs*	 *Hl*	 *Po*	 *Pm*	 *Mf*	SQUIRRELS *Rb*	 *Ra*	 *Cp*	 *Cn*	 *St*	 ?	BIRDS C	 Ps	 B	 P	 A	 E
Melastomataceae (cont.)																		
M. heteropleurum						F												
M. oleifolium	mimpis kulit besar											x						
Memecylon sp.	mimpis kulit kechil							F										
Pternandra capitellata	hembuyan daun kasar	F	F									x						
P. echinata	hembuyan daun halus	f	Ff					F										
	chenderai cherang	F																
Meliaceae																		
Aglaia (?) *aquea*	pelir tupai				F					F								
A. (?) *pseudolansium*	berberas	FL	L				F	F	F									
Aglaia sp.	pemanis gading	Ff	F		F	F						x						
Aglaia sp.	pemanis halus	F	F			F												
Amoora malaccensis	hempekak										b	x						
Amoora sp.	tengkuruk helang		F			F						x						
Chisocheton (?) *annulatus*	ganding bulu									s								
C. princeps	ganding betul									s								
Chisocheton sp.	ganding buah hitam																	
Chisocheton erythrocarpus	ganding buah merah			F														
Dysoxylum (?) *acutangulum*	kulim burung							s										
D. costulatum	lawen			F	F	F						x						
Dysoxylum sp.	mersindok			FL	FL	F	F	s							F			
Lansium domesticum	langsat hutan					F						x						
Prunus polystachys	pemanis halus			F	F	F						x						
Sandoricum koetjapi	ketapi			FL	F	F	F		F									
Walsura neuroides	mersindok		L		F							x						
	kayu cheneka	F	F															
	meresat											x						
	mersindok gading				F							x						
Moraceae																		
Antiaris toxicaria	ipoh					F	F											
Artocarpus integer	bankong											x						
A. lowii	mikor							F										
A. (?) *maingayi*	lidah kerbau			FL								x						
A. rigidus	perian	L		F	F	F			F									
A. scortechinii	terap					F						x						

SCIENTIFIC NAME Family, genus, species	LOCAL NAME	PRIMATES *Hs*	 *Hl*	 *Po*	 *Pm*	 *Mf*	SQUIRRELS *Rb*	 *Ra*	 *Cp*	 *Cn*	 *St*	 ?	BIRDS C	 Ps	 B	 P	 A	 E
Moraceae (cont.)																		
Artocarpus sp.	peradom	F	F	FL	F	F			F									
Artocarpus sp.	peradom halus	F	F									x						
Ficus (3 spp.)	ara	FL	FL	FL	FL	F	Fs	F	Fs	Fs			F	F	F	F		
Parartocarpus sp.	tunjung																	
Myristicaceae																		
Horsfieldia irya	pianggok			L	F	F						x						
H. sucosa	basong				F							x						
H. superba	basong bulu											x						
Horsfieldia	basong				F	F						x						
Knema cinerea	basong paya	F	F									x						
K. hookeriana	basong bulu	F	F									x						
Knema sp.	basong				F				F						F			F
Myristica maingayi	basong hitam											x						
Myristica sp.	basong				F				F						F			
Myristica sp.	penara											x						
Myrsinaceae																		
Ardisia colorata	semambu cherang	F				F						x						
Ardisia sp.																F		
Myrtaceae																		
Eugenia (at least 6 spp.)	kelat	FL	FL	L	FfL	FfL	F	F			s		F					
Olacaceae																		
Ochanostachys amentacea	petaling	L		L	F			F										
Scorodocarpus sp.	kulim											x						
Strombosia javanica	dedali			fL	fL							x						
Oxalidaceae																		
Sarcotheca griffithii	belimbing rot	FfL	FfL	L				F	F									
S. monophylla	penondok	L																

SCIENTIFIC NAME	LOCAL NAME	PRIMATES					SQUIRRELS						BIRDS					
Family, genus, species		*Hs*	*Hl*	*Po*	*Pm*	*Mf*	*Rb*	*Ra*	*Cp*	*Cn*	*St*	?	C	Ps	B	P	A	E
Polygalaceae																		
Xanthophyllum amoenum	teper		L	Ff	FL													
X. affine	bereher	L		FL	FfL	F		L										
X. rufum	minyak berok·	f	f									x						
X. stipitatum	teremok	fL	f															
Xanthophyllum	jenis teremok							F			s							
Xanthophyllum	teremok kasat																	
Protaceae																		
Heliciopsis velutina	dumpung bukit																	
Heliciopsis sp.	dumpung																	
Rhizophoraceae																		
Carallia brachiata	sebun gutu	Ff	Ff	L								x						
Gynotroches axillaris	hembuluh											x						
Gynotroches sp.	peretang																	
Rosaceae																		
Maranthes corymbosum	beretai	F	F	FL	F	Ff	Fs		s									
Parinari oblongifolium	kemalau	FL	L	L	L	L						x						
P. parva	hembalau	L																
Parinari sp.	merbatu																	
Rubiaceae																		
Canthium sp.	babi kurus			L														
Diplospora malaccensis	pelung		F															
Morinda citrifolia	mengkudu																	
Morinda sp.	mengkudu				F	Ff												
Nauclea and *Neonauclea* spp.	mengkal, mengkal paya, mengkunyet	F	F	FL	f													
Randia scortechinii	gula cheneka	F	F	F	F	F						x						
Rutaceae																		
Achronychia porteri	chenderoh daun satu				F													
Citrus	kumuning sangkan																	
Euodia glabra	yak chenderoh							s										

SCIENTIFIC NAME Family, genus, species	LOCAL NAME	PRIMATES *Hs*	*Hl*	*Po*	*Pm*	*Mf*	SQUIRRELS *Rb*	*Ra*	*Cp*	*Cn*	*St*	?	BIRDS C	Ps	B	P	A	E
Rutaceae (cont.)																		
Merrillia caloxylon	kumuning																	
Sapindaceae																		
Nephelium costatum	reming					F						x						
N. eriopetalum	gompal benang				FL							x						
N. mutabile	rambutan hutan				F	F			F									
N. ophioides	sanggol lutung											x						
Nephelium sp.	rambutan perningan											x						
Paranephelium macrophyllum	gesiar				F							x						
Pometia pinnata	kasai			F		F	F	F	F	F	b							
Xerospermum intermedium	kiki buntal buah gutu	F	F		F	FL		s	F	F	b							
X. wallichii	kiki buntal buah lichin	FL	F	F	F	F						x						
	chemba'ak																	
	gompal meyang bulu											x						
	riden											x						
Sapotaceae																		
Chrysophyllum lanceolatum	pulot getah		F									x						
Madhuca malaccensis	menjatoh cempelot			F		F						x						
Palaquium hispidum	palan		FL									x						
P. oxleyanum	getah taban											x						
Palaquium	palan halus											x						
Payena lucida	menjatoh	f	Ff	F	F	Ff		F		f								
P. maingayi	mentubak	fL	Ff		F	Ff	F											
	lancher																	
Simaroubaceae																		
Irvingia malayana	pauh	L		L		F	Fs	F	s							F	F	F
Sterculiaceae																		
Heriteria javanica	pu																	
H. simplicifolia	melima																	
Pterocymbium javanicum	mata lembu					f		b	b	b								
Pterospermum javanicum	bayur					F						x						

SCIENTIFIC NAME Family, genus, species	LOCAL NAME	PRIMATES *Hs*	*Hl*	*Po*	*Pm*	*Mf*	SQUIRRELS *Rb*	*Ra*	*Cp*	*Cn*	*St*	?	BIRDS C	Ps	B	P	A	E
Sterculiaceae (cont.)																		
Pterygota alata	benuang				F													
Scaphium linearicarpum	payang hantu																	
Sterculia parvifolia	kelompang			L							b							
Sterculia sp.	kelompang batu																	
Sterculia sp.	kelompang paya																	
Styracaceae																		
Styrax benzoin	keminyan					F		F										
Theaceae																		
Adinandra (?) *lamponga*	kelat paya	f			F													
Gardenia carinata	chempaka hutan											x						
Thymelaceae																		
Aquilaria malaccensis	kepang				F				s									
Gonystylus confusus	penggerak									F								
Tiliaceae																		
Grewia fibrocarpa	damak bulu	FL	F									x						
G. laurifolia	damak halus	F	F	F	F	FL				F								
G. tomentosa	chenderai tanjung					F						x						
Pentace floribunda	ba'ang					F						x						
P. triptera	pelimbang			f														
Schoutenia (?) *accrescens*	bayur cherang																	
S. (?) *corneri*	chenderai bekudaun																	
Ulmaceae																		
Celtis rigescens	semantit			L			b					x						
Gironnieria hirta	meskam						F	F										
G. nervosa	meskam											x						
G. parvifolia	meskam	Ff										x						
G. subaequalis	peretpong	F																

SCIENTIFIC NAME Family, genus, species	LOCAL NAME	PRIMATES *Hs*	*Hl*	*Po*	*Pm*	*Mf*	SQUIRRELS *Rb*	*Ra*	*Cp*	*Cn*	*St*	?	BIRDS C	Ps	B	P	A	E
Urticaceae																		
Sloetia elongata	tebakah	FfL	FfL	FfL	Ff	Ff	f	f		F								
Verbenaceae																		
Vitex pubescens	haleban	F	F		F	F						x						
V. trifoliata	berambong daun tiga	L	L									x						
Violaceae																		
Rinorea sp.	telur buaya				FL													

Data for primates comes from D.J. Chivers, J.J. Raemaekers, Sheila Curtin, F.P.G. Aldrich-Blake, J.R. and Kathy MacKinnon, Y.L. Mah and J.B. Payne; for squirrels from J.B. Payne and Kalang (see above), and for birds from D.R. Wells.

Frugivorous birds mostly eat figs, but there is great specialisation according to the size of the bird; large birds, such as hornbills, eat the largest figs, medium-sized birds, like parrots and pigeons eat the middle-sized figs, and the smaller birds, such as bulbuls and barbets, eat the smallest figs.

Dr David Wells submitted records for about 13 species of bird of six families, to which we have added observations for several more species including two more families, mostly by John Payne (see table 2.2, pp. 52-53, and table 9.1, pp. 263-265):

Columbidae (pigeons)	*Treron* spp. *Ducula aenea*
Psittacidae (parrots)	*Psittacula longicauda* *Psittinus cyanurus*
Bucerotidae (hornbills)	*Anorrhinus galeritus* *Anthracoceros malayanus* *A. convexus* *Rhyticeros undulatus* *Buceros rhinoceros* *Rhinoplax vigil*
Pycnonotidae (bulbuls)	*Pycnonotus simplex* *P. brunneus*
Aegithinidae (leaf birds)	*Irena puella*
Eurylaimidae (broadbills)	*Calyptomena viridis*
Capitonidae (barbets)	
Dicaeidae (flowerpeckers)	

We present incomplete information on birds, in the hope that it will stimulate further literature searches and field study; we recognise, however, that many of the fruit eaten by primates and squirrels are too large or thick-skinned to be eaten by most birds (see Chapter 2).

APPENDIX II

LONG-TERM OBSERVATIONS OF SIAMANG BEHAVIOUR

Background information on the quantitative analysis of dawn-to-dusk observations of the behaviour of the siamang group TS1 at Kuala Lompat on 487 days in 65 months from April 1969 to April 1979 inclusive will perhaps be of interest to readers who wish to pursue certain points. We clarify here the distribution of samples (fig. 8.15) and present mean values for each sample period for measures of activity period, environmental features and ranging and feeding behaviour; the percentages and indices used in the analyses (figs. 8.15-18) were calculated from these means.

Figures given are means for each period, except for CALLING where numbers of days on which calling occurred, and the number of bouts during the sample period, are presented. For FEEDING, differences between sums of food types (excluding figs) and mean daily feeding time represent differences in recording the start and stop of bouts and 10-min samples of major activity respectively, as well as unidentified feeding observations. CANOPY LEVEL is given as the percentage of 10-min samples during the activity period, and WEATHER was sampled at 10- or 15-min intervals from dawn to dusk for the sample period, with rainfall in mm. for the whole month in which the sample was taken.

Following the table, data from short sample periods (10-20 days) from 1969 to 1973 inclusive (Chivers *et al.*, 1975) are displayed to illustrate the extent of variation about the mean values used in this analysis. There are observations of the siamang group TS1 in the fruiting seasons of 1969, 1971 and 1972, and in the dry seasons of 1970 and 1973. They provide us with an opportunity to see how behaviour changes with time (see Chapter 8), how variable behaviour may be within a sample period, and how similar may be the behaviour of different, but adjacent groups in the same forest at the same time (since data are available for siamang group TS1a from 1971-73). The data presented here are concerned with different aspects of feeding and ranging behaviour.

The two siamang groups were very similar to each other in their ranging and feeding behaviour in each of the three observation

(continued on p. 338)

YEAR	MONTH	No. of days	Sun rise	Sun set h	ACTIVITY PERIOD start	stop h	duration min	CALLING days	bouts	RANGING day range	night shift m	no. of ha all	diff.	level % H	M	L	travel min/day	FEEDING feed min/day	fruit & fls	lvs	insects	figs min/day	WEATHER sun	cloud	rain %	rain mm
1969	Apr	10	0636	1829	0628	1610	582	0	0	1035	299	10.1	8.5				90	347	106	160	0	91	44	53	3	87
	May	9	0632	1847	0701	1629	568	0	0	784	289	7.3	6.3				84	384	140	224	0	14	50	48	2	266
	June	14	0634	1852	0637	1635	598	0	0	712	288	7.4	6.4				84	381	128	212	1	102	40	57	2	97
	July	14	0640	1857	0642	1857	600	1	1	767	288	8.3	7.3				79	350	203	147	0	45	49	49	2	104
	Aug	12	0642	1854	0707	1650	583	0	0	807	308	8.1	7.1				94	332	114	191	0	52	39	56	5	309
	Sept	10	0634	1843	0628	1648	620	0	0	765	217	6.1	5.3				74	400	190	203	0	181	56	41	3	80
	Oct	10	0627	1830	0635	1629	594	1	1	710	235	7.5	7.0	21	64	15	82	357	96	245	3	72	57	40	3	158
	Nov	10	0628	1827	0647	1611	564	1	1	746	307	7.5	6.8	22	69	9	83	388	129	210	0	94	25	69	6	236
	Dec	10	0639	1837	0653	1651	598	0	0	680	281	7.5	6.5	25	62	13	76	353	154	91	6	75	37	59	4	155
1970	Jan	10	0654	1851	0708	1658	590	0	0	723	246	7.0	6.3	31	62	7	74	393	122	246	0	103	48	49	3	
	Feb	10	0657	1858	0655	1858	635	0	0	886	363	8.7	8.2	34	57	9	84	394	145	220	1	89	67	33	0	78
	Mar	10	0649	1855	0629	1724	655	7	10	1506	268	13.1	11.4	25	61	14	133	395	139	240	3	106	55	42	3	175
	Apr	10	0638	1849	0610	1717	677	7	12	1865	448	15.9	15.0	35	60	5	141	397	181	139	50	103	61	37	2	59
	May	9	0632	1847	0611	1717	668	4	4	1811	276	16.0	13.9	21	71	9	174	359	152	167	11	85	45	51	4	122
1971	Sept	8	0634	1843	0634	1756	682	1	1	1264	278	12.3	11.1				96	290	139	128	0	29	62	36	3	138
	Oct	10	0627	1830	0655	1636	581	2	2	818	240	8.5	7.4				65	248	110	167	5	61	46	52	3	290
1972	Jy/Aug	10	0641	1855	0633	1742	669	4	4	1572	299	15.0	11.7	35	48	18	136	303	118	107	45	13	55	44	1	
	Se/Oct	10	0630	1828	0633	1735	662	5	5	1393	171	12.1	8.7				106	215					42	57	1	
1973	Jan	5	0655	1855	0638	1745	667	0	0	623	238	6.8	5.4					345	283	62	0	210				84
	Mar	5	0649	1855	0621	1726	665	1	1	741	215	8.0	6.4					354	157	188	9	164	55	39	5	14
	Apr	5	0638	1849	0624	1726	662	4	4	735	236	7.0	5.8					345	196	147	2	193	84	16	0	125
	May	7	0632	1847	0635	1722	647	3	3								88	380	120	246	13					330
	June	8	0634	1852	0632	1700	628	5	7								91	329	164	119	43					165
	July	4	0640	1857	0708	1713	605	3	3								70	363	190	128	42					99
1974	June	4	0634	1852	0646	1722	639	2	2	969	416	9.3	8.3	32	53	16	110	316	177	134	25	44	60	40	0	96
	Aug	4	0642	1854	0639	1734	655	3	3	1138	265	10.5	9.5	17	61	22	122	385	142	191	48	15	43	54	3	29
1974	Sept	9	0634	1843	0637	1726	674	7	7	928	333	9.9	8.6				131	271	137	95	63	58				166
	Oct	9	0627	1830	0629	1701	616	5	5	914	354	10.1	8.1				103	222	124	36	79	26				182
	Nov	10	0628	1828	0636	1704	627	4	6	817	289	9.3	7.8				126	302	191	81	50	10				240
		10	0630	1830	0648	1645	589	4	5	654	274	8.3	7.2				111	294	161	97	52	16				
	Dec	10	0635	1835	0650	1700	614	4	4	753	332	8.2	7.5				135	263	121	75	80	19				200

YEAR	MONTH	No. of days	Sun rise h	Sun set h	ACTIVITY PERIOD start h	ACTIVITY PERIOD stop h	ACTIVITY PERIOD duration min	CALLING days	CALLING bouts	RANGING day range	RANGING night shift m	no. of ha all	no. of ha diff.	level H	level M	level L %	travel min/day	FEEDING feed min/day	fruit & fls	lvs	insects	figs min/day	WEATHER sun	cloud	rain %	rain mm
1975	Apr	10	0638	1849	0631	1654	621	5	6	806	408	8.0	7.4	28	45	8	109	263	131	102	33	113	44	50	8	100
	May	10	0632	1847	0653	1651	598	2	2	805	392	8.6	7.7	44	56	5	137	283	152	93	35	74	43	51	6	145
	June	10	0634	1852	0635	1557	568	2	2	681	252	6.3	5.5	28	72	0	98	251	123	112	40	56	27	70	3	222
	July	10	0640	1857	0639	1615	576	0	0	555	254	5.4	4.9	48	51	1	107	292	100	174	19	55	39	58	3	130
	Aug	10	0642	1854	0658	1633	575	3	3	739	329	7.4	6.8	48	51	1	130	273	109	133	32	37	34	60	5	147
	Sept	10	0634	1843	0640	1639	600	3	3	755	325	7.5	7.2	36	62	2	145	293	120	144	30	74	42	57	0	103
	Oct	9	0627	1830	0641	1557	554	3	3	661	303	7.1	6.8	20	80	0	124	251	103	124	22	67	38	60	2	222
	Nov	10	0628	1827	0639	1553	555	2	2	495	200	5.1	4.6	24	76	0	95	262	123	109	32	119	25	61	14	333
	Dec	9	0639	1837	0641	1606	565	1	1	419	180	4.9	4.1	36	61	3	118	300	62	184	21	45	33	63	4	172
1976	Jan	10	0654	1851	0646	1626	580	4	4	667	279	7.7	6.8	71	29	0	120	291	83	132	92	55	35	65	0	47
	Feb	10	0657	1858	0640	1715	636	4	4	1039	405	11.6	11.0	29	60	11	161	315	145	60	116	51	64	36	0	9
	Mar	10	0649	1855	0642	1711	630	7	7	1180	487	12.5	10.8	56	43	1	168	298	147	90	60	60	43	54	3	63
1976	May	5	0632	1847	0626	1655	629	4	6								210	333	124	113	93					
	July	4	0640	1857	0640	1640	600	3	3	1374	336	12.3	11.8				160	180	87	53	33					170
	Aug/Se	6	0638	1848	0635	1649	614			1075	479	13.9					120	104	70	31	5	2				222
	Oct	3	0627	1830	0648	1620	572	3	3	1102	375	8.5	8.0				161	262	101	43	47					266
	Nov	4	0628	1827	0629	1607	578	1	1	874	316	8.5	7.8						167	58	51					250
1977	Jan	3	0654	1851	0705	1555	530	0	0	390	85	3.5	3.0													2
	Feb	4	0657	1858	0720	1604	524	0	0	373	179	4.8	4.4				75	219	30	169	37					80
	Apr	5	0638	1849	0641	1621	580	0	0	785	294	7.8	7.0				133	248	155	55	33					12
	May	3	0632	1847	0640	1540	540	0	0	567	317	6.0	6.0				122	274	139	126	8					184
	June	5	0634	1843	0636	1640	604	2	3	560	241	5.2	4.8				85	289	101	174	4					155
	Aug	5	0642	1854	0712	1711	599	0	0	790	240	6.6	6.0				114	288	156	101	5					134
	Sept	5	0634	1843				3	3	850	136	8.4	7.2				148	231	115	100	5					110
	Oct	5	0627	1830				1	1	520	224	5.0	4.8				81	408	315	106	4					399
	Nov	5	0628	1827				0	0	635	165	7.0	6.0				67	248	143	122	0					166
	Dec	5	0639	1837				1	1	548	312	4.8	4.6				54	269	93	197	2					108
1978	Jan	5	0654	1851				0	0	368	90	2.6	2.4				31	311	150	199	9					
	Mar	5	0649	1855	0705	1644	579	1	1	587	260	7.0	6.6				133	266	111	73	6					
	Apr	5	0638	1849	0652	1630	578	0	0	701	234	6.4	6.0				137	288	184	88	11					
	May	5	0632	1847	0652	1602	550	2	2	715	275	6.2	5.8				142	288	109	157	11					
1979	Apr	9	0638	1849	0645	1616	571	3	4	527	187	5.7	5.0				105	355	266	87	3					

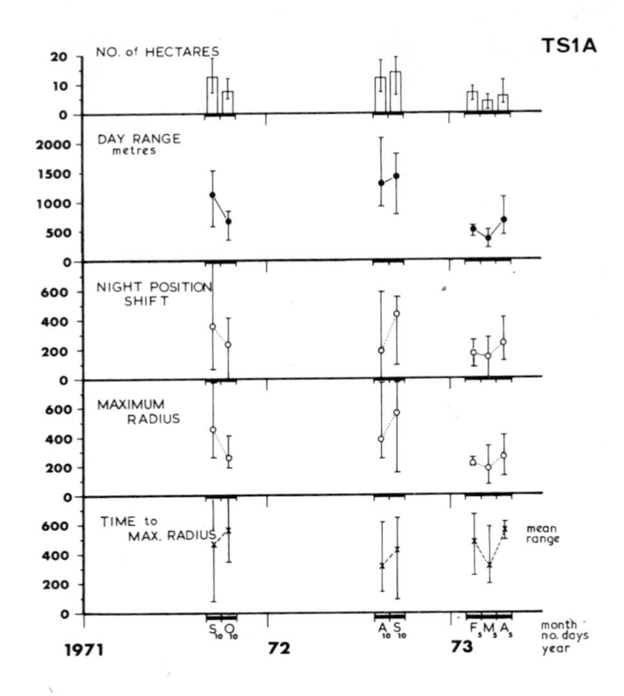
TS1A
NO. of HECTARES
DAY RANGE
metres
NIGHT POSITION
SHIFT
MAXIMUM
RADIUS
TIME to
MAX. RADIUS
mean
range
month
no. days
year
1971
72
73

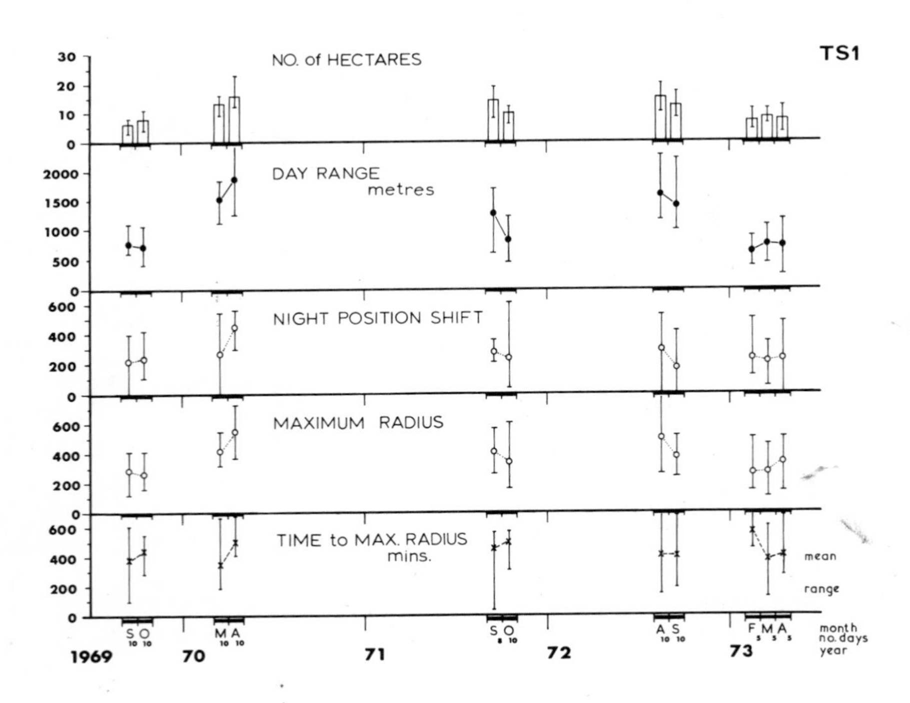
TS1
NO. of HECTARES
DAY RANGE
metres
NIGHT POSITION SHIFT
MAXIMUM RADIUS
TIME to MAX. RADIUS
mins.
mean
range
month
no. days
year
1969
70
71
72
73

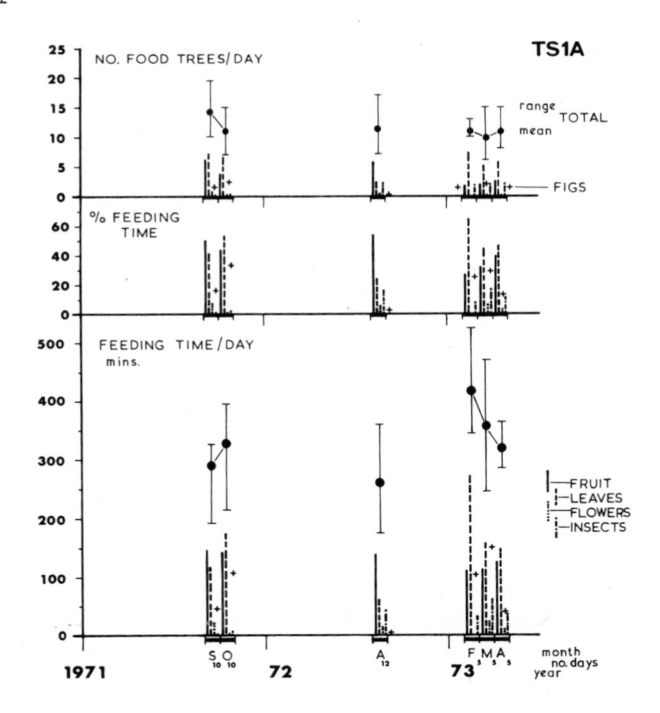
TS1A
NO. FOOD TREES/DAY
range
TOTAL
mean
FIGS
% FEEDING TIME
FEEDING TIME/DAY mins.
FRUIT
LEAVES
FLOWERS
INSECTS
S O
A
F M A
month
no. days
year
1971
72
73

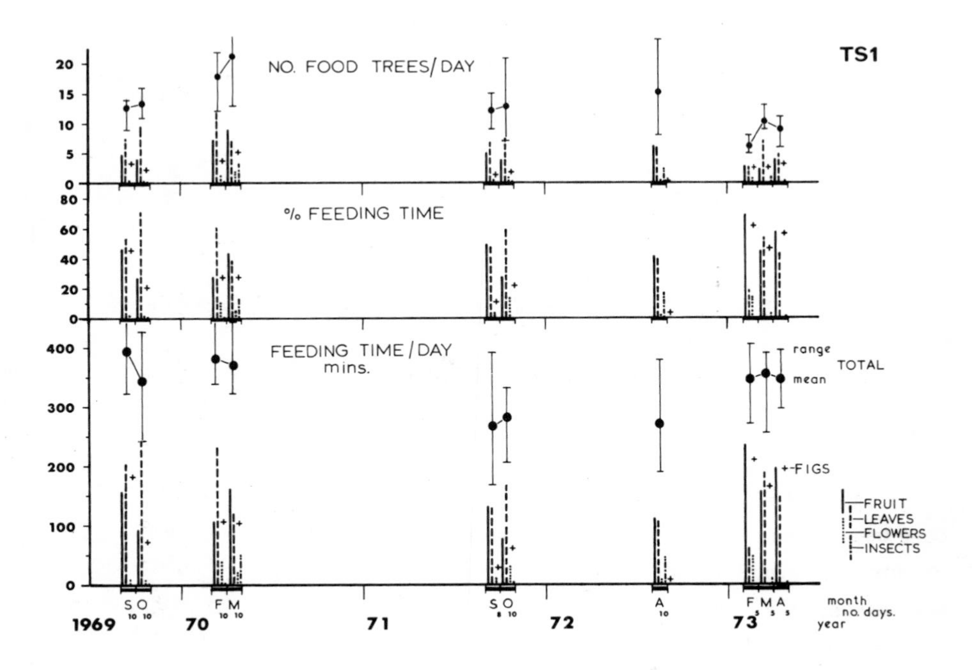
TS1
NO. FOOD TREES/DAY
% FEEDING TIME
FEEDING TIME/DAY mins.
range
TOTAL
mean
FIGS
FRUIT
LEAVES
FLOWERS
INSECTS
S O
F M
S O
A
F M A
month
no. days.
year
1969
70
71
72
73

periods pertaining to both groups. The only noticeable difference in ranging occurred in 1972, when TS1 was decreasing its ranging as the observation period progressed, but TS1a was increasing these values, especially those for night position shift and maximum radius (the farthest point reached from the previous sleeping position).

In 1973 TS1 maintained a fairly constant level of feeding (with high fig intake), while TS1a showed a steady decrease in feeding time about the TS1 level, as the lone male shifted from a predominance of leaf-eating to one of fruit-eating. Otherwise, levels and changes were very similar between the groups.

Day-to-day changes in behaviour can be more marked than changes between observation periods; nevertheless, these changes represent the unevenness of movement around the home range, as the group exploits food sources heavily in one part of the home range before moving on to another, and over a period of 5-10 days such behaviours average out at levels indicative of the circumstances of that period (see Chapter 8).

The map below shows the philopatry of each group: TS1 in particular visited that part of their home range marked as the 'core area' in each sample period, irrespective of whether they were mainly using the drier northern part, or the wetter south-east corner, of their range; TS1a also spent most of its time in certain parts of their home range. Group calls (marked) and night sleeping trees occur mostly in the area marked as the 'territory', which coincide with the usual limits of ranging in any sample period.

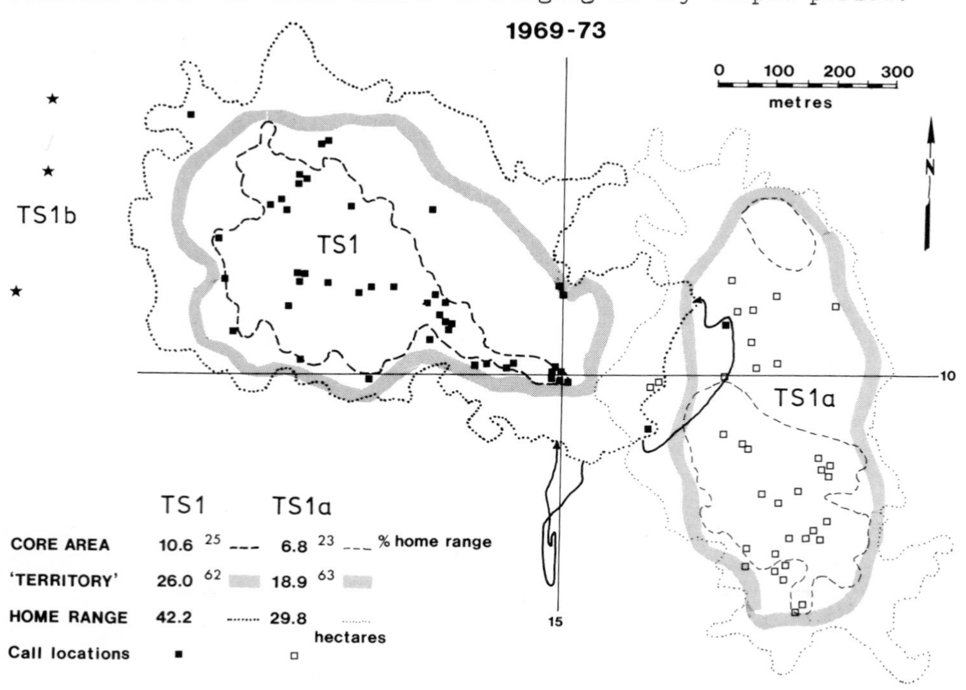

REFERENCES

Abe, T. (1975) Studies on the distribution and ecological role of termites in a lowland rain forest of West Malaysia. Unpubl. Ph.D. dissertation, University of Kyoto.

Aldrich-Blake, F.P.G. (1970) Problems in social structure in forest monkeys. In "Social Behaviour in Birds and Mammals" (J.H. Crook, ed.), pp. 79-101. Academic, London.

Aldrich-Blake, F.P.G. and Chivers, D.J. (1973) On the genesis of a group of siamang. *Amer. J. Phys. Anthrop.* **38**: 631-632.

Altmann, S.A. (1962) Social behaviour of anthropoid primates : analysis of recent concepts. In "Roots of Behaviour" (E.L. Bliss, ed.), pp. 277-285. Harper, New York.

Altmann, S.A., ed. (1967) "Social Communication among Primates". University of Chicago Press.

Andrews, P. and Groves, C.P. (1976) Gibbons and brachiation. In "Gibbon and Siamang" (D.M. Rumbaugh, ed.), vol. 4, pp. 167-218. Karger, Basel.

Angst, W. (1973) Pilot experiments to test group tolerance to a stranger in wild *Macaca fascicularis*. *Amer. J. Phys. Anthrop.* **38**: 625-630.

Angst, W. (1974) Das Ausdrucksverhalten des Javanafferen *Macaca fascicularis* Raffles 1821. *Fortschritte de Verhaltenforschung* **15**: 1-90 (*Zeit. f. Tierpsychol.* suppl.).

Angst, W. (1975) Basic data and concepts on the social organization of *Macaca fascicularis*. In "Primate Behaviour : developments in field and laboratory research" (L.A. Rosenblum, ed.), vol. 4, pp. 325-388. Academic, New York.

Angst, W. and Thommen, D. (1977) New data and a discussion of infant killing in Old World monkeys and apes. *Folia primatol.* **27**: 198-229.

Ashton, P.S. (1972) The Quaternary geomorphological history of western Malesia and lowland forest phytogeography. In "The Quaternary Era in Malesia" (P. and M. Ashton, eds.), pp. 35-49. Department of Geography, University of Hull. Transactions of the Second Aberdeen-Hull Symposium on Malesian Ecology.

Avis, V. (1962) Brachiation: the crucial issue for man's ancestry. *J. Anthrop.* **18**: 119-148.

Baldwin, L.A. and Teleki, G. (1974) Field research on gibbons, siamangs and orang-utans : an historical, geographical and bibliographical listing. *Primates* **15**: 365-376.

Baldwin, L.A. and Teleki, G. (1976) Patterns of gibbon behaviour on Hall's Island, Bermuda : A preliminary ethogram for *Hylobates lar*. In "Gibbon and Siamang" (D.M. Rumbaugh, ed.), vol. 4, pp. 21-105. Karger, Basel.

Bartels, E. (1964) On *Paradoxurus hermaphroditus*. *Beaufortia* **10**: 193-201.

Bauchop, T. (1971) Stomach microbiology of primates. *Ann. Rev. Microbiol.* **25**: 429-435.

Bauchop, T. (1978) Digestion of leaves in vertebrate arboreal folivores. In "The Ecology of Arboreal Folivores" (G.G. Montgomery, ed.), pp. 193-204. Smithsonian Institution, Washington D.C.

Bauchop, T. and Martucci, R.W. (1968) Ruminant-like digestion of the langur monkey. *Science* **161**: 698-700.

Bell, R.H.V. (1971) A grazing ecosystem in the Serengeti. *Scient. Amer.* **225**: 86-93.

Berkson, G., Ross, B.A. and Jatinandana, S. (1971) The social behaviour of gibbons in relation to a conservation program. In "Primate Behaviour: Development in Field and Laboratory Research" (L.A. Rosenblum, ed.), vol. 2, pp. 225-255. Academic, London.

Bernstein, I.S. (1967a) Intertaxa interactions in a Malayan primate community. *Folia primatol.* **7**: 198-207.

Bernstein, I.S. (1967b) A field study of the pig-tail monkey (*Macaca nemestrina*). *Primates* **8**: 217-228.

Bernstein, I.S. (1968a) The lutong of Kuala Selangor. *Behaviour* **32**: 1-16.

Bernstein, I.S. (1968b) Social status of two hybrids in a wild troop of *Macaca irus*. *Folia primatol.* **8**: 121-131.

Boulbet, M. (1979) Du climax forestier au paysage anthropique. Mimeo 35 pp., Mekong Committee, Bangkok.

Bourlière, F. and Hadley, M. (1970) The ecology of tropical savannas. *Ann. Rev. Ecol. System.* **1**: 125-152.

Brandon-Jones, D. (in prep.) Present primate distributions as a guide to Quaternary climatic change in Asia.

Brockelman, W.Y. (1975) Gibbon populations and their conservation in Thailand. *Nat. Hist. Bull. Siam Soc.* **26**: 133-157.

Brockelman, W.Y. (1978) Preliminary report on relations between the gibbons *Hylobates lar* and *Hylobates pileatus* in Thailand. In "Recent Advances in Primatology, vol. 3, Evolution" (D.J. Chivers and K.A. Joysey, eds.), pp. 315-318. Academic, London.

Brockelman, W.Y., Ross, B.A. and Pantuwatana, S. (1973) Social correlates of reproductive success in the gibbon colony on Ko Klet Kaeo, Thailand. *Amer. J. phys. Anthrop.* **38**: 637-640.

Brockelman, W.Y., Ross, B.A. and Pantuwatana, S. (1974) Social inter-actions of adult gibbons (*Hylobates lar*) in an experimental colony. In "Gibbon and Siamang" (D.M. Rumbaugh (ed.), vol. 3, pp. 137-156. Karger, Basel.

Brotoisworo, E. (1978) Nature conservation in Indonesia and its problems with special reference to primates. In "Recent Advances in Primatology, Vol. 2, Conservation" (D.J. Chivers and W. Lane-Petter, eds.), pp. 31-40. Academic, London.

Brown, J.L. (1964) The evolution of diversity in avian territorial systems. *Wilson Bull.* **76**:

Brown, J.L. (1975) "The Evolution of Behaviour". Norton, New York.

Burgess, P.F. (1971) Effect of logging on hill dipterocarp forest. *Malay. Nat. J.* **24**: 231-237.

Burt, W.H. (1943) Territoriality and home range concepts as applied to mammals. *J. Mammal.* **24**: 346-352.

Burton, G.J. (in prep.) Relationship between body and gonadal weight in the dusky leaf monkey (*Presbytis obscura*).

Caldecott, J.O. (1980) Habitat quality and populations of two sympatric gibbons (Hylobatidae) on a mountain in Malaya. *Folia primatol.* **33**:291-309.

Caldecott, J.O. and Bennett, E.L. (in prep.) Unexpected abundance: notes on the Lima Blas Estate Forest Reserve, Slim River, Perak.

Caraco, T. (1980) Time budgeting and group size: a test of theory. *Ecology* **60**: 618-627.

Carpenter, C.R. (1940) A field study in Siam of the behaviour and social relations of the gibbon, *Hylobates lar*. *Com. Psychol. Monogr.* **16**: 1-212.

Chalmers, N. (1968) Group composition, ecology and daily activities of free-living mangabeys in Uganda. *Folia primatol.* **8**: 247-262.

Charles-Dominique, P. (1977) "Ecology and Behaviour of Nocturnal Primates". Duckworth, London.

Chia, L.S. (1977) Seasonal rainfall distribution over West Malaysia. *Malay. Nat. J.* **31**: 11-39.

Chiang, M. (1968) The annual reproductive cycles of a free-living population of long-tailed macaques (*M. fascicularis*) in Singapore. Unpubl. M.Sc. dissertation, University of Singapore.

Chivers, D.J. (1969) On the daily behaviour and spacing of howling monkey groups. *Folia primatol.* **10**: 48-102.

Chivers, D.J. (1971) Spatial relations within the siamang family group. In "Proc. 3rd Int. Congr. Primat., Zurich 1970" (H. Kummer, ed.), vol. 3, pp. 14-21. Karger, Basel.

Chivers, D.J. (1972) The siamang and the gibbon in the Malay Peninsula. In "Gibbon and Siamang" (D.M. Rumbaugh, ed.), vol. 1, pp. 103-135. Karger, Basel.

Chivers, D.J. (1973) An introduction to the socio-ecology of Malayan forest primates. In "Comparative Ecology and Behaviour of Primates" (R.P. Michael and J.H. Crook, eds.), pp. 101-146. Academic, London.

Chivers, D.J. (1974) The Siamang in Malaya : a field study of a primate in tropical rain forest. *Contrib. Primatol.* **4**: 1-335. Karger, Basel.

Chivers, D.J. (1975) The behaviour of siamang in the Krau Game Reserve. *Malay. Nat. J.* **29**: 7-22.

Chivers, D.J. (1976) Communication within and between family groups of siamang, *Symphalangus syndactylus*. *Behaviour* **57**: 116-135.

Chivers, D.J. (1977) The lesser apes. In "Primate Conservation" (HSH Prince Rainier and G.H. Bourne, eds.), pp. 539-598. Academic, New York.

Chivers, D.J. (1978) The gibbons of Peninsular Malaysia. *Malay Nat. J.* **30**: 565-591.

Chivers, D.J. and Chivers, S.T. (1975) Events preceding and following the birth of a wild siamang. *Primates* **16**: 227-230.

Chivers, D.J. and Davies, A.G. (1979) Abundance of primates in the Krau Game Reserve, Peninsular Malaysia. In "The Abundance of Animals in Malesian Rain Forests" (A.G. Marshall, ed.), pp. 9-32. Miscell. Series, Dept. of Geography, University of Hull (Aberdeen-Hull Symposia on Malesian Ecology).

Chivers, D.J. and Herbert, J., eds. (1978) "Recent Advances in Primatology, vol. 1, Behaviour". Academic, London.

Chivers, D.J. and Bladik, C.M. (in press) Morphology of the gastro-intestinal tract in primates : comparison with other mammals in relation to diet. *J. Morph.*

Chivers, D.J. and MacKinnon, J.R. (1977) On the behaviour of siamang after the playback of their calls. *Primates* **18**: 943-948.

Chivers, D.J., Raemaekers, J.J. and Aldrich-Blake, F.P.G. (1975) Long-term observations of siamang behaviour. *Folia primatol.* **23**: 1-49.

Clutton-Brock, T.H. (1974) Activity patterns of red colobus (*Colobus badius tephrosceles*). *Folia primatol.* **21**: 161-187.

Clutton-Brock, T.H. (1975a) Ranging behaviour of red colobus (*Colobus badius tephrosceles*) in the Gombe National Park. *Anim. Behav.* **23**: 706-722.

Clutton-Brock, T.H. (1975b) Feeding behaviour of red colobus and black-and-white colobus in East Africa. *Folia primatol.* **23**: 165-207.

Clutton-Brock, T.H. ed. (1977a) "Primate Ecology : studies of feeding and ranging in lemurs, monkeys and apes". Academic, London.

Clutton-Brock, T.H. (1977b) Some aspects of intraspecific variation in feeding and ranging behaviour in primates. In "Primate Ecology" (T.H. Clutton-Brock, ed.), pp. 539-556. Academic, London.

Clutton-Brock, T.H. and Harvey, P.H. (1976) Evolutionary rules and primate societies. In "Growing Points in Ethology" (P.P.G. Bateson and R.A. Hinde, eds.), pp. 196-237. Cambridge University Press.

Clutton-Brock, T.H. and Harvey, P.H. (1977) Species differences in feeding ranging behaviour in primates. In "Primate Ecology" (T.H. Clutton-Brock, ed.), pp. 557-584. Academic, London.

Cody, M.L. (1974) "Competition and the Structure of Bird Communities". University Press, Princeton.

Cody, M.L. and Brown, J.H. (1969) Song asynchrony in neighbouring bird species. *Nature, Lond.* **222**: 778-780.

Conaway, C.H. and Sade, D.S. (1965) The seasonal spermatogenic cycle in free-ranging rhesus monkeys. *Folia primatol.* **3**: 1-12.

Corbet, G.B. and Southern, H.N. eds. (1977) "The Handbook of British Mammals". Blackwell, Oxford.

Corner, E.J.H. (1940) "Wayside Trees of Malaya". Government printer, Singapore.

Crook, J.H. (1965) The adaptive significance of avian social organisation. *Symp. Zool. Soc. Lond.* **14**: 181-218.

Crook, J.H. (1970) The socio-ecology of primates. In "Social Behaviour in Birds and Mammals" (J.H. Crook, ed.), pp. 103-166. Academic, London.

Crook, J.H. and Gartlan, J.S. (1966) Evolution of primate societies. *Nature, Lond.* **210**: 1200-1203.

Curtin, S.H. (1976a) Niche differentiation and social organization in sympatric Malaysian colobines. Unpubl. Ph.D. dissertation, University of California, Berkeley.

Curtin, S.H. (1976b) Niche separation in sympatric Malaysian leaf-monkeys (*Presbytis obscura* and *Presbytis melalophos*). *Yearb. of Phys. Anthrop.* **20**: 421-439.

Curtin, S.H. and Chivers, D.J. (1978) Leaf-eating primates of Peninsular Malaysia : the siamang and the dusky leaf-monkey. In "The Ecology of Arboreal Folivores" (G.G. Montgomery, ed.), pp. 441-464. Smithsonian Institution, Washington D.C.

Dale, W.L. (1959) The rainfall of Malaya, I. *J. trop. Geogr.* **13**: 23-27.

Dale, W.L. (1960) The rainfall of Malaya, II. *J. trop. Geogr.* **14**: 11-28.

Dale, W.L. (1964) Sunshine in Malaya. *J. trop. Geogr.* **19**: 20-26.

Dasmann, R. (1975) National Parks, nature conservation and "future primitive". *Ecologist* **6**: 164-167.

Davies, N.B. (1978) Territorial defence in the specialised wood butterfly (*Parage aegeria*): the resident always wins. *Anim. Behav.* **26**: 138-147.

Davies, N.B. and Halliday, T.R. (1979) Competitive male searching in male common toads, *Bufo bufo*. *Anim. Behav.* **27**: 1253-1267.

Dawkins, R. (1976) "The Selfish Gene". Oxford University Press.

Dawkins, R. and Krebs, J.R. (1978) Animal signals: information or manipulation? In "Behavioural Ecology: an evolutionary approach" (J.R. Krebs and N.B. Davies, eds.), pp. 282-309. Blackwell, Oxford.

Dawson, T.J. (1977) Kangaroos. *Scient. Amer.* **237** (2): 78-89.
DeVore, I. ed. (1965) "Primate Behaviour: field studies of monkeys and apes". Holt, Rinehart and Winston, New York.
Diamond, J.M. (1975) The island dilemma: lessons of modern biogeographic studies for the design of natural preserves. *Biol. Conserv.* **7**: 129-146.
Eckholm, E.F. (1975) "The Other Energy Crisis: Fuelwood". Worldwatch Institute, Washington.
Ehrlich, P.R. and Raven, P.H. (1964) Butterflies and plants: a study in coevolution. *Evolution* **18**: 586-608.
Eisenberg, J.F. (1966) The social organisation of mammals. *Handb. Zool. VIII* (39): 1-92.
Eisenberg, J.F., Muckenhirn, N.A. and Rudran, R. (1972) The relation between ecology and social structure in primates. *Science N.Y.* **176**: 863-874.
Eisenberg, J.F. and Thorington, R.W. (1973) A preliminary analysis of a Neotropical mammal fauna. *Biotropica* **5**: 150-161.
Ellefson, Judy (1967) Social communication in long-tailed macaques in Singapore. Unpubl. Ph.D. dissertation, University of California, Berkeley.
Ellefson, J.O. (1968) Territorial behaviour in the common white-handed gibbon, *Hylobates lar*. In "Primates: Studies in Adaptation and Variability" (P.C. Jay, ed.), pp. 180-199. Holt, Rinehart and Winston, New York.
Ellefson, J.O. (1974) A natural history of the white-handed gibbon in the Malayan peninsula. In "Gibbon and Siamang" (D.M. Rumbaugh, ed.), vol. 3, pp. 1-136. Karger, Basel.
Emlen, S.T. (1978) The evolution of cooperative breeding in birds. In "Behavioural Ecology" (J.R. Krebs and N.B. Davies, eds.), pp. 245-281. Blackwell, Oxford.
Emmons, L.H. (1975) Ecology and behaviour of African rain forest squirrels. Unpubl. Ph.D. dissertation, Cornell University.
Eudey, A.A. (1978) Earth-eating by macaques in Western Thailand: a preliminary report. In "Recent Advances in Primatology, Vol. 1, Behaviour" (D.J. Chivers and J. Herbert, eds.), pp. 351-353. Academic, London.
Eudey, A.A. (1979) Differentiation and dispersal of macaques (*Macaca* spp) in Asia. Unpubl. Ph.D. dissertation, University of Nevada.
F.A.O. (1976) "Development and Forest Resources in the Asia and Far East Region". FAO, Rome.
Fedorov, A.A. (1966) The structure of tropical rain forest and speciation in the humid tropics. *J. Ecol.* **54**: 1-11.
Feeny, P.P. (1975) Biochemical coevolution between plants and their insect herbivores. In "Coevolution of Animals and Plants" (L.E. Gilbert and P.H. Raven, eds.), pp. 3-19. University of Texas Press.
Ficken, R.W., Ficken, M.S. and Hailman, J.P. (1974) Temporal pattern shifts to avoid acoustic interference in singing birds. *Science, N.Y.* **183**: 762-763.

Fittinghoff, N.A. (1975) Riverine refuging in East Bornean *Macaca fascicularis*. *Amer. J. phys. Anthrop.* **42**: 300-301.

Fleagle, J.G. (1974) Dynamics of a brachiating siamang. *Nature, Lond.* **248**: 259-260.

Fleagle, J.G. (1976a) Locomotion, posture and comparative anatomy of Malaysian forest primates. Unpubl. Ph.D. dissertation, Harvard University.

Fleagle, J.G. (1976b) Locomotor behavior and skeletal anatomy of sympatric Malaysian leaf-monkeys (*Presbytis obscura Presbytis melalophos* . *Yearb. Phys. Anthrop.* **20**: 440-453.

Fleagle, J.G. (1976c) Locomotion and posture of the Malayan siamang and implications for hominoid evolution. *Folia primatol.* **26**: 245-269.

Fleagle, J.G. (1977) Locomotor behavior and muscular anatomy of sympatric Malaysian leaf-monkeys (*Presbytis obscura Presbytis melalophos*). *Amer. J. phys. Anthrop.* **46**: 297-308.

Fleagle, J.G. (1978) Locomotion, posture and habitat utilization in two sympatric, Malaysian leaf-monkeys (*Presbytis obscura* and *Presbytis melalophos*). In "The Ecology of Arboreal Folivores" (G.G. Montgomery, ed.), pp. 243-251. Smithsonian Institution, Washington D.C.

Fleagle, J.G. and Mittermeier, R.A. (1980) Locomotor behavior, body size and comparative ecology of seven Surinam monkeys. *Amer. J. phys. Anthrop.* **52**: 301-314.

Flenley, J. (1979) "The Equatorial Rain Forest: a Geological History". Butterworths, London.

Fooden, J. (1971) Report on primates collected in Western Thailand, January - April 1967.

Frankie, G.W., Baker, H.G. and Opler, P.A. (1974) Comparative phenological studies of trees in tropical wet and dry forests in the lowlands of Costa Rica. *J. Ecol.* **62**: 881-919.

Freeland, W.J. and Janzen, D.H. (1974) Strategies in herbivory by mammals: the role of plant secondary compounds. *Amer. Natur.* **108**: 269-289.

Frisch, J.E. (1971) Evolution of the siamang in South-east Asia during the Pleistocene. In "Proc. 3rd Int. Congr. Primat., Zurich 1970" (J. Biegert and W. Leutenegger, eds.), pp. 67-73. Karger, Basel.

Furuya, Y. (1961) The social life of silvered leaf-monkeys (*Trachypithecus cristatus*). *Primates* **3**: 41-60.

Gartlan, J.S., McKey, D.B. and Waterman, P.G. (1978) Soils, forest structure and feeding behaviour of primates in a Cameroon coastal rain-forest. In "Recent Advances in Primatology, Vol. 1, Behaviour" (D.J. Chivers and J. Herbert, eds.), pp. 259-267. Academic, London.

Gautier-Hion, A. (1978) Food niches and coexistence in sympatric primates in Gabon. In "Recent Advances in Primatology, Vol. 1, Behaviour" (D.J. Chivers and J. Herbert, eds.), pp. 269-286. Academic, London.

Gittins, S.P. (1978) The species range of the gibbon *Hylobates agilis*. In "Recent Advances in Primatology, Vol. 3, Evolution" (D.J. Chivers and K.A. Joysey, eds.), pp. 319-321. Academic, London.

Gittins, S.P. (1979) The behaviour and ecology of the agile gibbon (*Hylobates agilis*). Unpubl. Ph.D. dissertation, University of Cambridge.

Glander, K.E. (1978) Howling monkey feeding behaviour and plant secondary compounds: a study of strategies. In "Ecology of Arboreal Folivores" (G.G. Montgomery, ed.), pp. 561-574. Smithsonian Institution, Washington D.C.

Glick, B.B. (1979) Testicular size, testosterone level, and body weight in male *Macaca radiata*. Maturational and seasonal effects. *Folia primatol.* **32**: 268-289.

Grand, T. (1972) A biomechanical interpretation of terminal branch feeding. *J. Mammal.* **53**: 198-201.

Groves, C.P. (1972) Systematics and phylogeny of the gibbons. In "Gibbon and Siamang" (D.M. Rumbaugh, ed.), vol. 1, pp. 1-89. Karger, Basel.

Hafez, E.S.E., Schein, M.W. and Ewbank, R. (1969) The birth of cattle. In "The Behaviour of Domestic Mammals" (E.S.E. Hafez, ed.), pp. 235-295. Bailliere, Tindall and Cassell, London.

Haimoff, E.H. (in press) Video analysis of siamang (*Symphalangus syndactylus*) call bouts. *Behaviour*.

Hall, K.R.L. (1967) Social interactions of the adult male and adult females of a patas monkey group. In "Social Communication among Primates" (S.A. Altmann, ed.), pp. 261-280. Chicago University Press.

Harcourt, A.H. (1978) Activity periods and patterns of social interactions: a neglected problem. *Behaviour* **66**: 121-135.

Harrison, J.L. (1966) "An Introduction to Mammals of Singapore and Malaya". Malayan Nature Society, Singapore Branch.

Harrison, J.L. (1969) The abundance and population density of mammals in Malayan lowland forests. *Malay Nat. J.* **22**: 174-178.

Harrisson, B. (1973) "Conservation of Nonhuman Primates 1970". Karger, Basel.

Harrisson, T. (1962) Leaf monkeys at Fraser's Hill. *Malay. Nat. J.* **16**: 120-125.

Heekeren, H.R. van (1972) "The Stone Age of Indonesia". Martinus Nijhoff, The Hague.

Heinrich, B. (1978) The economics of insect sociality. In "Behavioural Ecology" (J.R. Krebs and N.B. Davies, eds.), pp. 97-128. Blackwell, Oxford.

Henwood, K. and Fabrick, A. (1979) A quantitative analysis of the dawn chorus: temporal selection for communicatory optimization. *Am. Nat.* **114**: 260-274.

Hinde, R.A. (1974) "Biological Bases of Human Social Behaviour". McGraw-Hill, New York.

Hladik, A. (1978) Phenology of leaf production in a rain forest of Gabon: distribution and composition of food for folivores. In "Ecology of Arboreal Folivores" (G.G. Montgomery, ed.), pp. 441-464. Smithsonian Institution, Washington D.C.

Hladik, A. and Hladik, C.M. (1969) Rapports trophiques entre vegetation et Primates dans la forêt de Barro Colorado (Panama). *Terre et Vie* **23**: 25-117.

Hladik, C.M. (1975) Ecology, diet and social patterning in Old and New World primates. In "Socio-ecology and Psychology of Primates" (R.H. Tuttle, ed.), pp. 3-36. Mouton, The Hague.

Hladik, C.M. (1977) A comparative study of the feeding strategies of two sympatric species of leaf monkeys: *Presbytis senex* and *Presbytis entellus*. In "Primate Ecology" (T.H. Clutton-Brock, ed.), pp. 324-354. Academic, London.

Hladik, C.M. (1978) Adaptive strategies of primates in relation to leaf-eating. In "Ecology of Arboreal Folivores (G.G. Montgomery, ed.), pp. 373-396. Smithsonian Institution, Washington D.C.

Hladik, C.M. and Chivers, D.J. (1978) Ecological factors and specific behavioural patterns determining primate diet. In "Recent Advances in Primatology, Vol. 1, Behaviour" (D.J. Chivers and J. Herbert, eds.), pp. 433-444. Academic, London.

Hladik, C.M., Hladik, A., Bousset, J., Valdebouze, P., Viroben, G. and Delort-Laval, J. (1971) Le regime alimentaire des Primates de l'île de Barro-Colorado (Panama). *Folia primatol.* **16**: 85-122.

Hooijer, D.A. (1960) Quaternary gibbons from the Malay archipelago. *Zool. Verh., Leiden,* **46**: 1-41.

Hrdy, S.B. (1974) Male-male competition and infanticide among the langurs (*Presbytis entellus*) of Abu, Rajasthan. *Folia primatol.* **22**: 19-58.

Hrdy, S.B. (1977) "The Langurs of Abu". Harvard University Press, Cambridge, Mass.

I.U.C.N. (1980) "Save the Rainforests". Bulletin **11** (5). Gland, Switzerland.

Janzen, D.H. (1970) Herbivores and the number of tree species in tropical forests. *Amer. Nat.* **104**: 501-528.

Janzen, D.H. (1971) Seed predation by animals. *Ann. Rev. Ecol. System.* **2**: 465-492.

Janzen, D.H. (1975) "Ecology of Plants in the Tropics". Arnold, London.

Janzen, D.H. (1978) Complications in interpreting the chemical defenses of trees against tropical arboreal plant-eating vertebrates. In "The Ecology of Arboreal Folivores" (G.G. Montgomery, ed.), pp. 73-84. Smithsonian Institution, Washington D.C.

Janzen, D.H. (1979) How to be a fig. *Ann. Rev. Ecol. System.* **10**: 13-51.

Jarman, P.J. (1974) The social organisation of antelope in relation to their ecology.

Jay, P.C. (1965) The common langur of North India. In "Primate Behaviour" (I. DeVore, ed.), pp. 197-249. Holt, Rinehart and Winston, New York.

Jay, P.C. ed. (1968) "Primates: studies in adaptation and variability". Holt, Rinehart and Winston, New York.

Jolly, A. (1966) "Lemur Behaviour". University of Chicago Press, Chicago.

Jolly, A. (1972) "The Evolution of Primate Behaviour". Macmillan, New York.

Jong, K., Stone, B.C. and Soepadmo, E. (1973) Malaysian tropical rain forest: an underexploited reservoir of edible-fruit tree species. *Proc. Symp. Biol. Res. and Nat. Dev.*: 113-121.

Kawabe, M. (1970) A preliminary study of the wild siamang gibbon, *Hylobates syndactylus*, at Fraser's Hill, Malaysia. *Primates* **11**: 285-291.

Kawai, M. (1965) Newly acquired pre-cultural behavior of the natural troop of Japanese monkeys on Koshima islet. *Primates* **6**: 1-30.

Kay, R.H. (1975) The functional adaptations of primate molar teeth. *Amer. J. Phys. Anthrop.* **43**: 195-216.

Kay, R.H. (1977) Diets of early Miocene African hominoids. *Nature, Lond.* **268**: 628-630.

King, B., Woodcock, M. and Dickinson, E.C. (1975) "A Field Guide to the Birds of South-East Asia". Collins, London.

Kleiber, M. (1961) "The Fire of Life". Wiley, New York.

Kleiman, D.G. (1977) Monogamy in mammals. *Quart. Rev. Biol.* **52**: 39-69.

Klein, L.L. and Klein, D.J. (1975) Social and ecological contrasts between four taxa of neotropical primates (*Ateles belzebuth, Alouatta seniculus, Saimiri sciureus, Cebus apella*). In "Socioecology and Psychology of Primates" (R.H. Tuttle, ed.), pp. 59-86. Mouton, The Hague.

Klein, L.L. and Klein, D.B. (1977) Feeding behaviour of the Colombian spider monkey. In "Primate Ecology" (T.H. Clutton-Brock, ed.), pp. 153-181. Academic, London.

Kohlbrugge, J.H.F. (1891) Versuch einer Anatomie des Genus *Hylobates*. In "Zoologisches Ergebnisse einer Reise in Niederlande Ost-Indien" (Weber, ed.), vol. 2, pp. 139-254. Leiden.

Koyama, N. (1971) Observations on the mating behaviour of wild siamang gibbons at Fraser's Hill, Malaysia. *Primates* **12**: 183-189.

Kuhn, H-J. (1964) Zur kenntnis von Ban und Funktion des Mageus der Schlankaffen (Colobinae). *Folia primatol.* **2**: 193-221.

Kummer, H. (1971) "Primate Societies: group techniques of ecological adaptation". Aldine, Chicago.

Kurland, J.A. (1973) A natural history of Kra macaques (*Macaca fascicularis* Raffles 1821) at the Kutai Reserve,

Kalimantan Timur, Indonesia. *Primates* **14**: 245-262.
Lack, D. (1968) "Ecological Adaptations for Breeding in Birds". Methuen, London.
Lack, D. (1971) "Ecological Isolation in Birds". Blackwell, Oxford.
Layne, J.N. (1954) The biology of the red squirrel, *Tamiasciurus hudsonicus loquax*, in central New York. *Ecol. Monogr.* **24**: 227-267.
Leigh, E.G. and Smythe, N. (1978) Leaf production, leaf consumption, and the regulation of folivory on Barro Colorado Island. In "The Ecology of Arboreal Folivores" (G.G. Montgomery, ed.), pp. 33-50. Smithsonian Institution, Washington D.C.
Lekagul, B. and McNeely, J.A. (1977) "Mammals of Thailand". Association for the Conservation of Wildlife, Bangkok.
Lekagul, B. and McNeely, J.A. (1978) Thailand launches extensive reafforestation programme. *Tigerpaper* **5**, 1: 9-13.
Leutenegger, W. (1978) Scaling of sexual dimorphism in body size and breeding systems in primates. *Nature, Lond.* **272**: 610-611.
Lucas, G. and Synge, H. (1978) "The IUCN Plant Red Data Book". IUCN, Morges, Switzerland.
MacKinnon, J.R. (1974) The behaviour and ecology of wild orang-utans (*Pongo pygmaeus* . *Anim. Behav.* **22**: 3-74.
MacKinnon, J.R. (1977) A comparative ecology of Asian apes. *Primates* **18**: 747-772.
MacKinnon, J.R. and MacKinnon, K. (1977) The formation of a new gibbon group. *Primates* **18**: 701-708.
MacKinnon, J.R. and MacKinnon, K.S. (1978) Comparative feeding ecology of six sympatric primates in West Malaysia. In "Recent Advances in Primatology, Vol. 1, Behaviour" (D.J. Chivers and J. Herbert, eds.), pp. 305-321. Academic, London.
MacKinnon, K.S. (1976) Home range, feeding ecology and social behaviour of the grey squirrel (*Sciurus carolinensis* Gmelin). Unpubl. D.Phil. dissertation, University of Oxford.
MacKinnon, K.S. (1978) Stratification and feeding differences among Malayan squirrels. *Malay. Nat. J.* **30**: 593-608.
Mah, Y.L. (in press) Ranging behaviour of *Macaca fascicularis* in two different habitats in Peninsular Malaysia. In "Proceedings of 5th International Symposium of Tropical Ecology, Kuala Lumpur 1979" (J.I. Furtado, ed.).
Mah, Y.L. (in prep.) Ecology and behaviour of the long-tailed macaque at two sites in Peninsular Malaysia. Unpubl. Ph.D. dissertation, Universiti Malaya.
Marshall, J.T. and Brockelman, W.Y. (in prep.) Natural hybrids between pileated and lar gibbons with implications for the origin of Bornean gibbons.

Marshall, J.T. and Marshall, E.R. (1976) Gibbons and their territorial songs. *Science, N.Y.* **193**: 235-237.

Marshall, J.T., Ross, B.A. and Chantharojvung, S. (1972) The species of gibbons in Thailand. *J. Mammal.* **53**: 479-486.

McCance, R.A. and Lawrence, R.D. (1929) The carbohydrate content of foods. *Spec. Rep. Ser. Med. Res. Coun. 135.*

McCann, C. (1933) Notes on the colouration and habits of the white-browed gibbon or hoolock (*Hylobates hoolock*). *J. Bombay Nat. Hist. Soc.* **36**: 395-405.

McClure, H.E. (1964) Primates in the dipterocarp forest of the Gombak valley. *Primates* **5**: 39-58.

McClure, H.E. (1966) Flowering, fruiting and animals in the canopy of a tropical rain forest. *Malay. Forester* **29**: 182-203.

McKey, D.B. (1978) Soils, vegetation and seed-eating by black colobus monkeys. In "Ecology of Arboreal Folivores" (G.G. Montgomery, ed.), pp. 423-437. Smithsonian Institution, Washington D.C.

McNab, B.K. (1963) Bioenergetics and the determination of home range size. *Amer. Nat.* **97**: 133-140.

Medway, Lord (1969) "The Wild Mammals of Malaya". Oxford University Press, Kuala Lumpur.

Medway, Lord (1970) The monkeys of Sundaland: ecology and systematics of the cercopithecids of a humid equatorial environment. In "Old World Monkeys: Evolution, Systematics, and Behaviour" (J.R. and P.H. Napier, eds.), pp. 513-553. Academic, London.

Medway, Lord (1972) Phenology of a tropical rain forest in Malaya. *Biol. J. Linn. Soc.* **4**: 117-146.

Medway, Lord and Wells, D.R. (1971) Diversity and density of birds and mammals at Kuala Lompat, Pahang. *Malay. Nat. J.* **24**: 238-247.

Medway, Lord and Wells, D.R. (1976) "The Birds of the Malay Peninsula. Volume 5: Conclusions, and survey of every species". Witherby, London.

Menzel, E.W. (1971) Communication about the environment in a group of young chimpanzees. *Folia primatol.* **15**: 220-232.

Michael, R.P. and Crook, J.H. eds. (1973) "Comparative Ecology and Behaviour of Primates". Academic, London.

Milton, K. and May, M.I. (1976) Body weight, diet and home range area in primates. *Nature, Lond.* **259**: 459-462.

Mitani, J.C. and Rodman, P.S. (1979) Territoriality. The relation of ranging pattern and home range size to defendability, with an analysis of territoriality among primate species. *Behav. Ecol. Sociobiol.* **5**: 241-251.

Mitchell, A. (1974) "A Field Guide to the Trees of Britain and Northern Europe". Collins, London.

Mohnot, S.M. (1978) The conservation of primates in India. In "Recent Advances in Primatology, Vol. 2, Conservation" (D.J. Chivers and W. Lane-Petter, eds.), pp. 47-53. Academic, London.

Montgomery, G.G. ed. (1978) "The Ecology of Arboreal Folivores". Smithsonian Institution, Washington D.C.

Montgomery, G.G. and Sunquist, M.E. (1975) Impact of sloths on neotropical forest energy flow and nutrient cycling. In "Tropical Ecological Systems: Trends in Terrestrial and Aquatic Research" (F.B. Golley and E. Medina, eds.), pp. 69-98. Springer, New York.

Moynihan, M. (1970) Some behaviour patterns of Platyrrhine monkeys *Saguinus geoffroyi* and some other tamarins. *Smithson. Contr. Zool.* **28**: 1-77.

Mukherjee, R.P. and Saha, S.S. (1974) The golden langurs (*Presbytis geei*, Khajuria, 1956) of Assam. *Primates* **15**: 327-340.

Muul, I. and Lim, B.L. (1978) Comparative morphology, food habits and ecology of some Malaysian arboreal rodents. In "The Ecology of Arboreal Folivores" (G.G. Montgomery, ed.), pp. 361-368. Smithsonian Institution, Washington D.C.

Myers, N. (1979) "The Sinking Ark: a new look at the problem of disappearing species". Pergamon, Oxford.

Napier, J.R. and Napier, P.H. (1967) "A Handbook of Living Primates". Academic, London.

Nice, M.M. (1941) The role of territory in bird life. *Am. Midl. Nat.* **26**: 441-487.

Ng, F.S.P. (1966) Age at first flowering in dipterocarps. *Malay. Forester* **29**: 290-295.

Ng, F.S.P. (1977) Gregarious flowering of dipterocarps in Kepong, 1976. *Malay. Forester* **40**: 126-137.

Ng, F.S.P. and Loh, H.S. (1974) Flowering to fruiting periods of Malayan trees. *Malay. Forester* **37**: 127-132.

Nieuwolt, S. (1965) Evaporation and water balances in Malaya. *J. Trop. Geogr.* **20**: 34-53.

Odum, E.P. (1971) "Fundamentals of Ecology". (3rd Ed.) Saunders, Philadelphia.

Opler, P.A. (1978) Interaction of plant life history components as related to arboreal herbivory. In "Ecology of Arboreal Folivores" (G.G. Montgomery, ed.), pp. 23-32. Smithsonian Institution, Washington D.C.

Orians, G.H. (1969) On the evolution of mating systems in birds and mammals. *Am. Nat.* **103**: 589-603.

Packer, C. (1979a) Inter-troop transfer and inbreeding avoidance in *Papio anubis*. *Anim. Behav.* **27**: 1-36.

Packer, C. (1979b) Male dominance and reproductive activity in *Papio anubis*. *Anim. Behav.* **27**: 37-45.

Pariente, G. (1974) Influence of light on the activity rhythms of two Malagasy lemurs, *Phaner furcifer* and *Lepilemur mustelinus*. In "Prosimian Biology" (R.D. Martin, G.A. Doyle and A.C. Walker, eds.), pp. 183-198. Duckworth, London.

Parra, R. (1978) Comparison of foregut and hindgut fermentation in herbivores. In "Ecology of Arboreal Folivores" (G.G. Montgomery, ed.), pp. 205-230. Smithsonian Institution,

Washington D.C.

Parsons, P.E. and Taylor, C.R. (1978) Energetics of brachiation versus walking: A comparison of a suspected and invested pendulum mechanism. *Physiol. Zool.* **50**: 182-188.

Payne, J.B. (1979a) Abundance of diurnal squirrels at the Kuala Lompat Post of the Krau Game Reserve, Peninsular Malaysia. In "The Abundance of Animals in Malesian Rain Forests" (A.G. Marshall, ed.), pp. 37-51. Department of Geography, University of Hull.

Payne, J.B. (1979b) Synecology of Malayan tree squirrels, with particular reference to the genus *Ratufa*. Unpubl. Ph.D. dissertation, University of Cambridge.

Petter, J-J. (1978) Ecological and physiological adaptations of five sympatric nocturnal lemurs to seasonal variations in food production. In "Recent Advances in Primatology, Vol. 1, Behaviour" (D.J. Chivers and J. Herbert, eds.), pp. 211-223. Academic, London.

Pocock, R.I. (1934) The monkeys of the genus *Pithecus* (or *Presbytis*) and *Pygathrix* found to the East of the Bay of Bengal. *Proc. Zool. Soc. Lond.*: 895-961.

Poirier, F.E. and Smith, E.O. (1974) The crab-eating macaques (*Macaca fascicularis*) of Angaur Island, Palau, Micronesia. *Folia primatol.* **22**: 258-306.

Poore, M.E.D. (1968) Studies in Malaysian rain forest. I. The forest on Triassic sediments in Jengka Forest Reserve. *J. Ecol.* **56**: 143-196.

Powell, G.V.N. (1974) Experimental analysis of the social value of flocking by starlings (*Sturnus vulgaris*) in relation to predation and foraging. *Anim. Behav.* **22**: 501-505.

Pringle, S.L. (1969) World supply and demand of hardwoods. In "Proceedings of Conference on Tropical Hardwoods, Syracuse".

Raemaekers, J.J. (1977) Gibbons and trees: Comparative ecology of the siamang and lar gibbons. Unpubl. Ph.D. dissertation, University of Cambridge.

Raemaekers, J.J. (1978a) The sharing of food sources between two gibbon species in the wild. *Malay. Nat. J.* **31**: 181-188.

Raemaekers, J.J. (1978b) Changes through the day in the food choice of wild gibbons. *Folia primatol.* **30**: 194-205.

Raemaekers, J.J. (1978c) Competition for food between lesser apes. In "Recent Advances in Primatology, Vol. 1, Behaviour" (D.J. Chivers and J. Herbert, eds.), pp. 327-330. Academic, London.

Raemaekers, J.J. (1979) Ecology of sympatric gibbons. *Folia primatol.* **31**: 227-245.

Raemaekers, J.J. (in press) Causes of variation between months in the distance travelled daily by gibbons. *Folia primatol.*

Rainier, HSH Prince and Bourne, G.H. eds. (1977) "Primate Conservation". Academic, New York.

Richard, A. (1974) Intra-specific variation in the sociology and ecology of *Propithecus verreauxi*. *Folia primatol.* **22**:

178-207.
Richards, D.G. and Wiley, R.H. (1980) Reverberations and amplitude fluctuations in the propagation of sound in a forest: implications for animal communication. *Am. Nat.* **115**: 381-389.
Richards, P.W. (1952) "The Tropical Rain Forest". Cambridge University Press.
Rijksen, H.D. (1978) "A field study on Sumatran orang-utans (*Pongo pygmaeus abelii* Lesson 1827)". Wageningen.
Ripley, S. (1970) Leaves and leaf monkeys: the social organisation of foraging in grey langurs, *Presbytis entellus thersites*. In "Old World Monkeys: Evolution, Systematics and Behaviour" (J.R. and P.H. Napier, eds.), pp. 481-509. Academic, London.
Rodman, P. (1973a) Synecology of Bornean primates. I. A test for interspecific interactions in spatial distribution of five species.
Rodman, P.S. (1973b) Population composition and adaptive organization among orang-utans of the Kutai Reserve. In "Comparative Ecology and Behaviour of Primates" (J.H. Crook and R.P. Michael, eds.), pp. 171-209. Academic, London.
Rodman, P.S. (1978) Diets, densities and distributions of Bornean primates. In "The Ecology of Arboreal Folivores" (G.G. Montgomery, ed.), pp. 465-478. Smithsonian Institution, Washington D.C.
Rodman, S. (1979) Skeletal differentiation of *Macaca fascicularis* and *Macaca nemestrina* in relation to arboreal and terrestrial quadrupedalism. *Amer. J. phys. Anthrop.* **51**: 51-62.
Rollet, B. (1969) Etudes quantitatives d'une forêt dense sempervirente de la Guiane vénézuelienne. Thése de Doctorat, Faculté de Toulouse.
Rothe, H. (1975) Some aspects of sexuality and reproduction in groups of captive marmosets (*Callithrix jacchus*). *Z. Tierpsychol.* **37**: 255-273.
Rowell, T.E. (1972) "The Social Behaviour of Monkeys". Penguin, Harmondsworth.
Rudran, R.R. (1973) The reproductive cycles of two sub species of purple-faced langurs (*Presbytis senex*) with relation to environmental factors. *Folia primatol.* **19**: 41-60.
Sade, D.S. (1964) Seasonal cycle in testes of free-ranging *Macaca mulatta*. *Folia primatol.* **2**: 171-180.
Schaller, G.B. (1963) "The Mountain Gorilla". Chicago University Press.
Schmidt, F.H. and Ferguson, J.H.A. (1951) "Rainfall types based on wet and dry period ratios for Indonesia with western New Guinea". Verh. Djawatan Met. dan Geofisik, Jakarta.
Schmidt-Nielsen, K. (1972) Locomotion: energy cost of swimming, flying and running. *Science* **177**: 222-228.

Schultz, A.H. (1933) Observations on the growth, classification and evolutionary specialization of gibbons and siamangs. *Hum. Biol.* **5**: 212-255, 385-428.
Schultz, A.H. (1974) The skeleton of the Hylobatidae and other observations on their morphology. In "Gibbon and Siamang" (D.M. Rumbaugh, ed.), vol. 3, pp. 1-54. Karger, Basel.
Smith, C.C. (1968) The adaptive nature of social organization in the genus of tree squirrels *Tamiasciurus*. *Ecol. Monogr.* **38** (1): 31-63.
Smythe, N. (1970) Relationships between fruiting seasons and seed dispersal methods in a neotropical forest. *Amer. Nat.* **104**: 25-35.
Snow, D.W. (1965) A possible selective factor in the evolution of fruiting seasons in a tropical forest. *Oikos* **15**: 274-281.
Snow, D.W. (1976) "The Web of Adaptation. Bird Studies in the American Tropics". Collins, London.
Southwick, C.H. and Cadigan, F.C. (1972) Population studies of Malaysian primates. *Primates* **13**: 1-19.
Spencer, C. (1975) Interband relations, leadership behaviour and the initiation of human-orientated behaviour in bands of semi-wild free-ranging *Macaca fascicularis*. *Malay. Nat. J.* **29**: 83-89.
Stern, J.T. and Oxnard, C.E. (1973) Primate locomotion: Some links with evolution and morphology. *Primatologia* **4** (11): 1-93.
Stevens, W.E. (1968) "The Conservation of Wildlife in West Malaysia". Federal Game Dept., Malaysia.
Stoel, T.B. (1980) Framework for IUCN programme on tropical rain forests. Mimeographed report.
Stott, K. and Selsor, C.J. (1961) Observations on the maroon leaf monkey in North Borneo. *Mammalia* **25**: 184-189.
Struhsaker, T.T. (1974) Correlates of ranging behaviour in a group of red colobus monkeys (*Colobus badius tephrosceles*). *Amer. Zool.* **14**: 177-184.
Struhsaker, T.T. (1975) "The Red Colobus". Chicago University Press.
Struhsaker, T.T. (1978) Food habits of five monkey species in the Kibale Forest, Uganda. In "Recent Advances in Primatology, Vol. 1, Behaviour" (D.J. Chivers and J. Herbert, eds.), pp. 225-248. Academic, London.
Struhsaker, T.T. and Oates, J.F. (1975) Comparison of the behaviour and ecology of the red colobus and black-and-white colobus monkeys in Uganda: a summary. In "Socio-ecology and Psychology of Primates" (R.H. Tuttle, ed.), pp. 103-123. Mouton, The Hague.
Sugiyama, Y. (1965) On the social change of hanuman langurs (*Presbytis entellus*) in their natural conditions. *Primates* **6**: 381-417.
Tenaza, R.R. (1975) Territory and monogamy among Kloss' gibbons (*Hylobates klossii*) in Siberut Island, Indonesia. *Folia*

primatol. **24**: 68-80.
Tenaza, R.R. (1976) Songs, choruses and countersinging of Kloss' gibbons (*Hylobates klossii*) in Siberut Island, Indonesia. *Z. Tierpsychol.* **40**: 37-52.
Tenaza, R.R. and Hamilton, W.J. (1971) Preliminary observations of the Mentawai Islands gibbon, *Hylobates klossii*. *Folia primatol.* **15**: 201-211.
Tenaza, R.R. and Tilson, R.L. (1977) Evolution of long distance alarm calls in Kloss' gibbon. *Nature, Lond.* **268**: 233-235.
Tilson, R.L. (1974) Man and Monkey: concepts in conservation for Mentawai Island primates. Ubpubl. ms.
Tilson, R.L. (in press a) Distribution of hoolock gibbons (*Hylobates hoolock*) and their behaviour in a seasonal environment. *J. Bombay Nat. Hist. Soc.*
Tilson, R.L. (in press b) Family formation strategies of Kloss gibbons. *Folia primatol.*
Tittensor, A.M. (1970) The red squirrel (*Sciurus vulgaris*) in relation to its food resources. Unpubl. Ph.D. dissertation, University of Edinburgh.
Trivers, R.L. (1972) Parental investment and sexual selection. In "Sexual Selection and the Descent of Man" (R.G. Campbell, ed.), pp. 136-179. Aldine, Chicago.
Vanzolini, P.E. (1978) Current problems of primate conservation in Brazil. In "Recent Advances in Primatology, Vol. 2, Conservation" (D.J. Chivers and W. Lane-Petter, eds.), pp. 15-25. Academic, London.
Verner, J. (1964) Evolution of polygamy in the long-billed marsh wren. *Evolution* **18**: 252-261.
Waser, P.M. and Homewood, K. (1979) Cost-benefit approaches to territoriality: a test with forest primates. *Behav. Ecol. Sociobiol.* **6**: 115-119.
Washburn, S.L. (1944) The genera of Malaysian langurs. *J. Mammal.* **25**: 289-294.
Wasserman, F.E. (1977) Intraspecific acoustical interference in the white-throated sparrow (*Zonotrichia albicollis*). *Anim. Behav.* **25**: 949-952.
Wells, D.R. (1971) Survival of the Malaysian bird fauna. *Malay. Nat. J.* **24**: 248-256.
Wheatley, B.P. (1976) The ecological strategy of the long-tailed macaque, *Macaca fascicularis*, in the Kutai Nature Reserve, Kalimantan. *Frontier* **5**: 27-32.
Wheatley, B.P. (1978) Foraging patterns in a group of long-tailed macaques in Kalimantan Timur, Indonesia. In "Recent Advances in Primatology, Vol. 1, Behaviour" (D.J. Chivers and J. Herbert, eds.), pp. 347-349. Academic, London.
Whitmore, T.C. (1975) "Tropical Rain Forests of the Far East". Clarendon, Oxford.
Whitten, A.J. (1980) The Kloss Gibbon in Siberut Rain Forest. Unpubl. Ph.D. dissertation, University of Cambridge.

Whitten, J.E.J. (1979) Ecological Isolation of an Indonesian Squirrel (*Sundasciurus lowii siberu*). Unpubl. M.Phil. dissertation, University of Cambridge.

Wiley, R.H. and Richards, D.G. (1978) Physical constraints on acoustic communication in the atmosphere: implications for the evolution of animal vocalizations. *Behav. Ecol. Sociobiol.* **3**: 69-94.

Wilson, C.C. and Wilson, W.L. (1973) Census of Sumatran primates. Final report, unpublished, to the Regional Primate Research Center, Seattle.

Wilson, C.C. and Wilson, W.L. (1975) The influence of selective logging on primates and some other animals in East Kalimantan. *Folia primatol.* **23**: 245-274.

Wilson, C.C. and Wilson, W.L. (1977) Behavioral and morphological variation among primate populations in Sumatra. *Yearb. Phys. Anthrop.* **20**: 207-233.

Wilson, E.O. (1975) "Sociobiology: The New Synthesis". Belknap Press, Cambridge, Mass.

Wolf, K. (in prep.) Social behaviour of silvered leaf monkeys at Kuala Selangor. Unpubl. Ph.D. dissertation, Yale University.

Wolf, K.E. and Fleagle, J.G. (1977) Adult male replacement in a group of silvered leaf monkeys (*Presbytis cristata*) at Kuala Selangor, Malaysia. *Primates* **18**: 949-956.

Wong, Y.K. (1967) Some indications of the total volume of wood per acre in lowland dipterocarp forest. *Malayan Forestry Dept. Research Pamphlet* no. 53.

World Population Data Sheet (1979) Population Reference Bureau, Inc., Washington.

Wrangham, R.W. (1979) On the evolution of ape social systems. *Social Science Information* **18**: 335-368.

Wyatt-Smith, J. (1953) Malayan forest types. *Malay. Nat. J.* **7**: 45-55, 91-98.

Wycherley, P.R. (1967) Rainfall probability tables for Malaysia. *R.R.I.M. Planting Manual* **12**: 1-85. Rubber Research Institute, Kuala Lumpur.

Wycherley, P.R. (1973) Phenology of plants in the humid tropics. *Micronesica* **9**: 75-96.

Yoda, K. (1974) Three-dimensional distribution of light intensity in a tropical rain forest in West Malaysia. *Jap. J. Ecol.* **24**: 247-254.

Yoshiba, K. (1968) Local and inter-troop variability in ecology and social behaviour of common Indian langurs. In "Primates" (P.C. Jay, ed.), pp. 217-242. Holt, Rinehart and Winston, New York.

AUTHOR INDEX

SUBJECT INDEX

(should be used in conjunction with List of Contents)